W9-BFQ-290

NORTH CAROLINA
STATE BOARD OF COMMUNITY COLLEGES
LIBRARIES
GASTON COLLEGE

Construction Paperwork:

An Efficient Management System

J. Edward Grimes

Construction Paperwork:

An Efficient Management System

J. Edward Grimes

Contributing Author: Mark Burgess

R.S. Means Company, Inc.

Morris Library
Gaston College
Dallas, NC 28034

R.S. MEANS COMPANY, INC.
CONSTRUCTION CONSULTANTS & PUBLISHERS
100 Construction Plaza
P.O. Box 800
Kingston, MA 02364-0800
(617) 585-7880

© 1989

In keeping with the general policy of R.S. Means Company, Inc., its authors, editors, and engineers apply diligence and judgment in locating and using reliable sources for the information published. However, no guarantee or warranty can be given, and all responsibility for loss or damage is hereby disclaimed by the authors, editors, engineers, and publisher of this publication with respect to the accuracy, correctness, value, and sufficiency of the data, methods, and other information contained herein as applied for any particular purpose or use.

No part of this publication may be reproduced, stored in a retrieval system, or transmitted in any form or by any means without prior written permission of R.S. Means Company, Inc.

This book was edited by Jeff Goldman and Mary Greene. Typesetting was supervised by Joan Marshman. The book and jacket were designed by Norman Forgit.

Printed in the United States of America

10 9 8 7 6 5 4 3
Library of Congress Catalog Number 89-189300

ISBN 0-87629-147-7

This book is dedicated to the men and women of the construction industry.

Table of Contents

Acknowledgments

I would like to acknowledge the special contribution made by Mark Burgess in the writing of this book. Without his efforts, advice, and writing assistance, this book would probably not have been written.

Through my association with the American Arbitration Association, I have been privileged to serve on many arbitration cases, and have learned a great deal from each case. The recurring effect of "lack of documentation" on the outcome of *all* of these cases has led to many of the conclusions and recommendations that appear in this book. I would, therefore, like to thank the Arbitration Association for the opportunity to serve, listen, and learn.

Norman Peterson, former President of the R.S. Means Company, Inc., urged me to write a book on Construction Paperwork. Ferol Breymann was very helpful in getting me started on this project. Appreciation is also due to Mary Greene and Jeffrey Goldman, who edited the manuscript.

Rod Wright, architect and President of HCH Partners, has given me advice and insight for the book from an architectural viewpoint. Leon Alderfer, President of Aldergraf Systems, Inc., assisted in the "Scheduling" chapter.

I would like to thank the many people in the construction industry who have helped me to learn my trade and, therefore, directly or indirectly contributed to this book. Among them are: Richard Bradshaw, Patricia Capen, Casie Carey, Mickey Carhart, Ronald Coover, Robert Drinkward, Harry Gallager, Steve Marble, Shayne McDaniel, S. F. Nielsen, Dale Pursel, Lester and Sheila Raikow, Astrid Shannon, James B. Still, and Joyce Von Graven.

Special thanks go to Mr. Fred Gibson, one of the best superintendents I have ever had the opportunity to work with on a construction project.

I would also like to thank the Associated General Contractors for granting permission to use their forms as samples in the book.

While I was acquiring experience and an education in the industry, my wife Frances has stuck by me through all of the ground-breakings and topping-outs of a few hundred different projects, and all of the emotional peaks and valleys in between. I can only say that without her support, I would neither have stood the test of time in this industry, nor been able to write this book. Thanks Frances.

Introduction

Buildings are born on paper: on plans, in specifications, and on the backs of napkins. Buildings also die on paper: in unpaid claims, rejected work, liquidated damages, arbitration, litigation, and in lots of red ink on the books.

The contract is the central, and perhaps biggest, single collection of paper required for a construction project. A contract is normally a written legal document that incorporates, by reference, other written documents. Items that are part of, or referenced in, the contract documents include, but are not limited to, the plans and specifications, the soils report, the Uniform Building Code, the construction price, the time to complete the project, and any penalties for not completing the work on time.

Once a contract has been signed, all work must comply with or modify that contract until the project is complete and the warranty period has ended. The contract should, and usually does, dictate the method and documentation required for modifying the contract. The methods specified apply to all parties to the contract. Even the word "contract" in the designation, *general contractor* or *subcontractor*, shows the legal significance placed on documentation. Building contractors, as well as in-house facilities departments and subcontractors, are all bound by the contract. Good communication documentation is recommended by everyone, but unfortunately, is not so universally achieved. The contractor who consistently uses good communication procedures has a better chance of receiving, in return, good written communications from the owner, the architect, and subcontractors. The better the documentation between all parties, the fewer the unknowns – which means reduced liability for all.

The communication requirements of a large general contractor, a small general contractor, a facilities management department, a machine shop, or an individual contractor, are functionally similar. There may be some unique individual requirements, but for the most part, documentation, processing, and completion of any project are much the same. The goal is to keep everyone involved in the project informed of what you are doing, what you intend to do, what you expect them to do, and when they should do it. Throughout this book, we use the term *manager* to refer to the individual who is given chief responsibility for the administration of the project, whether he or she is a project manager, facilities manager, owner's representative, or construction manager.

The purpose of this book is to show the importance of good communication documentation, and how to achieve it. Included are written letters, Submittals, Transmittals, and Requests for Information, as well as documentation of changes in scope, and cost quotes for those changes. Construction documentation control must be a concept that permeates the day-to-day aspects of doing business. Generally, the companies that are consistent in their documentation procedures have learned the value of careful documentation control the hard way—having lost money without it. The key reasons for establishing an effective communications and documentation program include *project management, liability, time and timely response, and marketing.*

Project Management

This book is, in a sense, a guide to total project management, since good project management starts and finishes with good communication. Effective construction communication keeps the project on track and everyone going in the proper direction. Many books and articles have been written for contractors on the separate subjects of bid documents, construction documents, estimating, accounting, contracts, scheduling, and supervision. In fact, each of these topics is only one part of construction project management. Recently, project scheduling has been touted by many as the cornerstone of project management. While project scheduling is extremely important, it is just one of the many tools to be used in the effective management of a project. Good project communication is the string that connects all of the parts. Only through proper documentation of the communications between the participants in a construction project can each contribute fully toward the success of that project.

Liability

Recent judicial decisions have set the contractor's exposure to legal claims from a minimum of ten years to almost forever if there is any negligence involved. This kind of liability demands a better job of documenting communications. This means documenting decisions, or the lack of decisions, of all parties as they affect the construction project, including, but not limited to, those of the owner and/or his representatives, the developer, the architect, the construction manager, the facilities manager, the general contractor, and his subcontractors and vendors.

Time and Timely Response

Proper documentation also includes the concept of timely response. Since "time is money" in construction, as in other businesses, we will present ideas on how to set up your documentation controls so that your responses are not only timely as required by the contract, but timely as required to increase profitability on a project. There is an old saying in the construction industry: "The job is complete when the money runs out." We will show you how to make sure all of your requests, claims, and changes in the scope of the project are documented and presented in time to make your claim (for additional funds for additional services rendered) before "all the money runs out."

Marketing

In the construction industry, you have only a brief opportunity to make a presentation to "sell" your services. Good written communication shows the client that you know what you are doing. A business-like image is extremely important in inspiring the confidence that owners, clients, and architects must have in the manager's ability to accomplish a given project. Good communication practices and procedures help to emphasize this professional image.

Summary

Good documentation cannot be generated today for an event that occurred in the past. It is part of the day-to-day, week-to-week activity that must take place in a standardized procedure. It is very difficult to create protective documentation after the fact.

Many builders, even with valid arguments and good cause, lose their legitimate claims due to poor documentation. Additionally, many, many more valid claims of the contractor never get a decent hearing, because they lack sufficient documentation for a presentable case.

We will show you the guidelines for good (proper and profitable) communication documentation, including the care and handling of communication to and from your office. If you maintain consistent procedures for issuing and receiving communications, they will go a long way towards keeping you on the tennis courts and out of the legal ones. Good communications, well-documented, will help your projects run more smoothly and help you maintain a profitable bottom line for your project. In this book, we will concentrate on the relationships, the interdependencies, and the functions of all of the participants in a construction project. We will emphasize the importance of establishing and controlling the flow of documentation on the project, and will show techniques you can use to manage paperwork, rather than it managing you.

Part One

Principles of Construction Communications

Chapter One

Principles of Construction Communications

All construction communications can be grouped into two categories, intended to:

1. provide information or directives to others
2. receive information or directives from others

The first category includes, among others: letters, advertisements, and other reference materials intended to convey information. It also includes notifications of events, such as the dates on which certain actions are scheduled. For example:

"Please start your work on the project on Monday, October 15."
"The man hoist will be removed on Friday, October 19."
"There will be a job meeting on Tuesday, October 16."

The second category is used to obtain information. These communications usually take the form of a question. They may request either an answer or approval. For example:

"What is the cost to install an extra door?"
"When will the elevators be operational?"

Or:

"Here is the quote for adding the new door. Please give me your approval to proceed as soon as possible."

Goals of Communications

The goals of effective communications in construction are similar to those for communications in other businesses. They are:

- To alleviate doubt
- To clarify
- To produce a record of events
- To reduce reliance on the spoken word
- To verify (document) verbal instructions
- To avoid the possibility of having to rely on selective memory

The purpose of construction communications is to exchange information about the project in order to build it according to the contract. The contract is a legally binding agreement and usually exists as a printed and signed document. Any modifications to the written contract for any purpose must be documented. Such changes include the many verbal "mini-contracts" made in the course of the work to allow the project to move ahead. These "mini-contracts" are often made as needed, over the telephone or at the site, and *must* be documented at some point, either in the project minutes or in a modification to the contract (in the form of a Field Order or Price Request). It is best to document these "mini-contracts" immediately, before one party forgets exactly what was decided, or that the conversation ever took place.

If the documentation procedure is not followed, the manager may be subjected to such "no win" client arguments as:

"I told you so!"

"I didn't tell you to do that!"

"You didn't tell me it was going to cost $2,000."

"I didn't give you approval."

"Your original schedule of completion was for November 15. You never told me you were not going to finish until January."

"Who told you to paint the rooms green? When? Why didn't you tell me?"

Effective construction communication requires that participants initiate, track, account for, follow up, and follow through on communications that require a response. For the most part, construction communication is a series of *questions and answers*. The communication is not complete until the question is answered, and the answer is understood by all involved parties.

Exchanging Information

It is imperative, and basic to effective communication, that each party to a construction contract is aware of the activities of the other parties. All parties must give each other an opportunity to point out a potential conflict with the proposed plan of action. Documentation is most effective for all parties when proposed decisions and plans are exchanged between participants *before* the decisions are finalized.

For example, the general contractor should advise the owner or architect of the cost of any change work desired before the change work is begun. This allows the owner to reject the change without spending money. Owners or contractors should never assume, "this won't cost very much," or "it won't take very long," when considering a change in the scope of the work. If the contractor seems pessimistic when submitting a time estimate, that pessimism is most likely based on previous, and sometimes bitter, experience.

The architect and owner should allow the contractor to have input into both the proposed schedule of completion and the scope of work. The contractor must have an opportunity to say, "no, this work cannot be done in the time allocated." The reverse also applies; the contractor should allow the architect and owner input on *their* schedules for a coordination "reality check," to ensure that the contractor understands the scope of the project and to allow for tentative furnishing and use dates.

Unrealistic or "quick" estimates of time or money should never be encouraged or provided simply because an owner feels that the scope of the change will be less costly or time-consuming than the contractor estimated. For example, first-time owners may try to tell experienced contractors what it costs to add a new door in a wall. The owner's version is usually about half of what the contractor thinks it will cost. Submitting unrealistic estimates to please the owner can only lead to eventual distrust and discouragement. The owner and architect must do their part to bring the contractor(s) into the decision process by considering his opinion a valid estimate. Otherwise, the project will end up over-budget, and will not be completed within the schedule.

Principles of Documentation

Typically, construction personnel know what has to be done to maintain good documentation and/or communication, but tend to neglect these measures. Looking back after the fact and wishing that communications had been well-documented is not enough. Regular documentation practices must be maintained among all parties. This includes secretaries, architect/engineers, estimators, managers, owners, facility users, subcontractors, and vendors.

In some cases, individuals do not have the knowledge or experience to document correctly. Too often, however, the cause is not lack of experience, but rather:

> *"It is just too much trouble."*
>
> *"There is a new person in the office, and they are already behind."*
>
> *"Who needs this information anyway? They already know about the problem."*
>
> *"There are other urgent problems at (another) job."*
>
> *"I will take care of it later, when I get the time. (Maybe it will go away.)"*

Each party to a construction project should document his or her own activities and keep a "score card" on the activities of the other players. That is the only way to maintain control of a project.

Traditionally, the manager (the general contractor's project manager, the facilities manager, or the construction manager) acts as a sort of "station master," assuming responsibility for monitoring the status of hundreds and hundreds of individual pieces of paper, quotes, contracts, subcontracts, and submittals that flow through the system. Decisions must be made for each step in the processing of these documents. Lack of decisions and the resulting confusion affect job completion by delaying the flow of paperwork. Job completion has a greater effect on the general contractor's profit and loss than on the fortunes of any of the other parties involved in the construction project. Since he has the most at risk, it makes sense that the general contractor should take a leading role in seeing that documentation is efficiently processed.

Functional Authority

The manager has *functional authority*, or the authority to use whatever means are at his disposal, in order to complete the project for which he is responsible. The manager has the power to hire and fire his own firm's employees who are in his charge for that project. However, the manager does not have this kind of influence or leverage over the employees of the project's architect, subcontractors, or vendors.

Lacking this direct control, the manager must find other means to ensure their cooperation. This is where an efficient management system for construction paperwork can make a tremendous difference. When a consistent, comprehensive communications system is used, all notifications, requests, etc., become part of the project documentation. Subcontractors and vendors thereby become responsible parties to the project, and realize that it is in their best interest to comply with their contractual obligations. Furthermore, it will be apparent to the architect that the manager is making every effort to fulfill his obligation to the owner.

Computer Use

Naturally, total communication and absolute understanding of all project issues are not realistic expectations. Communication and documentation are, however, easier, quicker, and, possibly more dependable, when using computers for generating documentation, as well as for follow-up and tracking that documentation. Parts II-IV of this book show effective formats for individual documents used in the course of the project. In each case, two options are shown: a pre-printed form filled out by hand, and a computer-generated version. Both of these documentation methods are valid. A computer system offers many advantages to firms who are willing to adapt to current technology.

Summary of Principles

- Document (write down and transmit) all decisions.
- Document all questions regarding the project and direct them to the individual responsible for answering them.
- Document all modifications or changes to the original contract.
- Keep all parties on the project aware of the direction, progress, problems, and timing of the project, thereby making the best use of their time and efforts to complete the project within the contract budget and schedule.

Chapter Two

The Participants: Exposure and Resources

It is important that all parties are aware of the relationships, interdependencies, functions, and exposures to profit/loss of all other participants within a construction project. On many projects, thousands of individuals are involved. Just think of all of the people in the owner's arena, the architect's offices, and all of the different contractor and field offices, including accounting, estimating, management, and field personnel. The possibilities for tragic miscommunication are astounding.

Each owner, architect, or contractor must emphasize the importance of establishing a system for managing and controlling the flow of documentation in the office as well as on the job site. Otherwise, chaos will reign, bringing inefficiencies that reduce profits and delay completion.

The Owner

The owner is the individual or organization who will be the user of the facilities. The owner orders the work performed for his use and pays all the fees and costs of a project.

An extension of construction time will reduce the owner's profits, due to lost use of the facility, but time overruns generally have a more significant effect on the general contractor. Delay costs affect the owner/developer most severely when interest rates rise, as occurred in the late 1970's and early 1980's.

The Architect

The architect is generally retained by the owner to design the new facilities to the owner's needs, and to transmit those requirements to the general contractor via plans and specifications. The architect usually works on a percentage fee based on the total value of the project. By the time a project is started, the architect and engineers have completed most of their work. If they continue to be involved in the project, they are generally paid additional fees for construction administration. Therefore, the duration of a construction project does not usually affect the architect's net profit significantly.

The General Contractor

The general contractor is responsible for taking the architect's plans and specifications and turning them into reality. He stands to make a profit based upon the fact that he also risks absorbing the biggest proportionate loss.

For a general contractor to make a profit, a project must finish close to the estimated completion time. A $1,000,000, six-month job that takes eight months to complete can possibly reduce the general contractor's anticipated net profit from two percent to nothing. By the same token, if the construction time can be reduced from six months to four, the profit can almost double. Figure 2.1 shows the effect of time on the general contractor's profit. The example in this figure assumes job site and other overhead costs to be 15%, and the estimated "net-net" profit, before taxes, to be 2%, which is approximately the national average.

The duration of a job, and whether it meets or exceeds the general contractor's estimated time of construction, exponentially affects the project's net profit. It is, therefore, in the contractor's best interest to obtain needed information as fast as possible– throughout the job.

Facility Manager

The facility manager's responsibilities are much the same as those of the general contractor. For the facility manager, good plans and specifications will make the difference between an in-house job completed within budget, and one that requires mountains of cost overrun justification.

Subcontractors

The subcontractor's function is similar to that of the general contractor, except that he or she is responsible for a smaller, specialized portion of the project. The potential for profit or loss is also similar but proportionally less. The subcontractor can generally shift forces back and forth, as necessary, between various jobs and, therefore, control direct job site overhead. The general contractor, in contrast, has a fixed job site overhead that cannot be adjusted easily. The longer the general contractor is involved with a project, the more profit the job site overhead consumes. On larger projects, the job site overhead versus time of completion factor affects the major subcontractors, as well as the general contractor.

Managers

A manager is any individual or organization responsible, by contract or job description, for the completion of a project. This includes the general contractor, construction managers, project managers, and facility managers, as well as subcontractors who manage their portions of work within the project. In some cases, the architect or engineer will serve as the overall project manager.

The concepts presented in this book can be applied to any company, group, or individual operating as "project manager," or to person(s) heading up a project or staff in charge of a project. Regardless of the size of the project, these basic concepts are necessary for effective management and for keeping project costs from mushrooming out of control.

Comparison of Actual Construction Time and Effect on Profit

	Original Job Estimate	Time (in months) past original completion date						
		4	5	6	7	8	9	10
Contract Amount	$1,000,000	$1,000,000	$1,000,000	$1,000,000	$1,000,000	$1,000,000	$1,000,000	$1,000,000
Cost of Job Direct Costs, of GC Work & Subcontracted Work	$810,000	$810,000	$810,000	$810,000	$810,000	$810,000	$810,000	$810,000
Monthly Job Overhead Office and Field	$25,000	$25,000	$25,000	$25,000	$25,000	$25,000	$25,000	$25,000
Estimated Time to Complete	6	4	5	6	7	8	9	10
Total Estimated Job Overhead	$150,000	$100,000	$125,000	$150,000	$175,000	$200,000	$225,000	$250,000
% Net Profit before Taxes	4.00%	9.00%	6.50%	4.00%	1.50%	-1.00%	-3.50%	-6.00%
Estimated Net Profit	$40,000	$90,000	$65,000	$40,000	$15,000	($10,000)	($35,000)	($60,000)
Percent Increase or Decrease from Original Estimate	0.00%	125.00%	62.50%	0.00%	-62.50%	-125.00%	-187.50%	-250.00%

Not usually considered are the "lost profits" of extended completion schedules—the amount that could have been earned by the same construction team on another project. The following table gives an example of this "lost profits" factor.

Calculation of Profits Including "Lost Profits" (40,000/6)	($6,667)	$90,000	$65,000	$40,000	$8,333	($23,333)	($55,000)	($86,667)
Expressed as a Percentage of Anticipated Profits	4.00%	9.00%	6.50%	4.00%	0.83%	-2.33%	-5.50%	-8.67%

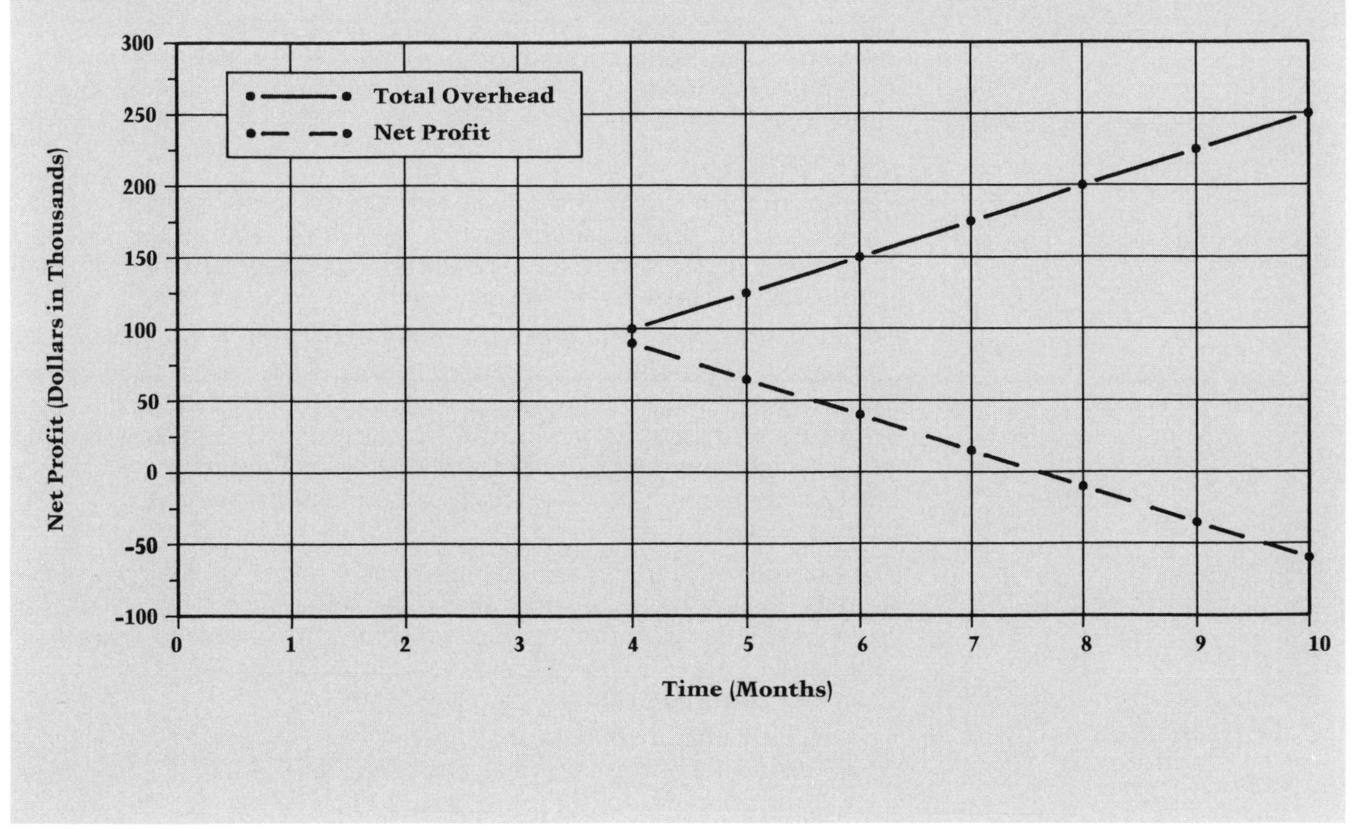

Figure 2.1

Optimizing Manpower Resources

The Problem

Construction is a highly paid and profitable industry–for those who survive. It is easy to determine whether an individual succeeds or fails, because there is little forgiveness in the system. Competent people are a valuable commodity; they are difficult to find and to keep within a firm.

Because of the uncertainty and time constraints involved in bidding, awarding, and starting jobs, there is generally not enough time to find, hire, and train new managers for each project. At the same time, the cost of good construction management people is high; therefore, it is too expensive to keep them on the "back burner," waiting for the right project for their skills.

One of the most time-demanding periods in a job is the first one-to-two months. Newly-hired people need to "hit the ground running," yet they have little knowledge of how the company likes to manage and supervise projects. The other most time-demanding period is at the end of a project. This is a point at which the manager of a construction team may falter if too many tasks vie for his time.

When, with little or no warning, a job priced six months ago is awarded, and the owner wants to start next week, few contractors would say: "No, let's not start until I find someone who can run the job." They usually say: "We're ready and waiting; in fact, why have you taken so long?"... and then go back to the office and find that all managers are busy completing their other jobs that may face liquidated damages claims if not completed on time. No one else is available, so the best manager is often saddled with more work than can be accomplished satisfactorily. The construction team cannot complete one job properly if their time is occupied by the new project. It is difficult enough to finish a project properly without the additional work of a new project competing for the available resources.

The Solution

The solution to this nightmare is to have a system–manual or computerized–that most effectively utilizes the good, experienced people, while delegating some of the routine, clerical aspects of the job to less experienced people.

It makes more sense to hire two good $10-per-hour clerical workers than one $30-per-hour or more senior person who must then get "up to speed" with the company's systems, procedures, and corporate structure, as well as doing the actual work. The goal is to get the most coverage out of experienced project managers, and give the less demanding detail work to personnel who are less experienced in construction.

For example, when preparing Submittal Requests, an experienced manager can review the specs and make some notes, and a less experienced construction person can write the Submittal Request letters and follow up on them. Given proper direction, even a newly-hired person can:

- Handle plan distribution
- Verify subcontractor prices once the job is awarded
- Confirm and identify items with long lead times
- Obtain insurance certificates
- Handle other necessary, but routine correspondence

Any project will start smoothly if an efficient system for delegating the workload is already in place. The in-house system should be efficient and flexible, to take full advantage of the capabilities and experience of the most valuable employees. With a small, efficient computer system, short-term help can be hired, and within a couple of hours, be up-to-speed and ready to start cranking out the communications.

Office Overhead

In the office, the benefits of manpower resource management are not always clear or evident as a solution to levelling out peak loads. Most companies try to function efficiently with a minimum of overhead. Low overhead is a reasonable goal, but at what cost to production? Cut it too low and you may find yourself always trying to catch up with paperwork and letting items "fall through the cracks"–which is what we are trying to avoid.

A good system, and additional clerical help, can produce a gain of as much as 30% in the work volume. Once again, try not to allow $30-per-hour people to do $10-per-hour work. For example:

> *Assume the company has a project manager on staff at $80,000/year (including fringes) who can handle $2,000,000 of work per year. With an efficient system, and clerical personnel (at a cost of an additional $20,000 per year), the combined total project management cost is $100,000, and the new gross volume made possible by that set-up should be $2,600,000.*

The increase in overhead cost versus the additional gross volume shows that the *functional ratio* (or percentage of project management overhead per construction dollar) was reduced, as work volume increased. This is, of course, desirable, but may not be the *most* important benefit of the increased productivity. The greatest benefit may be having a project management team that is better able to cover peak loads, while being flexible enough to pitch in to get a new project rolling until someone else is hired and/or trained.

This approach clearly makes sense, but is not practiced as frequently as it should be. Top construction management may more readily tolerate wasted effort in the field, because they cannot *see* people at the site all the time, and the cost is directly attributable to the job (out of sight–out of mind). On the other hand, top management sees the people in the office–at their less productive moments–a lot more often than they see inactivity at the job site. Consequently, there may be a reluctance to hire additional clerical staff because of a perceived image that office personnel are not working at full capacity.

It is also important to understand the different job functions and learning curves that apply to the office, as opposed to field personnel. A newly hired journeyman carpenter can immediately work at a high level of efficiency, despite being unfamiliar with the project. Hanging doors, constructing concrete forms, and other skills can be universally applied from one job to the next. It is more difficult to find a project manager, or assistant, who does not require some training in the procedures a particular company uses to manage projects and paperwork.

It may be best to view office project management positions and support staff as overhead, similar to that required of a fire department or the Army – the unit must be ready when needed, but will not always be working at full capacity. The project management team should be ready to "go" at all times–to handle additional new work, as well as finish the old. An efficient communication system can keep the management staff at a high level of performance. However, sufficient clerical help is needed to maintain a good system of documentation.

Summary

The key to successful manpower management is arriving at that fine line between too much and too little. Too much and the overhead erodes profit; too little and it will not allow for smooth-running projects, nor the capacity to take on unexpected work, again reducing profits. When new work cannot be handled properly, potential losses due to mismanagement of one job can far outweigh the annual cost of implementing an effective communication system and its necessary clerical support.

Part Two

General Communications

Chapter Three
Transmittals

A Transmittal is a short letter or memo. It may be used as a communication in and of itself, or as a cover letter for additional correspondence referred to, attached, or enclosed. The information referenced in the Transmittal may be already in the hands of the addressee, such as construction plans or specifications of a particular project.

There are two basic types of Transmittals.

- Transmittals for the addressee's information or action
- Transmittals for the addressee's attention and response

Purpose

To communicate, according to *Webster's Ninth New Collegiate Dictionary,* is *to convey* or *transmit information.* To many people in the construction industry, a Transmittal means a cover letter to which something is attached, like a roll of plans or material samples. However, the Transmittal format is also valid for letters. It can and should be used for transmitting information, or requesting information, whether it conveys the message itself or has other documents attached. The reason for assigning an identification number (for example, TR #018) to a Transmittal is to have a way of listing these documents, along with a short description. If an answer is expected, the identification number and a short description can be referenced and logged, making it easier to follow up.

For the general contractor, architect, engineer, subcontractor, and facility manager, the Transmittal concept is the same. Each of these parties is in the position of being responsible *to* someone as well as *for* someone else's work. For example:

The facilities department is responsible to the Vice President of Operations while charged with remodeling or building a new computer room. The facilities manager must contract with the architect, engineer, and general contractor (or directly with subcontractors). Using the Transmittal format and implementing the procedures recommended in this book, the facilities manager can document (and provide copies) to all concerned if and when the job is not done on time. The problem may have been indecision on the part of the client (computer department), or a shortage of material or design time. In this case, let us assume that the delay was a request for a change in the electrical characteristics for the switchgear supplying power to the mainframe computer.

If the request for information is not documented, the computer department may say: "We gave facilities all of the information they needed. They never told us they needed to know the electrical characteristics of the mainframe that we've got on order. The computer is going to be here in two weeks! What do you mean the switchgear won't be here for three months?"

Never mind that the order for the computer was not placed until yesterday. The facilities group is supposed to have the space ready next week. It happens all too often.

In this case, the facilities department requested the electrical characteristics several months in advance, but had not received it. The facilities manager should have followed up and forced the decision, but that, very often, is like trying to push a chain across the street. As a result, the project will not be completed on time. If facilities cannot get the information, they should at least document their attempts to do so.

Keeping the Records Together

Because the construction industry is job- or project-oriented (a specific contract to perform a specific scope of work), all records should be kept by the job, too. For example, the insurance records of various subcontractors who have worked on a variety of different projects should not be filed together in one folder. Each certificate should be filed according to the job to which it applies.

This does not mean that project information should be filed *on the job site.* It means that when information is collected, it should be identified and filed by the job. A search through five different file cabinets should not be necessary in order to find information about Job No. 321, Gibson Office Building. It often takes thirty minutes just to locate a piece of data and only five minutes to review it. Even worse is the case where the information about a situation is not reviewed at all, simply because it is *assumed* that the data cannot be found.

The suggestions presented in this book for logs, exception logs, and due logs can eliminate wasted time spent searching for an item. This will make the whole process quicker and easier, partly because people will be more likely to look something up if it is easier to find–data they may have ignored in a clumsier system. With so many pressures in the business, it is easy to let the items that are inefficient, and a pain in the neck, slip to some future date where they will probably never get done.

Transmittals for Special Functions

Unique Transmittals used for specific functions are named something other than *Transmittal.* They are intended for one special function category only, and are generally named according to that special function. The following special function Transmittals are covered in detail in other chapters of this book.

- Requests for Information
- Submittal Requests
- Submittal to Architect
- Returned Submittals
- Price Requests, and Price Request Quote to Owner
- Field Order, and Field Order Quote to Owner
- Change Log, and Change Quote to Owner Log
- Request for Change, and Request for Change Quote to Owner

Each of these unique Transmittal types has its own set of logs and reports. Each represents a specific type of communication; therefore, they should not be combined. The remainder of this chapter is concerned with the "universal" type of Transmittal, used for a multitude of communication tasks.

General Uses of Transmittals

Often, a short letter or memo must be sent to a number of different people involved in a project. There might even be occasion to send a Transmittal to the insurance company (for example, regarding a job site accident). Everyone involved in a construction project should be aware of the possibility that they may be called upon to provide evidence for a case many years later. It is, therefore, important to keep all of the records that concern a job *with the Job File*. For instance, do not keep the accident records in a Safety file only. Also keep a copy with the job records. Keep information on safety meetings, accidents, and witnesses with the job records so that 15 years from now when the case goes to trial, the information is in one place and easy to find. Remember, in the case of negligence, there may be no statute of limitations on liability.

The "Smoking Gun"

Be careful of the "smoking gun" in record keeping. The discovery of a smoking gun seems to indicate someone must have been shot. However, such a conclusion, based on incomplete information, can be very dangerous. The same applies to record keeping, where one document might refer vaguely to the possibility of damaging evidence in another document that cannot be found. All relevant information should be kept within one physical area. In fact, having no information available may be better than having just part of the information, although that approach is never recommended. If the system is computerized, all correspondence, Transmittals, Job Minutes, and Submittals will be stored on media such as "hard" or "soft" disks, or tape, as well as the printed, or "hard" copies. Duplicates should be made, and one set kept off-site, after the project is completed.

Those firms that are not planning to be in business five years hence need not worry about careful record keeping. However, those that are planning to stay in business long-term should review all record keeping systems–not just the accounting records–and make sure their procedures are adequate.

Use of Transmittal Copies

When sending an insurance company a Transmittal regarding a job site accident, indicate on the original the other parties who are receiving copies. In this case, the letter would be addressed to the insurance carrier, and copies should be distributed to:

- The insurance file (all accident reports)
- The job file (accidents only on that specific job)
- The route file sent to "need to know" individuals in the organization
- The field superintendent

If the data base of construction correspondence is computerized, the insurance agency can be a name in that data base. The stored, computerized data will be in one place (with the job) and it will show that both Insurance Company X and the insurance file received a copy. All the Transmittal text, as well as the record of copies sent, will be stored permanently.

The examples of general topic Transmittals in Figures 3.1a through 3.1d show the flexibility of the Transmittal. Figures 3.1a and 3.1c are computer-generated, while 3.1b and 3.1d are pre-printed forms filled out by hand. These examples are intended to show the appropriate items to include in the Transmittal.

Transmittals as Cover Letters

Between the very specific and very general uses of a Transmittal are many other situations that benefit from the Transmittal format. These documents do not otherwise have any specific place in the system for recording when, where, or to whom they were sent. More importantly, without the Transmittal, there is no easy way of tracking whether or not a response was ever returned. The advantages of an otherwise careful documentation tracking system can be lost if these letters and memos cannot be followed through.

At some point, a decision must be made to draw the line on the number of specific letter and log types to maintain. The following factors should influence this decision.

- How the company or department is structured
- The complexity of the job
- The type of job
- The type of work performed by the company or department

The same correspondence structure should be used for each job. A 500 S.F. office remodel may not require a large number of Transmittals for each of the special functions to be accomplished, such as Submittals, Field Orders, etc. However, each functional *type* of correspondence will be required. For instance, there may not be a large number of Price Requests, but there may be quite a few Field Orders issued. Using the same system on all jobs not only ensures consistency, but also provides a reliable system that is especially helpful when a small job grows into a large one. In any case, whether the job is large or small, proper documentation is the key to avoiding legal and other problems.

On a very large or complex job, subsidiary logs or lists are useful. The Transmittal function can and should be used to keep track of due dates and other pertinent information in the text of the Transmittal. Contracts, purchase orders, and backcharges are typical categories of Transmittal correspondence that refer to specific types of records and subsidiary lists or logs.

One of the principles of effective communications is to be able to determine who has *not* done which tasks. Utilize the Transmittal format to keep track of the information sent out as well as who has not responded. Two examples of correspondence for which the Transmittal would serve as a cover letter for a "subsidiary log" are contracts to subcontractors, and Change Orders to subcontractors and material suppliers. Contracts, purchase orders, and backcharges are reviewed later in this chapter.

Transmittals contain all the subject matter of a normal letter. They are generally formatted for short text messages and do not lend themselves to multi-paragraphed, or multi-paged, documents. This is true for manual (handwritten or typed) Transmittal letters, as well as for computerized printouts.

DELTA CONSTRUCTION COMPANY

6065 Mission Gorge Road, #193
San Diego, CA 92120
(619) 582-5829

*** TRANSMITTAL ***

Date: 2/18/90

Job Number: 321
Transmittal Number: 5

TO: STRONG STRUCTURAL STEEL
4567 IRON STREET
SAN DIEGO, CA 92312

ATTENTION: ROBERT STRONG

JOB NUMBER: 321 GIBSON OFFICE BUILDING

FROM: J. EDWARD GRIMES

SUBJECT: TRANSMITTAL NUMBER: 5
CPM SCHEDULE

We are transmitting to you the following which is AS LISTED BELOW for your RESPONSE REQUIRED by: 2/27/90.

I am preparing the construction CPM schedule for this project. Please advise me of any long lead items that may affect the schedule. I'm planning to present the schedule to the client no later than 3/2, so I will need your input no later than 2/27. Give me a call if you need to discuss anything. If you don't tell me your problems now after 3/2 it will be too late.

Please contact me as soon as possible if you need any further information or if you have any questions.

Sincerely,

J. Edward Grimes

Copies were sent to the following:

QUICK PLUMBING COMPANY FOR RESPONSE (ORIGINAL)
COOL AIR CONDITIONING FOR RESPONSE (ORIGINAL)
CLEAR GLASS COMPANY FOR RESPONSE (ORIGINAL)

dp: 2/18/90 16:55 **PaperWorks**, a Construction Software Program

Figure 3.1a

Means Forms

LETTER OF TRANSMITTAL

Delta Construction Company
6678 Mission Road
San Diego, CA 92120
(619) 581-6315

FROM: J. Edward Grimes
Project Manager

DATE 2/18/90
PROJECT Gibson Office Bldg. #321
LOCATION San Diego, Ca.
ATTENTION Robert Strong
RE: CPM Schedule

TO: Strong Structural Steel
4567 Iron St.
San Diego, Ca. 92312

TR 321.05

Gentlemen:

WE ARE SENDING YOU ☐ HEREWITH ☐ DELIVERED BY HAND ☐ UNDER SEPARATE COVER

VIA ______ THE FOLLOWING ITEMS:

☐ PLANS ☐ PRINTS ☐ SHOP DRAWINGS ☐ SAMPLES ☐ SPECIFICATIONS

☐ ESTIMATES ☐ COPY OF LETTER ☐ ______

COPIES	DATE OR NO.	DESCRIPTION
CC		Quick Plumbing
CC		Cool Air Conditioning
CC		Clear Glass Company

THESE ARE TRANSMITTED AS INDICATED BELOW

☐ FOR YOUR USE ☐ APPROVED AS NOTED ☐ RETURN ______ CORRECTED PRINTS

☐ FOR APPROVAL ☐ APPROVED FOR CONSTRUCTION ☐ SUBMIT ______ COPIES FOR ______

☐ AS REQUESTED ☐ RETURNED FOR CORRECTIONS ☐ RESUBMIT ______ COPIES FOR ______

☒ FOR REVIEW AND COMMENT ☐ RETURNED AFTER LOAN TO US ☐ ~~FOR BIDS~~ DUE 2/27

☐ ______

REMARKS: I am preparing the construction CPM schedule for this project. Please advise me of any long lead time items that may affect the schedule. I'm planning to present the schedule to the client no later than 3/2, so I'll need your input no later than 2/27. Give me a call if you need to discuss anything. If you don't tell me your problems now, after 3/2 it will be too late.

IF ENCLOSURES ARE NOT AS INDICATED, PLEASE NOTIFY US AT ONCE.

SIGNED: J Edward Grimes

Figure 3.1b

DELTA CONSTRUCTION COMPANY

6065 Mission Gorge Road, #193
San Diego, CA 92120
(619) 582-5829

*** TRANSMITTAL ***

Date: 2/18/90

Job Number: 321
Transmittal Number: 6

TO: STRONG STRUCTURAL STEEL
4567 IRON STREET
SAN DIEGO, CA 92312

ATTENTION: ROBERT STRONG

JOB NUMBER: 321 GIBSON OFFICE BUILDING

FROM: J. EDWARD GRIMES

SUBJECT: TRANSMITTAL NUMBER: 6
ANCHOR BOLTS

We are transmitting to you the following which is AS LISTED BELOW for your ACTION.

Bob, we're going to need those anchor bolts no later than March 3. Make sure that they're on the job by that time, we can't get behind at this point.
Call me if you have a problem.

Please contact me as soon as possible if you need any further information or if you have any questions.

Sincerely,

J. Edward Grimes

Figure 3.1c

Means Forms

LETTER OF TRANSMITTAL

Delta Construction Company
6678 Mission Road
San Diego, CA 92120
(619) 581-6315

FROM: J. Edward Grimes
Project Manager

DATE 2/18/90
PROJECT Gibson Office Bldg. #321
LOCATION San Diego, Ca.
ATTENTION Robert Strong
RE: Anchor Bolts

TR- 321.06

TO: Strong Structural Steel
4567 Iron Street
San Diego, Ca. 92312

Gentlemen:

WE ARE SENDING YOU ☐ HEREWITH ☐ DELIVERED BY HAND ☐ UNDER SEPARATE COVER

VIA ______ THE FOLLOWING ITEMS:

☐ PLANS ☐ PRINTS ☐ SHOP DRAWINGS ☐ SAMPLES ☐ SPECIFICATIONS

☐ ESTIMATES ☐ COPY OF LETTER ☐ ______

COPIES	DATE OR NO.	DESCRIPTION

THESE ARE TRANSMITTED AS INDICATED BELOW

☒ FOR YOUR USE ☐ APPROVED AS NOTED ☐ RETURN ______ CORRECTED PRINTS

☐ FOR APPROVAL ☐ APPROVED FOR CONSTRUCTION ☐ SUBMIT ______ COPIES FOR ______

☐ AS REQUESTED ☐ RETURNED FOR CORRECTIONS ☐ RESUBMIT ______ COPIES FOR ______

☐ FOR REVIEW AND COMMENT ☐ RETURNED AFTER LOAN TO US ☐ FOR BIDS DUE ______

☒ Action

REMARKS: We're going to need those anchor bolts no later than March 3 - Make sure they're on the job by that time; we can't get behind at this point.

Call me if you have a problem.

IF ENCLOSURES ARE NOT AS INDICATED, PLEASE NOTIFY US AT ONCE.

SIGNED: J Edward Grimes

Figure 3.1d

The following are the primary reasons to issue a Transmittal cover letter in place of a letter that could stand on its own, such as a report, the job minutes, etc.

- A stand-alone letter has no identification other than the subject matter or date.
- If a Transmittal referencing other documentation is used, there is a place in the system of logs for tracking the communication.
- There is a secondary source of documentation of where, what, when, why, and how the original letter was sent.
- The logs can be carried to the job site to be used as a convenient reference to information about the job.

In the not-too-distant future, all of this data will probably be carried in a portable, lap-top computer for almost instant access. The computer technology exists and the software is available. It is just a matter of getting the general contractor, architect, engineer, and consultants to realize what a benefit this tool can be.

Transmittals Used as Memos, Letters, or Telephone Calls

An easy-to-use Transmittal form and procedure will be used more consistently and will provide better documentation than either conversations or telephone calls. Memos or letters, in Transmittal format, may also be more effective. A Transmittal does not allow the recipient to experience "selective memory," as it covers, *in writing*, all relevant issues. Used properly, the submittal advises the addressee of a potential problem or subject in a timely, clear way.

If the electrical contractor in the example (Figures 3.2a and 3.2b) fails to comply with the rules set forth by the general contractor, this could be grounds for termination of the subcontract or other disciplinary action.

Please note that while Transmittals have great potential to improve project management, an *easy-to-use* Transmittal will be used far more and will provide better control and documentation. Keep this in mind when drawing up the format for Transmittals.

Written Transmittal vs. a Telephone Call

In most cases, it is better to use a written Transmittal than a telephone call for several reasons. The most obvious is that the subject matter is documented and it can be proved that notification was given and when it was given. Another advantage is that, in many cases, a Transmittal is significantly less expensive than a telephone call. Many telephone conversations easily drift away from the original discussion (the redesign of the pond, for example) to other subjects, such as why the gas and electric company is demanding the right-of-way, or how the redesign problem can be solved in detail. This represents a lot of time wasted in speculation and/or design questions that should be resolved by the landscape architect.

DELTA CONSTRUCTION COMPANY

6065 Mission Gorge Road, #193
San Diego, CA 92120
(619) 582-5829

*** TRANSMITTAL ***

Date: 2/11/90

Job Number: 321
Transmittal Number: 3

TO: SPARKS ELECTRIC COMPANY
7654 WIRE STREET
SAN DIEGO, CA 92343

ATTENTION: ED LIGHT

JOB NUMBER: 321 GIBSON OFFICE BUILDING

FROM: J. EDWARD GRIMES

SUBJECT: TRANSMITTAL NUMBER: 3
DRINKING ON THE JOB

We are transmitting to you the following which is AS LISTED BELOW for your ACTION.

> There will be no drinking allowed on the job site. This includes the parking lot across the street which we are renting. Neither your insurance nor ours will allow such activity. Make sure that you inform all men working on the job.

Please contact me as soon as possible if you need any further information or if you have any questions.

Sincerely,

J. Edward Grimes

Copies were sent to the following:

QUICK PLUMBING COMPANY	FOR ACTION (ORIGINAL)
LOT'S OF REBAR	FOR ACTION (ORIGINAL)
FRANCES GIBSON	FOR INFORMATION (ORIGINAL)

dp: 2/11/90 16:54 **PaperWorks**, a Construction Software Program

Figure 3.2a

Means Forms

LETTER OF TRANSMITTAL

Delta Construction Company
6678 Mission Road
San Diego, CA 92120
(619) 581-6315

FROM: J. Edward Grimes
Project Manager

TO: Sparks Electric Company
7654 Wire Street
San Diego, Ca. 92343

DATE 2/11/90
PROJECT Gibson Office Bldg. #321
LOCATION San Diego, Ca.
ATTENTION Ed Light
RE: drinking on the job

TR 321.3

Gentlemen:

WE ARE SENDING YOU ☐ HEREWITH ☐ DELIVERED BY HAND ☐ UNDER SEPARATE COVER

VIA ________________ THE FOLLOWING ITEMS:

☐ PLANS ☐ PRINTS ☐ SHOP DRAWINGS ☐ SAMPLES ☐ SPECIFICATIONS

☐ ESTIMATES ☐ COPY OF LETTER ☐ ________________

COPIES	DATE OR NO.	DESCRIPTION
CC		Quick Plumbing
CC		Lot's of Rebar
CC		Owner

THESE ARE TRANSMITTED AS INDICATED BELOW

☒ FOR YOUR USE ☐ APPROVED AS NOTED ☐ RETURN ________ CORRECTED PRINTS

☐ FOR APPROVAL ☐ APPROVED FOR CONSTRUCTION ☐ SUBMIT ________ COPIES FOR ________

☐ AS REQUESTED ☐ RETURNED FOR CORRECTIONS ☐ RESUBMIT ________ COPIES FOR ________

☐ FOR REVIEW AND COMMENT ☐ RETURNED AFTER LOAN TO US ☐ FOR BIDS DUE ________

☒ Action

REMARKS: There will be no drinking allowed on the job site. This includes the parking lot across the street that we are renting. Neither your insurance nor ours will allow such activity. Make sure that you inform all men working on the job.

IF ENCLOSURES ARE NOT AS INDICATED, PLEASE NOTIFY US AT ONCE.

SIGNED: J Edward Grimes

Figure 3.2b

The written Transmittal, on the other hand, provides the other party with documentation that can be used to formulate a plan of action. Figure 3.3 compares the direct cost of a telephone call to that of a written Transmittal, using a correspondence program.

The advantages of the Transmittal form over a telephone call are significant, providing a system exists in-house that can get a Transmittal out and into the mail within one day. Naturally, if it takes three days to get the Transmittal out of the building, then the impact and relevance are lost. This is where a computerized system of construction correspondence can really pay off. It makes it easy to get those Transmittals written, out of the office, and into the hands of the recipient. The Transmittal's advantages are summarized below.

- There is a direct time savings on the part of the manager.
- The manager does not get caught up in "telephone tag" (i.e., "I called you last, therefore, you are "it" and must now call me.") Telephone tag can waste so much time that the subject of the call may be forgotten by the time the connection is finally made.
- There is a dollar savings in the actual method used to convey the information.
- It allows the manager to do other things. In the example, there is a time savings of nine minutes, which the manager can use to accomplish other necessary tasks.
- The manager can write the Transmittal during "off" hours, i.e., early in the morning, late at night, or even dictating the message while out of the office.
- Documentation is created.
- If required, follow-up may be tracked in the form of logs, and is not left to memory alone.

Transmittals as Memos or Letters

Figure 3.4 shows some typical subjects for Transmittals used as a memo or a letter. This figure shows the addressee, the sender, and subject matter, without the "stock" part of the Transmittal.

Comparison of Mailed Transmittal Versus Phone Call

Description	Phone Call	Transmittal
Manager's Time (12 min/3 min)	$ 6.00	$ 1.50
Clerical Time (0 min/12 min)	$ .00	$ 2.00
Phone Call/Stamps/Supplies	$ .40	$ .35
	$ 6.40	**$ 3.85**

Figure 3.3

****

To: AAA Steel Fabricators
From: Goodenchepe Construction Co.
Subject: Anchor bolts for footings
Date: February 27, 1990

We're going to need those anchor bolts for footing types F-2, F-3, F-4, 5, 6, 10, no later than 3/15. Make sure that they're on the job by then. We've got concrete scheduled for 3/17 and that only gives us a day to set templates."

cc: job site

****

To: I. Tower
From: Bildenhope Construction, Inc.
Subject: Notice of occupancy
Date: October 23, 1989

As of October 18, 1989, all permits have been signed off and the city of San Diego has cleared the building to be used for the purpose designed. A copy of the signed-off building permit is enclosed for your records.

cc: Architect
cc: Lender

****

To: M. Greene Assoc.
From: Gauley Construction
Subject: Beginning of the warranty period
Date: October 3, 1989

As of October 1, 1989, all equipment has been tested and inspected. Your forces have used the HVAC, elevator, and other equipment since that date. For warranty purposes, the starting date will be October 1, 1989, even though the Notice of occupancy has not been obtained.

cc: Architect
cc: Affected subcontractors

Figure 3.4

```
****************************************************

To:          Breymann Decorators
From:        Moyland Contractors
Subject:     Punch list dated September 15, 1989
Date:        September 18, 1989

Enclosed is the punch list of this project dated 9/15/89. Please
correct all items that affect your work and advise either the job
superintendent or this office when the work is complete. The
remedial work must be accomplished not later than October 1, 1989.

****************************************************

To:          N. Smit
From:        Chandler Construction
Subject:     Possible Labor Dispute
Date:        November 6, 1989

We want to confirm that we are encountering delays to the project
due to strike conditions with the Plumber's Notice to Owner of
labor disputes. This strike is impacting our construction schedule.
As soon as the labor dispute is settled, we will inform you of the
estimated impact on the completion schedule.

****************************************************

To:          Tonsa Rebar Co.
From:        G. Adams Contractors
Subject:     Delivery of rebar to job site
Date:        November 22, 1989

You are 2 to 3 days late getting the rebar from your fab shop to
the job in order to get the rebar in place. Please improve your
delivery relative to our agreed-on schedule. Right now you are
holding us up.

cc: Job site
```

Figure 3.4 (cont.)

****

To: Broecken Glass Co.
From: Fawdoun Construction
Subject: Delivery of glass to job site
Date: October 13, 1989

Please confirm delivery of high performance glass for no later than Nov. 17, 1989. The delivery of the glass and your installation performance is critical to the completion of this job. If you want to get paid on time, please have supplier confirm delivery in writing. Also make sure you have insurance on stored materials; the bank will ask for it.

cc: Job site
cc: Accounts payable (in house)

****

To: Moss Melman Assoc.
From: Gallo Construction
Subject: Finish Schedules
Date: April 9, 1990

Lack of selection of the finish colors for all materials will impact our construction schedule if we do not have the information as shown on the schedule by next week, April 16. This delay may impact our overhead costs and might require air freight or other means to complete the project on schedule.

cc: Owner
cc: Job superintendent

Figure 3.4 (cont.)

Computer Applications

By maintaining a computer data base of Transmittals, two functions occur without the conscious participation of the clerical staff. First, the data in the Transmittal now exists in more than one place: on the desks of the recipients and in the computer data base of Transmittals for this job. Second, merely by using the computerized Transmittal system, the Job Record and Job Logs may be constructed. The computer tracks the date sent, the subject, and whether or not a response was requested and required.

Items to be Included in all Transmittals

The following items should be included in *all* Transmittals. Therefore, this same information, in abbreviated form, should be included in the Transmittal Summary Log.

- Date
- To
- From
- Job identification
- Transmittal number and identification
- Text of Transmittal
- Date response required (if any)
- Names of those to whom copies were sent

Subcontracts and Purchase Orders Transmittals

Contracts to subcontractors, consultants, and vendors should be tracked and followed up to verify that the contract has been returned. Changes to the subcontract must also be signed and returned. The general contracting firm can get into serious difficulty if either the original contract or the changes to the contract are not signed and/or initialled and returned.

The Transmittal as a cover letter (with a due date for return or change to subcontract) is an excellent tracking device, even though a subsidiary list or log may also be kept. A more complete description of the Transmittal as a cover letter for subcontractors and purchase orders is presented in Chapter 10, "Subcontracts."

Backcharges Transmittals

Backcharges are an unending source of problems. In fact, there would not be a backcharge if there was not a problem in the first place. The whole matter is complicated by the fact that the timing, as well as the scope, of the backcharge, tends to be interpreted differently by each involved party.

A backcharge may be defined as: "a charge against an individual or a firm that did not perform the work required by their contract or subcontract. The work was, therefore, necessarily accomplished by a third party, and the cost of that work attributed to the original contract or subcontract."

The following are examples of typical backcharges.

- A subcontractor does not clean up adequately; therefore, the subsequent trade cannot proceed with work in the same area. Hired personnel must, therefore, do the work in order for the project to continue on schedule.
- Punch list items are not completed. Consequently, the general contractor must use its own labor forces or hire someone else to perform or complete the work.

- The general contractor "loans" labor to the subcontractor to unload a delivery when the subcontractor is not on the site to receive it or does not have adequate labor.

Since the backcharge is usually issued on the basis of the "golden rule" (in this case, "he who has the gold, gets to rule"), backcharges are open to the sort of argument, discussion, and threats that go along with one-sided decisions. Many backcharges made in the field are made in a fit of anger, and cannot be substantiated. Valid backcharges should be documented and sent with a Transmittal to the individual or company being charged as soon as possible.

On some projects, all backcharges are left to the very end of the job. Usually, the company being backcharged does not find out about the backcharge until it is withheld from the final payment. In some cases, this approach may be effective for the owner because of poor documentation on the part of the contractor. However, if it goes to court, the backcharge may not withstand scrutiny, primarily because it was not timely, was arbitrary, and the company being backcharged had no chance to properly present their side of the case and defend themselves. Again, it is always important to give the other party a chance to say "No." If they do not object or argue, then the backcharge is assumed to be valid and can be included in the accounting process.

Sometimes, when backcharges are saved until the end of the job, it may be discovered that a particular backcharge was not valid and should have instead been the subject of a Field Order to convey a change in scope of work to the owner. By processing the backcharges when they occur, two things are accomplished: (1) the backcharge is brought out into the open and the subcontractor is made aware of it, and (2) the accuracy of the backcharge can be verified prior to the completion of the job accounting.

When all of the backcharges are "saved" until the end of the job, both the valid backcharges and the invalid backcharges may not be processed because of a failure to notify the involved parties. It is very important that the party being backcharged be notified promptly of the backcharge, complete with all relevant details and an accounting of the dollars spent. This notification is easily accomplished using a Transmittal as a cover letter documenting the backcharge (which may or may not have a numerical designation). By using the Transmittal (as illustrated in Figures 3.5a and 3.5b) to convey the backcharge information, the following is recorded.

- The recipient
- The date it was sent
- The major topic
- The date an answer is due to be returned
- Additional information included in the standard Transmittal form

DELTA CONSTRUCTION COMPANY

6065 Mission Gorge Road, #193
San Diego, CA 92120
(619) 582-5829

*** TRANSMITTAL ***

Date: 2/28/90

Job Number: 321
Transmittal Number: 19

TO: QUICK PLUMBING COMPANY
7654 32ND STREET
SAN DIEGO, CA 92110

ATTENTION: ED WATERHOUSE

JOB NUMBER: 321 GIBSON OFFICE BUILDING

FROM: J. EDWARD GRIMES

SUBJECT: TRANSMITTAL NUMBER: 19
BACKCHARGE FOR CLEAN UP

We are transmitting to you the following which is ENCLOSED/ATTACHED for your ACTION.

Backcharge #321.011, signed by your field representative is for cleaning up your work area on the first floor. The pricing detail appears on the enclosed backcharge form in the amount of $599.00. This amount will be deducted from your account.

Please contact me as soon as possible if you need any further information or if you have any questions.

Sincerely,

J. Edward Grimes

dp: 2/28/90 11:14 **PaperWorks**, a Construction Software Program

Figure 3.5a

Means Forms

LETTER OF TRANSMITTAL

Delta Construction Company
6678 Mission Road
San Diego, CA 92120
(619) 581-6315

FROM: J. Edward Grimes
Project Manager

DATE 2/28/90
PROJECT Gibson Office Bldg. #321
LOCATION San Diego, Ca.
ATTENTION Ed Waterhouse
RE: Backcharge for clean up

TO: Quick Plumbing Co.
7654 32nd St.
San Diego, Ca. 92110

TR 321.19

Gentlemen:

WE ARE SENDING YOU ☒ HEREWITH ☐ DELIVERED BY HAND ☐ UNDER SEPARATE COVER

VIA ______ THE FOLLOWING ITEMS:

☐ PLANS ☐ PRINTS ☐ SHOP DRAWINGS ☐ SAMPLES ☐ SPECIFICATIONS

☐ ESTIMATES ☐ COPY OF LETTER ☒ Backcharge

COPIES	DATE OR NO.	DESCRIPTION
1	321.011	Back charge

THESE ARE TRANSMITTED AS INDICATED BELOW

☒ FOR YOUR USE ☐ APPROVED AS NOTED ☐ RETURN ______ CORRECTED PRINTS

☐ FOR APPROVAL ☐ APPROVED FOR CONSTRUCTION ☐ SUBMIT ______ COPIES FOR ______

☐ AS REQUESTED ☐ RETURNED FOR CORRECTIONS ☐ RESUBMIT ______ COPIES FOR ______

☐ FOR REVIEW AND COMMENT ☐ RETURNED AFTER LOAN TO US ☐ FOR BIDS DUE ______

☐ ______

REMARKS: Backcharge #321.011 signed by your field representative, is for cleaning up your work area on the first floor. The pricing detail appears on the enclosed backcharge form in the amount of $599.00. This amount will be deducted from your account.

SIGNED: J Edward Grimes

IF ENCLOSURES ARE NOT AS INDICATED, PLEASE NOTIFY US AT ONCE.

Figure 3.5b

Insurance Transmittals

It seems that each individual organization in the construction industry has a different method for organizing and filing its own insurance records and the certificates of insurance of contractors, subcontractors, consultants, and architects. It does not really matter what record keeping method is used, or who does it; the important thing is that each person or group on the job – the consultant, the contractor, the subcontractor, or anyone who works on the job– must be able to verify that he or she has adequate insurance. Lack of appropriate insurance can be a disaster. Each member of the management team should verify that the companies or individuals they contract or allow to work on the site have the appropriate insurance coverage.

The Transmittal letter is an ideal medium for requesting the appropriate insurance information. By keeping this data listed with the job, the request and return can be noted automatically. In addition to the "normal" insurance requirements dictated by the contract, make sure that the insurance company includes the Certificate of Insurance "notice of cancellation" clause. The following is a breakdown of the responsibilities of the owner, architect, engineer, facilities management department, and general contractor regarding maintenance of insurance coverage and documentation.

The **owner** should make sure that the architect, engineer, and general contractor have adequate insurance coverage. The architect and engineer should have public liability and property damage insurance, as well as errors and omissions coverage.

The **facilities management department** should have a valid insurance certificate for any work that is contracted and for anyone else who actually performs any work on the site. This includes public liability and property damage, auto, and Workers' Compensation Insurance. If the facilities management department contracts for design work, they will also need insurance on the architect, engineer, and general contractor, as required.

The **architect and engineer** must have on file a certificate of insurance for Personal Liability and Property Damage and, most importantly, a certification of the errors and omissions coverage of the consultants who will be working on the project. It may sound a bit excessive, but the architect/engineer should also ask for certification of the consultant's state registration.

The **general contractor** may choose to keep a master insurance file of all subcontractors. He can then request insurance certificates only for those who do not have one on file or who are close to the end of their policy period. It is recommended that an individual certificate of insurance be kept for each subcontractor on each job. *Make sure that the insurance company is obligated to provide notification if the policy is cancelled.*

In the frenetic atmosphere of a new job starting up, it is all too easy to forget to check the insurance file in order to find out if the subcontractor's coverage is up to date and will remain so throughout the course of the job. It is actually simpler (as well as more efficient and accurate) to have someone send out a Transmittal requesting a certificate of insurance for each job, as shown in Figures 3.6a and 3.6b.

DELTA CONSTRUCTION COMPANY

6065 Mission Gorge Road, #193
San Diego, CA 92120
(619) 582-5829

*** TRANSMITTAL ***

Date: 2/06/90

Job Number: 321
Transmittal Number: 1

TO: COOL AIR CONDITIONING
7776 MISSION GORGE PLACE
SAN DIEGO, CA 92344

ATTENTION: LARRY ICEBERG

JOB NUMBER: 321 GIBSON OFFICE BUILDING

FROM: J. EDWARD GRIMES

SUBJECT: TRANSMITTAL NUMBER: 1
INSURANCE REQUIREMENTS

We are transmitting to you the following which is AS LISTED BELOW for your RESPONSE REQUIRED by: 2/20/90.

> You must have insurance certificates on file with our office prior to starting work on the job site. NO EXCEPTIONS. The minimum amount and types of insurance required are summarized on the attached Insurance Information Sheet. You must name Delta Construction Company as additionally insured.

Please contact me as soon as possible if you need any further information or if you have any questions.

Sincerely,

J. Edward Grimes

Copies were sent to the following:

LOT'S OF REBAR	FOR RESPONSE (ORIGINAL)
QUICK PLUMBING COMPANY	FOR RESPONSE (ORIGINAL)
STRONG STRUCTURAL STEEL	FOR RESPONSE (ORIGINAL)
FRANCES GIBSON	FOR INFORMATION (ORIGINAL)

dp: 2/06/90 16:52 **PaperWorks**, a Construction Software Program

Figure 3.6a

Means Forms

LETTER OF TRANSMITTAL

Delta Construction Company
6678 Mission Road
San Diego, CA 92120
(619) 581-6315

FROM: J. Edward Grimes
Project Manager

DATE 2/06/90
PROJECT Gibson Office Bldg. #321
LOCATION
ATTENTION Ed Waterhouse
RE: Insurance Certificate

TO: Quick Plumbing Company
7654 32nd Street
San Diego, Ca. 92110

TR 321.01

Gentlemen:

WE ARE SENDING YOU ☒ HEREWITH ☐ DELIVERED BY HAND ☐ UNDER SEPARATE COVER

VIA ______ THE FOLLOWING ITEMS:

☐ PLANS ☐ PRINTS ☐ SHOP DRAWINGS ☐ SAMPLES ☐ SPECIFICATIONS

☐ ESTIMATES ☐ COPY OF LETTER ☐ ______

COPIES	DATE OR NO.	DESCRIPTION
1		Specification regarding insurance requirements

THESE ARE TRANSMITTED AS INDICATED BELOW

☒ FOR YOUR USE ☐ APPROVED AS NOTED ☐ RETURN ______ CORRECTED PRINTS

☐ FOR APPROVAL ☐ APPROVED FOR CONSTRUCTION ☐ SUBMIT ______ COPIES FOR ______

☐ AS REQUESTED ☐ RETURNED FOR CORRECTIONS ☐ RESUBMIT ______ COPIES FOR ______

☒ FOR REVIEW AND COMMENT ☐ RETURNED AFTER LOAN TO US ☐ FOR BIDS DUE ______

☐ ______

REMARKS: You must have insurance certificates on file in our office prior to starting work on the job site. NO EXCEPTIONS. The attached specs indicate the limits necessary. Please name us and the owner as being additionally insured on your policy and include a "30 day notice of termination or cancellation" clause

IF ENCLOSURES ARE NOT AS INDICATED, PLEASE NOTIFY US AT ONCE.

SIGNED: J Edward Grimes

Figure 3.6b

The certificates of insurance can be evaluated, and the length of the insurance coverage verified, at the time the certificate is received. All of these certificates can now be filed in one place (and kept there) with the job, and then archived with the other job records when the project is complete.

General Information Transmittals

The Transmittal is the ideal format for getting general information to the companies involved on a project. The use of the Transmittal format provides a record of the recipient of this information and the date sent. Normally, no "due date" is placed on the general "for your information" type of correspondence. Figure 3.7 shows some examples (subject matter only) of Transmittals used to relay project information.

Sample Computer Logs for Transmittals

Figures 3.8a, b, and c are pages from typical computer program report logs for Transmittals. Note that the log shown in Figure 3.8b records only the Transmittals that require responses, while the report shown in Figure 3.8a is from a log that shows all the detail of all the Transmittals ever sent out on the project. Figure 3.8c is a report of those Transmittals for which responses are delinquent.

Figure 3.8d is an example of a manual log showing the same information that is included in the computer-type logs shown in Figures 3.8a and 3.8b. The manual log is adequate if maintained on a regular basis and if there are not a lot of Transmittals to make the report 10-15 pages long. When the manual Transmittal log exceeds a few pages, it becomes easy to overlook the fact that some Transmittals may not have the required responses returned. For this reason alone, we recommend a computerized system of correspondence control.

To: Owner, Architect, Subcontractors, job site for posting
From: Gauley Construction
Subject: Holiday Schedule
Date: January 2, 1989

The following holiday schedule will be observed on this job:

1/15: Martin Luther King Jr. Birthday
2/19: President's Day
5/28: Memorial Day,
7/4: Independence Day
9/3: Labor Day

****

To: Subcontractors
From: Smit Bros. Contractors
Subject: Lien Information
Date: February 5, 1990

Attached is the information on the Owner, Lender, and Contractor for your use for the filing of your preliminary lien notice.

****

To: All subcontractors
From: Grant Construction, Inc.
Subject: Monthly Invoicing
Date: February 15, 1990

The general conditions and supplemental General Conditions require that you supply particular information when sending us your invoice for processing. A recap of all the requirements are enclosed. Please review and comply. We don't have a problem paying you...if you follow the rules.

****

To: Everyone Connected with Project
From: D & H Contractors
Subject: General Job Information
Date: Janaury 15, 1990

Enclosed is a job list with the name of the company, the contact, telephone number, address, trade, and info about the architect, owner and engineers, job site, telephone, etc.

Figure 3.7

```
                        DELTA CONSTRUCTION COMPANY
=========================================================PaperWorks Version 1.23F==
                              TRANSMITTAL LOG                                .TR1
            Job: 321        GIBSON OFFICE BUILDING
Date: 03/09/90                                                          Page:    2
==================================================================================
Transmittal     Name of Addressee           Date       Date      Date         Status
 Number                                     Sent       Due       Received
----------------------------------------------------------------------------------
    4                                     02/11/90    N/A
  subj: HOLIDAY SCHEDULE
        The following holidays will be observed on this project:
            Presidents Day                      Memorial Day
            Independence Day                    Labor Day
  SPARKS ELECTRIC COMPANY         for: INFORMATION                 N/A     Printed
  QUICK PLUMBING COMPANY          for: INFORMATION                 N/A     Printed
  STRONG STRUCTURAL STEEL         for: INFORMATION                 N/A     Printed
  LOT'S OF REBAR                  for: INFORMATION                 N/A     Printed

----------------------------------------------------------------------------------

    5                                     02/18/90   02/27/90
  subj: CPM SCHEDULE
        I am preparing the construction CPM schedule for this project.
        Please advise me of any long lead items that may affect the
        schedule.  I'm planning to present the schedule to the client no
        later than 3/2, so I will need your input no later than 2/27.
        Give me a call if you need to discuss anything.  If you don't
        tell me your problems now after 3/2 it will be too late.
  STRONG STRUCTURAL STEEL         for: RESPONSE REQ              02/27/90 Returned
  QUICK PLUMBING COMPANY          for: RESPONSE REQ              02/27/90 Returned
  COOL AIR CONDITIONING           for: RESPONSE REQ              NOT RET. Printed
  CLEAR GLASS COMPANY             for: RESPONSE REQ              NOT RET. Printed

----------------------------------------------------------------------------------

    6                                     02/18/90    N/A
  subj: ANCHOR BOLTS
        Bob, we're going to need those anchor bolts no later than
        March 3.  Make sure that they're on the job by that time, we
        can't get behind at this point.
        Call me if you have a problem.
  STRONG STRUCTURAL STEEL         for: ACTION                      N/A     Printed

----------------------------------------------------------------------------------

    7                                     02/27/90    N/A
  subj: JOB MINUTES OF MEETING #1
        Please review the job minutes and make sure that you agree with
        the conclusions and action items.

        Next job meeting will be March 1, 9:00 at the job site.
  ONE LINE DESIGN                 for: INFORMATION                 N/A     Printed
  QUICK PLUMBING COMPANY          for: INFORMATION                 N/A     Printed
  STRONG STRUCTURAL STEEL         for: INFORMATION                 N/A     Printed
  QUALITY MASONRY                 for: INFORMATION                 N/A     Printed
  LOT'S OF REBAR                  for: INFORMATION                 N/A     Printed
```

Figure 3.8a

```
                    DELTA CONSTRUCTION COMPANY
=========================================PaperWorks Version 1.23F==
                         TRANSMITTAL DUE LOG                      .TR3
            Job: 321     GIBSON OFFICE BUILDING
Date: 03/09/90                                              Page:    1
=======================================================================
Transmittal    Name of Addressee        Date      Date      Date          Status
 Number                                 Sent      Due       Received
-----------------------------------------------------------------------
   1                                  02/06/90  02/20/90
  subj: INSURANCE REQUIREMENTS
        You must have insurance certificates on file with our office
        prior to starting work on the job site.  NO EXCEPTIONS.  The
        minimum amount and types of insurance required are summarized on
        the attached Insurance Information Sheet.  You must name Delta
        Construction Company as additionally insured.
    cc: LOT'S OF REBAR                  for: RESPONSE REQ          Printed
    cc: STRONG STRUCTURAL STEEL         for: RESPONSE REQ          Printed

-----------------------------------------------------------------------

   5                                  02/18/90  02/27/90
  subj: CPM SCHEDULE
        I am preparing the construction CPM schedule for this project.
        Please advise me of any long lead items that may affect the
        schedule.  I'm planning to present the schedule to the client no
        later than 3/2, so I will need your input no later than 2/27.
        Give me a call if you need to discuss anything.  If you don't
        tell me your problems now after 3/2 it will be too late.
    cc: COOL AIR CONDITIONING           for: RESPONSE REQ          Printed
    cc: CLEAR GLASS COMPANY             for: RESPONSE REQ          Printed

-----------------------------------------------------------------------

   9                                  02/27/90  03/13/90
  subj: SUBCONTRACT AGREEMENT
        Please return two copies of the signed subcontract agreement in
        the amount of $300,000 for this project.  Your will not be
        allowed to proceed with work on the project without a signed
        subcontract on file in our offices
    cc: COOL AIR CONDITIONING           for: RESPONSE REQ          Printed

-----------------------------------------------------------------------

  12                                  02/27/90  03/13/90
  subj: SUBCONTRACT AGREEMENT
        Please return two copies of the signed subcontract agreement in
        the amount of $415,000 for this project.  You will not be
        allowed to proceed with work on this project without a signed
        subcontract on file in this office
    cc: CLEAR GLASS COMPANY             for: RESPONSE REQ          Printed

-----------------------------------------------------------------------

  13                                  02/27/90  03/13/90
  subj: SUBCONTRACT AGREEMENT
        Please sign and return two copies of the signed subcontract
        agreement in the amount of $67,000 for this project.  You will
        not be allowed to work on this project without a signed
        subcontract agreement on file in our offices.
    cc: APPLIED ROOFING COMPANY         for: RESPONSE REQ          Printed

-----------------------------------------------------------------------
```

Figure 3.8b

```
                    DELTA CONSTRUCTION COMPANY
=========================================PaperWorks Version 1.23F==
              TRANSMITTAL DELINQUENT LOG--AS OF 03/09/90           .TR5
          Job: 321       GIBSON OFFICE BUILDING
Date: 03/09/90                                                 Page:   1
=========================================================================
Transmittal    Name of Addressee        Date      Date      Date         Status
 Number                                 Sent      Due       Received
-------------------------------------------------------------------------
    1                                 02/06/90   02/20/90
  subj: INSURANCE REQUIREMENTS
        You must have insurance certificates on file with our office
        prior to starting work on the job site.  NO EXCEPTIONS.  The
        minimum amount and types of insurance required are summarized on
        the attached Insurance Information Sheet.  You must name Delta
        Construction Company as additionally insured.
    cc: LOT'S OF REBAR                   for: RESPONSE REQ          Printed
    cc: STRONG STRUCTURAL STEEL          for: RESPONSE REQ          Printed
-------------------------------------------------------------------------

    5                                 02/18/90   02/27/90
  subj: CPM SCHEDULE
        I am preparing the construction CPM schedule for this project.
        Please advise me of any long lead items that may affect the
        schedule.  I'm planning to present the schedule to the client no
        later than 3/2, so I will need your input no later than 2/27.
        Give me a call if you need to discuss anything.  If you don't
        tell me your problems now after 3/2 it will be too late.
    cc: COOL AIR CONDITIONING            for: RESPONSE REQ          Printed
    cc: CLEAR GLASS COMPANY              for: RESPONSE REQ          Printed
-------------------------------------------------------------------------
```

Figure 3.8c

Delta Construction Co., Job #321 - Gibson Office Building
TRANSMITTAL LOG

#	TO	SUBJECT	DATE SENT	DATE DUE	DATE REC'D	COMMENTS
4	Sparks Electric	Holiday Schedule	2/11/90	N/A		for your information
	Quick Plumbing		2/11/90	N/A		
	Strong Steel		2/11/90	N/A		
	Lot's of Rebar		2/11/90	N/A		
5	Strong Steel	CPM Schedule	2/18/90	2/27/90	2/27	call subs if not
	Quick Plumbing		2/18/90	2/27/90	2/27	returned by 2/27!
	Cool Air Conditioning		2/18/90	2/27/90		Important
	Clear Glass		2/18/90	2/27/90		
6	Strong Steel	Anchor Bolts	2/18/90	N/A		
7	One Line Design	Job Mtg. Minutes	2/27/90	N/A		
	Quick Plumbing		2/27/90	N/A		
	Strong Steel		2/27/90	N/A		
	Quality Masonry		2/27/90	N/A		
	Lot's of Rebar		2/27/90	N/A		

Figure 3.8d

Chapter Four

Requests for Information

The Request for Information (RFI) is a specific type of written communication to document questions and record the resulting answers. The subjects addressed by an RFI can and will affect the progress of a job in one way or another, and generally equate to dollars in or out of someone's pocket. An RFI may address critical issues with immediate time constraints, or it may relay requests that are not quite so urgent.

The Purpose

The primary purpose of the RFI is to keep the job moving by getting questions answered. Information conveyed promptly assists the project immensely. By the same token, lack of information hinders the project by creating doubt, indecision, frustration, and general dissatisfaction. The following axioms apply:

> *"Lacking direction, people will make decisions."*
>
> *"Lacking direction or answers, individuals and/or companies will choose the path that maximizes their own profit ..... not yours!!"*

Participants in the project should make decisions throughout the course of the job. The challenge is to get them to base these decisions on the best interests of the project rather than their own immediate benefit. Favorable responses are much more likely if the information they need is available.

The purpose of an RFI is to obtain accurate, reasonable, well-thought-out answers to questions so you can build the project to the best of your ability and maximize your profitability. When the general contractor or subcontractor puts the question in writing, the chances of receiving a well-thought-out, intelligent answer are increased dramatically. By documenting these questions in an RFI, you help to ensure three things:

1. Obtaining an answer to your question does not depend solely on someone remembering that you asked.
2. Documenting your questions today, no matter how insignificant they may seem at the moment, will help to protect your position in the future when that item affects the progress of the whole job.
3. You will be assured that your actions are correct per the instructions of the architect, owner, or user.

Without documentation, the participants proceed with a solution based on hearsay or someone's interpretation of the user's intent. The item in question may be installed or built incorrectly.

Consequences of Failing To Use an RFI

Even if the plans and specifications are incorrect, both the architect and the contractor have an obligation to the project owner to use their "construction expertise" to point out the problems on the owner's behalf. If the contractor does not follow up on the questionable items, then it will be his word (that he asked "the question") against someone else's word (that he did not ask). When the time comes to assign the blame and or the cost, many people suddenly develop a "selective memory." Selective memory is cured by writing out the questions in an RFI and then following up to obtain clear and correct answers to your questions.

One often hears an owner or architect/engineer say:

"I didn't know it was important," or

"You never asked us that," or

"You didn't give me enough time."

Problems arising from unanswered RFI's are compounded when the architect/owner must contact a third party for information. By documenting the question in an RFI and then tracking and following up on that document, both the contractor and the job profit, and everyone is saved from unnecessary future problems.

Sample RFI

Note that the format of the RFI is not that of a letter or a memo. The sample RFI letter in Figure 4.1a is computer-generated. Figure 4.1b is a manual RFI (Means Transmittal Form from *Means Forms for Building Construction Professionals*).

DELTA CONSTRUCTION COMPANY

6065 Mission Gorge Road, #193
San Diego, CA 92120
(619) 582-5829

*** REQUEST FOR INFORMATION ***

Date: 2/11/90

Job Number: 321
R.F.I. Number: 1

TO: ONE LINE DESIGN
2001 DREAM STREET
SAN DIEGO, CA 92100

ATTENTION: ROD WRIGHT

JOB NUMBER: 321 GIBSON OFFICE BUILDING

FROM: J. EDWARD GRIMES

SUBJECT: R.F.I. NUMBER: 1
GRADING PLAN INFORMATION

We are requesting information from you which is AS LISTED BELOW for your RESPONSE REQUIRED by: 2/25/90.

> Please clarify and provide dimensions for the driveway to the garage. Where is the pivot point for the curve? Additional information is needed by the grading contractor.

Please contact me as soon as possible if you need any further information or if you have any questions.

Sincerely,

J. Edward Grimes

Copies were sent to the following:

BOB MEAN	COPY
FRANCES GIBSON	FOR INFORMATION (ORIGINAL)

dp: 2/11/90 16:29 **PaperWorks**, a Construction Software Program

Figure 4.1a

Means Forms

LETTER OF TRANSMITTAL

Delta Construction Company

6678 Mission Road
San Diego, CA 92120
(619) 581-6315

FROM: J. Edward Grimes
Project Manager

DATE 2/11/90
PROJECT Gibson Office Bldg. #321
LOCATION
ATTENTION Rod Wright
RE: slope of garage entry

TO: One Line Design
2001 Dream St.
San Diego, Ca. 92100

RFI 321.01

Gentlemen:

WE ARE SENDING YOU ☒ HEREWITH ☐ DELIVERED BY HAND ☐ UNDER SEPARATE COVER

VIA ____________ THE FOLLOWING ITEMS:

☐ PLANS ☐ PRINTS ☐ SHOP DRAWINGS ☐ SAMPLES ☐ SPECIFICATIONS

☐ ESTIMATES ☐ COPY OF LETTER ☒ R.F.I.

COPIES	DATE OR NO.	DESCRIPTION

THESE ARE TRANSMITTED AS INDICATED BELOW

☐ FOR YOUR USE ☐ APPROVED AS NOTED ☐ RETURN ______ CORRECTED PRINTS

☐ FOR APPROVAL ☐ APPROVED FOR CONSTRUCTION ☐ SUBMIT ______ COPIES FOR ______

☐ AS REQUESTED ☐ RETURNED FOR CORRECTIONS ☐ RESUBMIT ______ COPIES FOR ______

☐ FOR REVIEW AND COMMENT ☐ RETURNED AFTER LOAN TO US ☒ ~~FOR BIDS~~ DUE 2/25/90

☐ ______

REMARKS: Please clarify and provide dimensions for the driveway to the garage. Where is the pivot point for the curve? Additional information is needed by the grading contractor.

IF ENCLOSURES ARE NOT AS INDICATED, PLEASE NOTIFY US AT ONCE.

SIGNED: J Edward Grimes

Figure 4.1b

Items to be Included In All RFI's

The following information should be included in all Requests For Information. The sample RFI letters (Figures 4.1a and 4.1b) include each of these items at least once. The identification title and number of each RFI identifies it as a unique correspondence.

By adding the due dates, it is now possible to create a Correspondence Receivable list or a log similar to that maintained by the firm's accounting department for accounts receivable and accounts payable. The identification information for the RFI log is as follows:

1. FROM:
2. IDENTIFICATION TYPE: in this case, "Request For Information"
3. IDENTIFICATION NUMBER: e.g., "RFI #3"
4. DATE OF RFI LETTER:
5. TO:
6. FROM:
7. SUBJECT: e.g., "Slope of Garage Entry Ramp."
8. RESPONSE DUE BY:
9. TEXT: The question, or reason the RFI is being sent.
10. REFER TO STATEMENT: Include in the RFI instructions for the addressee to refer to the appropriate Job Number and RFI identification and number, e.g., **Job #321, RFI #3.** This system makes it easier to identify a particular job when a firm is dealing with a number of jobs.
11. COPIES: Sent to whom? This standard item can be an important tool. For example, if a copy is sent to the owner, he or she is alerted to the possibility that the subject of the RFI may cost some money. The owner should, as a result, be less surprised by the extra expense later on. Sending copies also makes field personnel and selected subcontractors aware that you have a question (RFI) in to the architect.

Any project correspondence or documentation that leaves your office should include this information. It does not have to be in the exact format shown in our RFI examples, but the information must be included, as it enables the recipient to make a proper and complete response to your correspondence.

By including a response due date, it is now possible to create a Correspondence Receivable list or a log similar to that maintained by the firm's accounting department for accounts receivable and accounts payable.

General Rules for RFI's

Keep it Simple, Limit the Number of Questions

The number of questions addressed in one RFI should be minimized so the answers are not delayed until *all* of the information for *all* of the questions can be obtained. In fact, it is best to limit each RFI to one topic. This does not mean that eight separate RFI's should be sent to address eight different questions about finish hardware. However, individual RFI's should be sent if they pose questions on separate subjects, such as structural steel and plumbing.

Separate RFI's should also be sent even if they are on the same subject–if one of the questions must be answered immediately, and the other one can wait several weeks. Again, it is easier and faster to obtain an answer to one item rather than waiting for a long series of questions involving several aspects of the work to be researched. It may take weeks to obtain answers to all of the questions when the most important one could have been answered in a day.

Be Specific

An RFI should address a specific item or group of items such as:

> **RFI Example 1:** "What is the dimension of the hallway No. 101 as shown on page A-5? It scales to 4'-0", but the given dimension is 4'-8". Please respond ASAP as we are laying out this area."

Note that this example begins with **the** question. The question to be answered is right up-front with no room for ambiguity. Example No. 1 also briefly explains why a *quick* answer is requested. "ASAP" indicates the urgency of the situation. This term should never be used, however, unless you mean it. Otherwise, the phrase becomes just another part of the form.

In Example No. 1, the answer most likely will be that the corridor was supposed to be 4'-0" and the drawing was simply dimensioned incorrectly. This may be a simple error, but it should still be documented.

As a side note, it makes no difference that the drawings specify that the printed dimensions govern and that the drawings should not be scaled. In this case, if the contractor had not scaled the drawings, he would have been working with the wrong dimensions. Good communication often means good cooperation.

> **RFI Example 2:** "Should the door hardware called for at door #1001 have panic hardware, rather than a standard lock set?"

In Example No. 2, the contractor, in reviewing the hardware schedule, found a door that, based on his experience, would seem to require panic hardware. Or, perhaps it should be hung to swing from right to left instead of as shown. Such issues must be brought to the attention of the architect. In this example, the more strongly the contractor feels he is right, the more important it is to get these questions down on paper.

> **RFI Example 3:** "In corridor number 100, the door schedule and door legend call for doors 101, 102, and 103 to be 20-minute ratings, while doors 104 and 105 call for 45-minute ratings. I think that all of the doors should have the same rating. Please advise."

Example No. 3 illustrates a number of specific questions contained within the same RFI. Note that the subject matter is all the same, and asks for specific information regarding the fire rating on the doors in question, due to an apparent lack of specification continuity in the construction documents.

> **RFI Example 4:** "There is insufficient space for the HVAC duct to pass under the steel beam. The mechanical HVAC drawing shows a 24 × 36 duct passing under/through the 26WF beam at the first floor ceiling at column lines F-5 and F-6. Please advise, and issue the appropriate paperwork."

Example No. 4 is almost sure to cause a change in scope and cost. In this case, the architect and mechanical engineer should answer the RFI and also issue a Price Request or a Field Order to effect the work required.

Many architects try to issue this type of change in scope as a "Clarification," claiming it should not increase costs because they do not want the owner aware that the original plans were not totally correct. However, the owner did not pay for a duct transition in the first place. It is now the obligation of the general contractor and the architect/engineer to see that the HVAC subcontractor does not overprice the cost for the duct transition.

The RFI performs the service that its title suggests: it requests information. A standard RFI format, such as the computer-generated example shown in Figure 4.1a, contains all of the information necessary to make the RFI unique and to establish due dates and other needed data. Alternate RFI phrasing for Example No. 2 (for use with a manual RFI letter such as the modified Transmittal form shown in Figure 4.1b) would be as follows:

> "Should door No. 1001 have panic hardware? Please advise. If additional or different hardware is required, please issue a Price Request or Field Order with the appropriate specifications, instructions, and documentation. Note that the hardware for this project is already on order. Changes to that order should be made no later than 10/28/89, if different hardware is required. Please respond no later than 10/18/89."

These statements have established the specific nature of the RFI. No one has to guess the real meaning behind it. They know the specific question, why it is asked, what they must do and when, and what the effects will be.

Unfortunately, most construction managers, project managers, facility managers, and superintendents believe that if a problem is obvious to them, then everyone else will see it too. This is not always the case. The couple standing on the bow of the Titanic must have assumed that nobody could possibly miss the giant iceberg looming out of the mist. The fact is, the more simple a situation is, the more likely something will be overlooked. Always ask, document your questions, and get the answers in writing.

Additional RFI examples are shown in the sample logs at the end of the chapter.

Make it Easy

A clear, straightforward question conveyed in any RFI can be reviewed and then passed along by the architect to a junior architect or staff person. This junior person may have more time to research the question and respond to an RFI. The architect's subordinate may also be less likely to respond with an over-the-shoulder "yes" or "no," or "figure it out yourself; everything is shown on the plans!"

Once such a request is documented, the architect or user is responsible for determining what must be done and then documenting the response. This exchange builds good job communication and clarity, which is what RFI's are all about.

Inadequate Response to RFI's

Even if the architect/user receives an RFI properly documenting the situation, he or she may simply telephone the sender and say, "you were right, why don't you go ahead and . . . .?" This oral response is misuse of an RFI in that it represents improper follow-up.

Avoid taking any action on an oral response (with no written back-up). If there is a scope and cost change, for example, do not proceed without a Field Order or authorization to proceed with the work; otherwise, the three-fold purpose of an RFI is defeated.

Do not Forget the Purpose

The first purpose of an RFI is *to document the question.* The second purpose of an RFI is *to get the other party to write out and document an answer,* so it can be distributed to the appropriate people. Insist that the other party respond in writing to your RFI.

At this point, some positive encouragement may be needed. Proper follow-up means more than just keeping track of what happens. It also means receiving and distributing timely and reasonable responses. If necessary, include the RFI response in the regular job meetings, where it will be documented in the minutes.

The third purpose of an RFI is to *make sure that the appropriate pricing documentation is issued by the architect in case of a scope/cost change.* The appropriate documentation is one of these two pricing documents: Price Request or Field Order. The use of each of these documents is discussed in detail in Chapters 19 and 20.

The architect may argue that because the contractor noticed the hardware problem (or another subject of an RFI), it must be included in his bid. The contract may even specify that the contractor must provide all items according to local codes. However, by using the RFI to document job-related questions and sticking to the philosophy "we build 'em, we don't design 'em," you have a good chance of coming out ahead of the game.

Once the architect understands that it is his or her obligation to supply the right information (i.e., to properly respond to RFI's), the contractor can build the project to code, and then start the paperwork for getting paid for the changes.

Do Not Use RFI's for Pricing

Many construction firms use the RFI as a pricing vehicle. This is wrong. The RFI is not a pricing vehicle. It is an easy, clear method for documenting questions and answers relating to the construction project. This is the primary purpose of the Request For Information.

Most RFI's issued on the job do not involve any significant sum of money, although ignoring them may cause significant time delay and expense to all parties later in the project.

Once the RFI is properly processed, the architect should issue a Field Order (or Price Request if there are scope changes or an increase or decrease in cost to the project). If the architect does not issue one of these pricing documents, then the contractor should initiate a Request For Change. (See Chapters 18-22 for detailed coverage of these documents.)

Send the RFI to the Right Person

This step is crucial to the success of the document. Send the RFI to the right person. Do not automatically send it to the chief architect, for example, or to the architect's secretary. Nor should it be automatically sent to the same person who answered your last question. Address the RFI to the one person with the responsibility and the knowledge to answer the question. The pre-construction meeting is a good opportunity to confirm who should send and receive these documents.

RFI questions regarding a structural detail may need to be sent to the architect of record and then forwarded by this person to the structural engineer. A good system is to send the RFI to the architect, and send a copy of the RFI to the structural engineer. In this way, a week need not be lost in transit, either in the mail or on someone's desk. Clarify who is to get what, because sending correspondence to the wrong person leads to needless delay and confusion. Another useful strategy for getting a response is to let a third party know that the question has been asked, then indicate on the original that copies have been sent to the project owner. Use discretion with this technique, however, as it can easily make enemies. Nevertheless, there are occasions when immediate information is absolutely needed. Construction project management is no beauty pageant, and there is no Congeniality award.

Follow-up, Follow-up, Follow-up

A specific, unique tracking number should be assigned to all RFI's. The best numbering system is a sequential one. If RFI's are initiated in the office and in the field on the same job, then let the office have numbers 101 to 200, and the field have RFI numbers 001 to 100. This will ensure that duplicate RFI numbers are not issued. This system makes it easier to find and reference data in correspondence between contractor and architect.

Sample RFI Logs

Figure 4.2a is an example of a computer-generated RFI log. Figure 4.2b is a computer-generated due date report for RFI's, Figure 4.2c is a computer-generated delinquent RFI Report, and Figure 4.2d is a manually-generated log for RFI's. Remember, in addition to writing the RFI, it is equally important to track and follow up the RFI to obtain the information requested. Computer systems can make this process much less painful. Most useful are systems that provide a report of the items that have not been answered. An RFI system without adequate tracking is like a car without wheels–you can sit in it and look good, but you are not going anywhere.

Summary

RFI's have a specific role in construction communications. Their sole purpose is to ask a question and elicit an answer. In the process, both the question and the answer (or lack of an answer) are documented. Do not misuse the RFI format. It is as its title indicates–a Request For Information.

```
                        DELTA CONSTRUCTION COMPANY
=============================================PaperWorks Version 1.23F==
                              R.F.I.  LOG                          .RF1
          Job: 321      GIBSON OFFICE BUILDING
Date: 03/11/90                                                 Page:    1
========================================================================
 R.F.I          Name of Addressee        Date      Date      Date        Status
 Number                                  Sent      Due       Received
------------------------------------------------------------------------
   1                                   02/11/90  02/25/90
  subj: GRADING PLAN INFORMATION
        Please clarify and provide dimensions for the driveway to the
        garage.  Where is the pivot point for the curve?  Additional
        information is needed by the grading contractor.
  ONE LINE DESIGN                 for: RESPONSE REQ        NOT RET. Printed
  BOB MEAN                        for: COPY                  N/A    Printed
  FRANCES GIBSON                  for: INFORMATION           N/A    Printed

------------------------------------------------------------------------
   2                                   02/11/90  02/25/90
  subj: SLOPE OF GARAGE ENTRY RAMP
        Please review the slope of the garage entry ramp.  It appears
        to us that the grade is too steep and cars could hang up on the
        hump, or the big ones would drag at the bottom of the ramp.
        Please issue the necessary documents as required.
  ONE LINE DESIGN                 for: RESPONSE REQ        NOT RET. Printed
  BOB MEAN                        for: INFORMATION           N/A    Printed

------------------------------------------------------------------------
   3                                   02/18/90  03/04/90
  subj: DETAILS FOR RETAINING WALL FOOTING
        The Civil Drawings indicate a retaining wall at the north end of
        the parking lot.  There are no details on the type of footing.
        It appears that there should be an engineered footing for this
        retaining wall.  Please issue the necessary documents.
  ONE LINE DESIGN                 for: RESPONSE REQ        NOT RET. Printed
  QUALITY MASONRY                 for: INFORMATION           N/A    Printed

------------------------------------------------------------------------
   4                                   02/18/90  03/04/90
  subj: HOLLOW METAL FRAME DETAILS
        There are several different conditions where the HM frames
        connect to wood, steel studs, concrete and block.  Your details
        only show the wood stud connection.  Please provide us with the
        appropriate details.
  ONE LINE DESIGN                 for: RESPONSE REQ        NOT RET. Printed
------------------------------------------------------------------------
   5                                   02/27/90  03/11/90
  subj: HARDWARE GROUP INFORMATION
        Please provide additional information on Hardware Groups:
        2, 3, 7, 14, 33, 41, 59 per Cont Hdwe letter attached.  If you
        have any questions, please call John direct at 777-6666.  Please
        make sure that any corrective information is return back though
        us, i.e. normal channels, and issue a Field Order or
        Clarification.
  ONE LINE DESIGN                 for: RESPONSE REQ        NOT RET. Printed
  CONTINENTAL HARDWARE          for: INFORMATION           N/A    Printed

------------------------------------------------------------------------
```

Figure 4.2a

```
                      DELTA CONSTRUCTION COMPANY
=====================================================PaperWorks Version 1.23F==
                         R.F.I.  D U E   L O G                          .RF3
             Job: 321      GIBSON OFFICE BUILDING
Date: 03/11/90                                                      Page:    1
================================================================================
 R.F.I          Name of Addressee          Date      Date      Date         Status
 Number                                    Sent      Due       Received
--------------------------------------------------------------------------------

    2                                    02/11/90  02/25/90
  subj: SLOPE OF GARAGE ENTRY RAMP
        Please review the slope of the garage entry ramp.  It appears
        to us that the grade is too steep and cars could hang up on the
        hump, or the big ones would drag at the bottom of the ramp.
        Please issue the necessary documents as required.
    cc: ONE LINE DESIGN                    for: RESPONSE REQ           Printed
--------------------------------------------------------------------------------

    4                                    02/18/90  03/04/90
  subj: HOLLOW METAL FRAME DETAILS
        There are several different conditions where the HM frames
        connect to wood, steel studs, concrete and block.  Your details
        only show the wood stud connection.  Please provide us with the
        appropriate details.
    cc: ONE LINE DESIGN                    for: RESPONSE REQ           Printed
--------------------------------------------------------------------------------

    5                                    02/27/90  03/11/90
  subj: HARDWARE GROUP INFORMATION
        Please provide additional information on Hardware Groups:
        2, 3, 7, 14, 33, 41, 59 per Cont Hdwe letter attached.  If you
        have any questions, please call John direct at 777-6666.  Please
        make sure that any corrective information is return back though
        us, i.e. normal channels, and issue a Field Order or
        Clarification.
    cc: ONE LINE DESIGN                    for: RESPONSE REQ           Printed
--------------------------------------------------------------------------------

    6                                    02/27/90  03/04/90
  subj: ELEVATOR PIT AND LANDING DETAILS AT GARAGE LEVEL
        We would like to keep the elevator pit no more than 3'-11" deep
        in order not to have to provide an access ladder.  Please advise
        us of your decision.
        Please review the marble landing detail at the entry from the
        lobby to the elevator.  It does not appear to have enough
        and may need some misc iron.
    cc: ONE LINE DESIGN                    for: RESPONSE REQ           Printed
--------------------------------------------------------------------------------

    7                                    03/04/90  03/04/90
  subj: ELECTRICAL INFORMATION FOR ELEVATOR
        Please review the attached information request from Up and Down
        Elevator.  This info is a must prior to the ordering of the
        elevator motors and switches.  I'm sending Sparks Elec. a copy
        of this letter so he can review his shop drawings to ensure
        proper coordination.
    cc: ONE LINE DESIGN                    for: RESPONSE REQ           Printed
--------------------------------------------------------------------------------
```

Figure 4.2b

DELTA CONSTRUCTION COMPANY

PaperWorks Version 1.23F

R.F.I. DELINQUENT LOG--AS OF 03/10/90 .RF5

Job: 321 GIBSON OFFICE BUILDING

Date: 03/11/90

Page: 1

R.F.I Number	Name of Addressee	Date Sent	Date Due	Date Received	Status
2		02/11/90	02/25/90		

subj: SLOPE OF GARAGE ENTRY RAMP
Please review the slope of the garage entry ramp. It appears to us that the grade is too steep and cars could hang up on the hump, or the big ones would drag at the bottom of the ramp. Please issue the necessary documents as required.
cc: ONE LINE DESIGN — for: RESPONSE REQ — Printed

R.F.I Number	Name of Addressee	Date Sent	Date Due	Date Received	Status
4		02/18/90	03/04/90		

subj: HOLLOW METAL FRAME DETAILS
There are several different conditions where the HM frames connect to wood, steel studs, concrete and block. Your details only show the wood stud connection. Please provide us with the appropriate details.
cc: ONE LINE DESIGN — for: RESPONSE REQ — Printed

R.F.I Number	Name of Addressee	Date Sent	Date Due	Date Received	Status
6		02/27/90	03/04/90		

subj: ELEVATOR PIT AND LANDING DETAILS AT GARAGE LEVEL
We would like to keep the elevator pit no more than 3'-11" deep in order not to have to provide an access ladder. Please advise us of your decision.
Please review the marble landing detail at the entry from the lobby to the elevator. It does not appear to have enough and may need some misc iron.
cc: ONE LINE DESIGN — for: RESPONSE REQ — Printed

R.F.I Number	Name of Addressee	Date Sent	Date Due	Date Received	Status
7		03/04/90	03/04/90		

subj: ELECTRICAL INFORMATION FOR ELEVATOR
Please review the attached information request from Up and Down Elevator. This info is a must prior to the ordering of the elevator motors and switches. I'm sending Sparks Elec. a copy of this letter so he can review his shop drawings to ensure proper coordination.
cc: ONE LINE DESIGN — for: RESPONSE REQ — Printed

Figure 4.2c

Delta Construction Co., Job #321 - Gibson Office Building
RFI LOG

#	TO	SUBJECT	DATE SENT	DATE DUE	DATE REC'D	COMMENTS
1	One Line Design Frances Gibson	Grading Plan	2/11/90	2/25/90		
2	One Line Design	Slope of garage entry	2/11/90	2/25/90		
3	One Line Design Quality Masonry	Ret. wall ftg.	2/18/90	3/4/90		
4	One Line Design	Hollow Mtl. frame details	2/18/90	3/4/90		
5	One Line Design	Hardware Group Info.	2/27/90	3/11/90		
6	One Line Design	Elev. Pit & Landing det. at Gar.	2/27/90	3/4/90		
7	One Line Design	Elec. Info. for Elevator	3/4/90	3/4/90		

Figure 4.2d

Chapter Five

Logs

The concept of keeping logs is a lot like buying a shovel: you do not buy a shovel because you want a shovel. You buy a shovel because you want a hole. By the same token, you do not keep logs because you want all that paper. You keep logs because you want to track what is sent, whether a reply is necessary, and when that reply is due. Without this tracking of the information flow, important facts may slip through the cracks and ultimately cost you time and/or money.

Traditionally, the manager (general contractor, facility manager, or construction manager) monitors hundreds of individual pieces of paperwork that flow through the project. With overall project responsibility, the manager must ensure that non-performance by one subcontractor on one item will not slow or stop the whole project. Any variance to the original time or cost estimate usually affects the general contractor's pocketbook more directly than any other contractor on the job. While the other participants do not share the same financial exposure, all have the same need to log and track their communications.

Each piece of paper usually represents a series of events. Therefore, the more paperwork, the greater number of events that must be tracked. The more complex the communication, the more important record keeping becomes. A subcontractor's log of communications with the general contractor is as important to his financial health as the general contractor's more complex log of subcontractors on the job.

Logs represent the history of the communications on a project–a written memory of how things got to where they are today. While meeting minutes reflect a history of *decisions*, logs reflect the history of *activity*. For example:

> *The concrete contractor pours a slab for the utilities building. The electrical contractor moves in the following week to rough in the wiring, only to discover that the slab was poured without block-outs. The project manager marches back to the office, opens up the logs, and reviews the communications concerning that utilities building. The story behind this problem is recorded right there.*
>
> *The concrete contractor had requested permission to pour that slab out of sequence four weeks earlier. Transmittals were sent notifying everyone on the project that the slab would be poured early, and requesting a description of the projected impact. The electrical contractor responded (seven days later) with a letter stating that there was no impact–omitting the fact that shop drawings had not yet been*

submitted, as they were not yet required by the original schedule. However, the electrical contractor had fallen behind schedule and pulled workers from the utilities building for other work. The concrete contractor assumed that the original go-ahead sent by the general contractor ten days after the original request was still valid. The logs show a letter sent to the electrical contractor requesting an action plan for getting back on schedule and a letter to the concrete contractor with a list of items affected by the electrical delays. The logs show no response from either party. The project manager now knows how a seemingly obvious mistake was made. Without that log, reconstructing those events might have taken much longer, and may not even have been possible.

Purpose

Logs have three basic functions:

1. To track the location of an item
2. To monitor information due dates
3. To give notice when a late item starts to impede job progress

One of the less confrontational ways of calling to attention those who fail to act or respond is to get into the habit, at the regularly scheduled job meetings, of giving the architect a copy of the logs, showing what has not been done. This is another instance where a computerized system (and selective reports) can be helpful. From those logs, a list of pressing, incomplete items can be determined.

With the exception log, the architect, owner, and/or subcontractors should be aware that each item on the list requires a response. Without a response, the project may be delayed, or in the case of subcontractors, the invoices will not be paid on time.

The log system offers other advantages as well. For example, it is important that the job atmosphere does not become confrontational. After all, the general contractor must work with the owner, architect, and subcontractors throughout the project. A documented list of incomplete work, like a log, keeps project communications on a business level–not a personal one. Distributing the list prevents finger-pointing at meetings. The responsible parties can calmly solve the problem later.

It is easy to review logs and determine what has not been done, thereby minimizing surprises downstream. Of course, some surprises cannot be avoided, but try to eliminate as many as possible by getting the loose items "tied down." Maintaining comprehensive and up-to-date logs will help avoid the possibility that small but important items will be overlooked.

Uses of Logs

Logs can, and should, be kept for almost every type of activity or correspondence. Whether manually- or computer-generated, a log should be maintained for every type of correspondence produced by the architect or contractor. We will cover most of them in this book. (Examples of manual and computerized logs are included in all applicable chapters.) Keep a log on purchase orders, equipment on the job, backcharges, and signed subcontracts, in addition to the normal correspondence.

Information Usually Contained in Logs

A unique tracking number and a date are examples of essential basic information to be included in a log. Names of those involved, contract document references, and other specific peripheral data are examples of specialized information. It is crucial that the quantity of information be kept to a minimum and the KISS (Keep It Simple, Silly) principle liberally applied.

One project manager's assistant kept a very involved coding system of different colored pencil checks, dots, and lines, plus a few highlighted areas. The trouble was that different colors do not reproduce on copy machines; and the assistant was the only one who could remember the coding system. To identify those parties who have failed to respond properly, it is necessary to have a simple log system that everyone can understand.

Types of Logs

Not every company requires the same identification system for logs and correspondence. For that matter, not every job needs all of the logs. Every job is unique and, therefore, has different requirements. Some of the classifications may be combined for certain job situations, but, again, keep it simple. It is better to have three logs that are understood easily, than one log that no one can understand. The possible types of logs include, but are not limited to:

- Submittal Requests
- Submittals to Architects
- Returned Submittals to Subcontractors/Vendors
- Transmittals
- Requests for Information
- Price Requests (sent to subcontractors)
- Price Request Quotations
- Field Orders
- Field Order Quotations
- Clarification
- Clarification Quotations
- Requests for Change and Quotations
- Subcontract Logs
- Purchase Order Logs
- Plan Distributions
- Change Order Logs
- Insurance Coverage
- Backcharges

Remember: try to keep people informed of what is happening (or not happening, as the case may be). The key participants should be given copies of some of the contractor's logs to review.

Log Entries

1. Identification Number

Each piece of correspondence should be assigned a unique identification number. Each type of correspondence should also have a unique numbering system. For example, Submittal Request numbers should start with "SR," followed by a number, just as Price Requests should start with "PR," and so on. This will make it much easier not only for the manager, but for all other project participants, to recognize what the correspondence refers to. It also makes it easier to log and file. Incoming correspondence should, where applicable, refer to an identification number when it is in response to an inquiry.

2. Date Received

It is vitally important to date-stamp every piece of paper that comes into a firm. Whether it is a set of plans, specifications, bids, proposals, or other correspondence items, put a date on it the day you receive it. Do not depend on the date printed on the correspondence by the sender. Many contested claims are determined based on the sequence of events and information received or not received. By date-stamping the documents yourself, you apply a record keeping tool of your own.

3. Date Correspondence Goes Out

It is important to be conscientious about accurately dating (and logging) outgoing correspondence. There are two reasons for making sure this practice is carried out. First, your reputation for putting the correct date on a piece of correspondence supports you in negotiations and other areas of question on the job. A reputation for back-dating memorandums or otherwise manipulating correspondence will only make your case more difficult to support when the true dates really are on your side. In such cases, people will assume you misrepresented the date, even when they have no proof. Second, accurate dates will give you a reliable history of the way events really occurred in order to evaluate your own performance.

4. Short Description

This entry should be what it says–brief, and should simply convey the subject matter included in the correspondence.

5. From and To

In most logs, it is necessary to identify *only* a "from" or a "to," but not both. It really depends on who originates the correspondence. If you are the originator, then you will always know the "from," but not necessarily the "to."

For instance: a Submittal Request might be from the general contractor to a *specific subcontractor.* In other cases, however, the correspondence may originate from the subcontractor in the form of a Submittal. In this instance, you know the "to" is you, but the "from" must be identified.

Pricing correspondence needs further clarification in that the document is issued by the architect and then passes through the general contractor to the subcontractors. Now the process reverses. A number of subcontractors submit their responses; these are gathered together by the general contractor and then sent on to the architect/owner. At this point, a new log is started for change orders.

6. Due Date

A crucial element in the management of a construction project is the assignment of "due dates" (and the associated follow-up dates). There should be a due date on every piece of correspondence–especially if there is a response required. The due date should appear clearly in both the correspondence and the logs.

7. Date Actually Received

The date a document is received should always be recorded by immediately date-stamping it. Often, we are so glad to see the information that we forget to document how late we received it. By consistently recording the date the document is actually received, and by logging it, you will have a record of the subcontractor's, the architect's, and/or the owner's promptness for reference, should there be a problem down the line.

Dedicated Computer Program

A dedicated computer program designed specifically for construction can provide the required types of logs with conveniences and efficiencies not easily attainable in a manual system. The Paperworks construction software program has been used throughout this book to create example illustrations of computer-generated letters, Transmittals, and logs.

Not only does a computerized system make it easy to produce correspondence with proper identification automatically attached, but it can also produce the following logs.

- **Summary Log:** Shows all of the information that was included in the original letter sent from your office.
- **Summary Log, abbreviated:** Same as above, but it does not include the text of the letter that was sent out. It will show due dates, date sent, subject covered.
- **Open Log:** Shows all of the letters sent that require a response, and have not yet received one.
- **Open Log, abbreviated:** Same as the Open Log, but it does not include the letter or item detail.
- **Delinquent Log:** A log that shows only those letters and items that are past due according to the due date stated in the letter.
- **Delinquent Log, abbreviated:** Same as the Delinquent Log, showing all of the letters and items that are past due, but not the detail of the letter.

Remember: one of the advantages of a computerized system is that the basic information can be input once and then easily reproduced in different useful formats. In a manual system, a handwritten or typed document exists in only one place and one form, unless further steps are taken.

The Summary Logs act as reference reports, providing adequate detail for quick reference.

The Open Logs are "exception" reports. They show what needs to be done, either tomorrow, yesterday, or a year from now.

The Delinquent Log is both an exception report and a management report. It shows only what has not been done and is past due. If the due dates have been set with reasonable care, this is the report to watch carefully. This is where the trouble starts and, for the most part, if you can keep your correspondence items off this list, you should have a smooth-running project. Take care of the little problems, and there will be more time to resolve any big issues that may arise on any construction project.

Sample Logs

Each of the specific document chapters in this book contains examples of typical logs completed in a manual system, as well as examples of computer-generated logs compiled from the original correspondence documents. These chapters (and logs) cover most of the various types of construction correspondence. See Chapter 6, "Computer Applications," for more about the advantages of computerized correspondence programs.

Examples of the logs can be found in their appropriate chapters as follows:

Abbreviation	Subject Covered	Chapter
TR	Transmittals	3
RFI	Requests for Information	4
SR	Submittal Requests	11
SA	Submittal to Architect	12
RS	Returned Submittals	13
PR	Price Requests	19
PQ	Price Request Quote to Owner	19
FO	Field Order	20
FQ	Field Order Quote to Owner	20
CL	Clarification	21
CQ	Clarification Quote to Owner	21
RC	Request for Change	22
RQ	Request for Change Quote to Owner	22

Chapter Six
Computer Applications

Introduction

Making the most productive use of experienced personnel over as many projects as possible is the most important reason for installing a computer program dedicated to construction correspondence. A dedicated system helps personnel not only to work more efficiently, but to respond with increased timeliness and accuracy. Another benefit is the reduction in clerical costs.

The axiom, "People do well, what the boss checks," certainly is applicable to the construction industry. One of the most effective applications of the computer is to quickly and easily identify what has not been done.

A computer can help utilize personnel more efficiently by providing:

1. A systematic method for storing information
2. Automatic processing of information, retaining the vital elements

The first feature, systematic information storage, allows less experienced personnel to do a more effective job and even to take over functions that normally require the attention of more experienced staff members. The second feature, automatic processing, relieves everyone of tedious work, such as cross-filing documents, filling out secondary tracking logs, and date-stamping every transaction.

Improvement Requires Change

Any company working on any project must deal with a variety of people and entities, and the assignment of "traditional duties" within the company structure. Integrating each of those players into a program to manage correspondence may require some realignment of duties, reeducation, and changes. Such changes, however, need not be earth-shaking, and will increase the efficiency, accuracy, and, therefore, the profitability of the organization.

In today's marketplace, a consistent system for tracking communications becomes very important. It cannot be done one way on one job and not at all on another job. This is one of the reasons to install a computerized system: it makes consistency easier to achieve.

A computer correspondence program benefits managers in the following ways:

- It places emphasis on getting the important things done, rather than trying to find out what the important things are.
- It provides timely information. All logs are updated as soon as the letter is written.
- Standardized reports allow a quick review of all data pertaining to a project.
- It makes possible quick, on-screen reviews, and saves the user from paper clutter.
- Special reports allow in-depth review of a particular job.
- Accuracy is improved.
- A standardized correspondence format creates consistency.
- Correspondence can be logged quickly and accurately.
- All correspondence can be reviewed and sent out more quickly, resulting in:
  - Quicker responses
  - Quicker processing of changes, thereby allowing for billing sooner.
- Easier, more accurate training of new personnel enables them to be more productive sooner.
- The number of items that "slip through the cracks" is significantly reduced.
- All correspondence and scope-of-work processing is done in the same way for every job.

Give Your High-Priced Help Some Help

Construction is, in general, a highly paid industry–for those who remain in it. It is fairly easy to determine whether an individual succeeds or fails, or is just marginally effective. Therefore, only the capable people are able to stay in the business over the long haul.

As a result, good people are becoming harder to find and to keep within an organization. The time constraints on the way jobs are bid, awarded, and started do not allow sufficient time to find, hire, and train new managers. Furthermore, due to the relatively high cost of maintaining construction management personnel on staff, it is not economical to keep too many "waiting in the wings."

The most time-intensive period in a new project is the first one-to-two months. New and existing personnel will be required "to hit the ground running," without much training or knowledge of the company's procedures. The second most time-consuming phase of a project is at the project's end, when the manager of a construction team can get into trouble by not documenting, reviewing, and seeing things properly completed. To make matters worse, that same manager may be involved in the start-up of a new project, so that both busy periods may demand his attention at the same time.

Generally, there is no warning when a job priced six months ago is suddenly awarded and the owner wants to start next week. Few companies will decline the opportunity to accept such work, even if they are involved in other projects already. Very often, the key project managers of the company are locked into completing another project. Removing a key manager or diluting his efforts in these critical last weeks may cause the project to finish late, thereby risking liquidated damages, among other considerations.

Because of the pressures of time, there is an urgency to assign someone to the new job in order to get started. These pressures may contribute to some of the worst hiring and training possible, with a resulting negative impact on project profits. The terrible part is that even a good project manager, recently hired, will be blamed for bad results, when it really was a failure of management to provide adequate, trained personnel on the project.

An efficient correspondence control system can help to eliminate some of these pressures and the resulting problems by utilizing the experience of trained personnel, and transferring some of the clerical work to less experienced, or junior, persons.

For example, when it comes to processing a Submittal Request, a good manager begins by reviewing the specifications and making some notes. A junior construction person can then use the computer system to create the Submittal Request letters and get that part of the procedure under way (see Chapter 11 for coverage of Submittal Requests.) A still less experienced person can handle plan distribution–provided proper direction is given.

Since the job was bid some months ago, the subcontractor prices must be verified. A junior person can handle this task, provided he or she is properly instructed (see Chapter 3, "Transmittals").

If the system provides easily-retrievable logs showing what has not been done and who is responsible, then a junior person can call around to find out why this work has not been completed.

Items with a long lead time must be confirmed and insurance certificates obtained before work commences (see Chapter 10, "Transmittals"). An efficient, established system can get this process started with a minimum of pain. Using a construction correspondence computer system, new or temporary help can be up-to-speed on what needs to be done within a couple of hours. The in-house system should be simple, efficient, and flexible enough to help fill the void caused by too much work and not enough skilled people. It should allow experienced personnel to direct their talents where they are needed most. The system should also have a good "exception" listing program to make senior management aware of what has not been done.

It is clearly more cost-effective to hire two capable, but less experienced $10.00-per-hour persons rather than one good $30.00-per-hour senior person (provided such a resource is available). Not only is the senior person paid more, but he or she must take the time to become familiar with the company's systems, procedures, and corporate structure.

Detail work can be delegated to the less experienced personnel in order to get the most coverage from senior personnel. This concept is as valid for existing projects as it is for new, surprise projects. A field supervisor would not hesitate to recommend that a laborer be hired if that would make 25% more production possible from three existing skilled craftsman. In the office, the benefits may not be evident so immediately. Generally, most firms try to "get by" with as low an overhead as possible. Although low overhead is good, up to 30% more volume is possible on a job with a good system and good clerical help.

Suppose only a dollar-for-dollar increase in productivity is possible. Assume the project manager's cost to the company is $80,000/year for handling $20 million of work. Assume that a correspondence system and clerical staff cost an additional $20,000 per year for a total Project Management cost of $100,000. With the implementation of a good system, it is quite conceivable that these same people could be responsible for a gross volume of a possible $26 million. The ratio of increased volume to increased cost demonstrates an actual reduction in functional overhead per construction dollar and, therefore, an increase in profit per project.

Reductions in cost are one side of the coin. Smoother project starts with less miscommunication are the other side. The new project team will be able to cover peak loads better and could even help get a new job rolling until someone else is hired and properly trained.

Unfortunately, the scenario above is not put into practice often enough, even though it makes perfect sense. Management tends to tolerate loss of efficiency in the field more easily than an extra clerk in the office or a modification of their paperwork flow "because we've always done it this way." Job site waste is not as visible and is usually charged to the project. The people in the office are often viewed as undesirable overhead–no matter what real benefit they bring to the operation.

It is important to understand the different functions of the field and the office. A journeyman carpenter hired off of the street will be pretty efficient. He knows how to hang a door, or construct concrete forms. It is hard to find a project manager or assistant who does not have to receive some degree of training. It takes six months to find out if a new project manager is going to "cut it," and three months for an assistant. It is no more possible to run out to a "Rent-a-PM" firm than it would be practical for a major city to rely on "Rent-a-Fireman." With fires and with construction, experienced personnel are required, often very suddenly.

A key ingredient in good management is finding the fine line between too big a staff and too little. Too much and the overhead erodes profit. Too little and unexpected work is lost or bungled, which also erodes profit. Many times, the cost of an effective communication system can be lost on *one job* poorly handled.

Computerized Correspondence

A computer-generated correspondence program is designed to include all information that should go into a business letter. The system should print all letters in a standardized format so they are consistently well-organized. It also "saves" a copy of each letter and makes it easy to set target dates for various types of letters issued. It is not intended to be an accounting program, but in a way, it keeps account of the communications flow by tracking the location and status of all construction correspondence.

The major types of construction correspondence are discussed in Chapters 9 through 12, and 14 through 17. Each of these chapters includes an example document prepared by an in-house computer correspondence system, as well as a manual (pre-printed form filled out by hand) version. (See the Table of Contents for a complete listing.)

Project correspondence information is stored in a computerized data base,where almost any kind of report can be generated in minutes to solve a particular problem or answer a particular question. No exceptional skill level is required to run the system because most of the procedures (document numbering, filing, etc.) are menu-driven and programmed to run automatically.

A correspondence program can also generate special reports such as: *Summary Reports, Exception Logs, Special Exception Logs, Subcontractors by Trade, Owners and Architects by Category or Name, Insurance Status of Subcontractors or Vendors,* and many more. The computer can produce these specific reports through its ability to isolate data within a given set of parameters. Some of the reports that can be generated are described below.

Summary Reports are reference reports that show all data about the particular piece of correspondence.

Exception Logs show all outstanding correspondence that requires a response. Correspondence that has either received a response, or did not require one, will not show up on this log. The Exception Report/Log is a valuable tool for identifying what needs to be done, rather than what has been done.

Special Exception Logs are a further type of Exception Log. This type of log lists all the items of correspondence for which a response is required by a given date, but has not been responded to by the due date. A correspondence item does not show up on this report until that due date has arrived and the response becomes delinquent. This is the "bad guys" list.

If the due date originally assigned is reasonable, management should find out why there has not been a response. In practice, it is very easy to scan the "delinquent report" and circle or highlight the "past due" critical items. A junior employee can then be assigned to call the subject party and ask why they have not responded. This is called *follow-up,* and *follow-through.* With the sorting capability of a good computer program, much follow-up can be done by junior personnel.

One warning worth repeating: In the same way a computerized scheduling system can be made to distort the truth about a project's time frame, a computerized correspondence program only reflects the project based on the information input. Therefore, internal procedures must be established. A messy job correspondence and filing system input to a computer is simply a *faster* messy system.

Summary

More and more contractors are automating their accounting systems because they recognize the utility of the computer for tracking that information. Similarly, a computerized correspondence control system can be thought of as Communication Accounting. Properly established within the organization, a good correspondence "accounting" system will account for the location, due dates, and past due dates of all correspondence. This accounting of the job correspondence helps to prevent problems and lost items, and contributes to a well-managed and profitable project.

Part Three
Special Communications

Chapter Seven
The Bidding Process

To show a profit, you must first get the job. That point is so obvious, it is often forgotten. Keeping the potential contract "pipeline" full is an often neglected part of a firm's activity. There should always be enough work starting to offset the jobs that are being finished. An ideal work ratio is about 50% negotiated work and 50% "hard bid" contract work. Whatever the firm's work ratio, or product mix, it is important to have a clear understanding of what is to be priced or bid.

No contract should be entered into unless there is a firm and substantiated belief that the company will make money on the job. The Contract Documents are the basic communication vehicle to ensure that the framework for a good project is established properly. A *good* project is one that makes the owner proud and the contractor profitable.

Special Considerations for the Facility Manager

For the facility manager, good plans and specifications will make the difference between an in-house job completed within budget, and one that requires mountains of cost overrun justification. In-house projects that are not completed within budget or on schedule tend to weaken the confidence of the user departments. As a result, those users will be requesting that an outside contractor be hired for future projects. They might even be more cooperative in providing information to an outside contractor than they were with their own in-house contractor. In order to overcome such effects of interdepartmental competition, one must be very professional, require information in a timely manner, and create an overall image of competence.

Since the role of the facility manager is, in a sense, the same as that of the general contractor, the guidelines in this book as directed to the general contractor also apply to the facility manager. In fact, since the facility manager may have to cope with interdepartmental competitiveness, it may be even more important that he or she document all requests, questions, and answers from the various departments served.

Coverage

Of paramount importance to any contractor or facility manager is *coverage.* This is the number of qualified individual subcontractor bids received and reviewed for each element of the project. As a rule of thumb, there should *never* be fewer than three qualified bids for each phase of the work. If possible, obtain five-to-eight separate, *qualified* bidders for each subtrade.

A bidder call list is a useful tool for ensuring adequate coverage. A sample bidder call list is shown in Figure 7.1. Follow-up and follow-through are essential to obtaining adequate subcontractor coverage.

Requests for Information

Good plans make for a good job. If the plans and specifications are not well-prepared, then it is the obligation of the contractor or facility manager to ask the right questions and obtain the appropriate answers *in writing* prior to making a price commitment. In other words, make sure the necessary information is known and available before making a commitment.

The Request for Information (RFI) form is an effective tool in the bidding period. (It is also an important device during the course of construction; its use at that time will be discussed in Chapter 4). The following excerpts from the contractor's part of the conversation with the architect show how *not* to obtain information.

> *"What do you mean by the steel beam detail on page A.3?*
>
> *Oh? You mean the beam is supposed to be a glu-lam and not steel?*
>
> *Well, that changes the picture, doesn't it? Ok, I'll figure it as a glu-lam beam."*

If there is time available, the contractor should send an RFI to obtain this information, and then make sure that a written response is received. If there is not enough time to process an RFI, then the following procedure should be used.

The Transmittal as a Confirming Letter

If there is no time to write up and send an RFI and await a response, the general contractor, facility manager, or subcontractor should make an inquiring phone call, but follow it up with a confirming Transmittal. Refer to Chapter 3, "Transmittals," for further information about Transmittals and Transmittal logs.

If a Transmittal confirming a particular detail is sent, and the architect/engineer does not respond by saying, "no, that is not a correct interpretation of the plans or specifications," it is correct to assume that the Transmittal was correct. Figure 7.2 is an example of a confirming Transmittal of a telephone conversation. The Transmittal also could convey an "assumption," or "basis," of the estimate and bid.

On those projects with "fuzzy" plans and specifications, include a copy of all confirming Transmittals as a part of the bid submittal. A recap of each Transmittal could also be submitted. In either case, the qualifying assumptions should accompany the bid for the project.

By taking the time to clarify project information, one could miss out on a job opportunity. However, not getting a job is preferable to losing money on it later. Always be concerned when the drawings appear incomplete or vague. This may indicate that someone does not know what they want or what they are doing.

Means® Forms

BIDDER CALL LIST

DATE 1/17/90

BY JEG

PROJECT #321 Gibson Office Building

ITEM Plumbing System

DATE	RESPONSE: BIDDING Y/N	RESPONSE: NO ANSWER	RESPONSE: CALL BACK	RESPONSE: WILL CALL	PHONE	SUBCONTRACTOR/MATERIAL SUPPLIER	COMMENTS
1/17	Y			✓	555-1234	Candu Plumbing	
1/17	Y			✓	555-6789	Dondatube Plumbing	
1/17	N				555-1357	Broecken Piping	Mech/HVAC only
1/18			1/20		555-2468	I. Flushem & Sons	Answer machine: call back
1/18	Y			✓	555-4321	Watt A. Rieleaf Co.	
1/18	Y			✓	555-5555	I. Gottago and Daughter Co.	

Figure 7.1

Means Forms

LETTER OF TRANSMITTAL

Delta Construction Company
6678 Mission Road
San Diego, CA 92120
(619) 581-6315

FROM: J. E. Grimes
Project Manager

TO: One Line Design
2001 Dream Street
San Diego, Ca. 92100

DATE 2/21/90
PROJECT Gibson Office Bldg. #321
LOCATION
ATTENTION Rod Wright
RE:
Beam Detail - A.3.

TR 321.29

Gentlemen:

WE ARE SENDING YOU ☒ HEREWITH ☐ DELIVERED BY HAND ☐ UNDER SEPARATE COVER

VIA ______ THE FOLLOWING ITEMS:

☐ PLANS ☐ PRINTS ☐ SHOP DRAWINGS ☐ SAMPLES ☐ SPECIFICATIONS

☐ ESTIMATES ☐ COPY OF LETTER ☒ confirmation of information

COPIES	DATE OR NO.	DESCRIPTION

THESE ARE TRANSMITTED AS INDICATED BELOW

☐ FOR YOUR USE ☐ APPROVED AS NOTED ☐ RETURN ______ CORRECTED PRINTS

☐ FOR APPROVAL ☐ APPROVED FOR CONSTRUCTION ☐ SUBMIT ______ COPIES FOR ______

☐ AS REQUESTED ☐ RETURNED FOR CORRECTIONS ☐ RESUBMIT ______ COPIES FOR ______

☒ FOR REVIEW AND COMMENT ☐ RETURNED AFTER LOAN TO US ☐ FOR BIDS DUE ______

☐ ______

REMARKS: Per our conversation this morning, 2/21/90, we will figure the beam, as detailed on drawing A.3, as a glu-lam member instead of as a steel beam.

IF ENCLOSURES ARE NOT AS INDICATED, PLEASE NOTIFY US AT ONCE.

SIGNED: J Edward Grimes

Figure 7.2

Taking Telephone Bids

After the plans and specifications have been clarified and the assumptions verified, the bid will depend on the qualified subcontractor coverage already requested and verified by someone in the company or department. A majority of these subcontractor bids will be received by telephone. It is, therefore, necessary and important to record these communications accurately and uniformly.

The individual who is responsible for, or most familiar with, the plans should prepare a sheet listing items to be included or questions to be asked of any company submitting a telephone bid. An example bid question sheet is shown in Figure 7.3. These questions can be asked of each of the telephone bidders.

Telephone Bid Form

By using the subcontractor question sheet and a standard telephone bid form, the bid responses can be recorded in a uniform way, for easier reference. The bids can also be more easily compared, one to another. This is especially helpful when the bid is due in 15 minutes and none of the subcontractor bidders can be reached.

When filling in telephone bid forms, it is important to list one trade or specification item per page. Frequently, when a bidder is quoting on a number of different specification sections, such as "Toilet Partitions, Toilet Accessories, Mirrors, and Finish Hardware," the various sections cannot be compared to another bidder who may bid only "Finish Hardware" or "Mirrors." Figure 7.4 is a standard Means Telephone Quotation Form (from *Means Forms for Building Construction Professionals*). The choice of form used is not crucial, as long as all the information is recorded for every quotation, complete with the information requested in the Subcontractor Bid Question Sheet.

Make sure that all exceptions noted by the bidder are clearly shown. Always have bidders confirm their telephone quotes. Have them state in writing all exceptions, variations, inclusions, and exclusions (plus inevitable assumptions). Also ask where money can be saved. (Use their experience!)

A spreadsheet (such as that shown in Figure 7.5) can be used in conjunction with the telephone bid forms to keep a current record of information. Across the top of this ledger-type spreadsheet in the boxes provided are the names of the expected bidders (allow spaces for additional bidders). Below is a space for the "base bid."

One can list (along the side space for "Description") all items called for in the specifications, as well as some that are not actually spelled out, but should be. This form can be used as a checklist when the bid is being called in. If the subcontractor says that he has included the item, simply put a check mark or circle the box in which a dollar figure can be recorded. In this way, an equal comparison can be made between contractors at the "bottom line" of the page.

The example shown in Figure 7.5 is a simplified version, but it shows how such a system can work.

Means® Forms

CONDENSED
Bid SUMMARY

SHEET NO. 1/1

PROJECT Gibson Office Bldg. #321

ESTIMATE NO. 615-90

LOCATION

TOTAL AREA/VOLUME

DATE 1/27/90

ARCHITECT One Line Design

COST PER S.F./C.F.

NO. OF STORIES

Bid Section: 03-Concrete

EXTENSIONS BY:

CHECKED BY:

	ABC Conc.	Krackup & Brache	Miracle Works Inc	Bildenhope Co.	Budget
Contact	K. Kong	Pete Sakes	Jack nJille	JuanMotime	
Phone #	555-1928	555-2189	555-3214	555-4912	
Base Bid					
Does bid include: Conc Forms Reinf.					
A Bolts - Material Labor					
Finish Cure Protect					
Special Seal @ Lab					
Joints					
Driveway approach aprons					
Topping at Mezzanine					
Precast floor/deck					
Flagpole Bases					
Sidewalks					
Ramps					
Curbs					
Stairs/Fill pans					
All taxes, fees					
Per plans/specs					
permits					

Figure 7.3

Means® Forms

TELEPHONE QUOTATION

	DATE
PROJECT	TIME
FIRM QUOTING	PHONE ( )
ADDRESS	BY
ITEM QUOTED	RECEIVED BY

WORK INCLUDED	AMOUNT OF QUOTATION
DELIVERY TIME **TOTAL BID**	
DOES QUOTATION INCLUDE THE FOLLOWING: If ☐ NO is checked, determine the following:	
STATE & LOCAL SALES TAXES ☐ YES ☐ NO MATERIAL VALUE	
DELIVERY TO THE JOB SITE ☐ YES ☐ NO WEIGHT	
COMPLETE INSTALLATION ☐ YES ☐ NO QUANTITY	
COMPLETE SECTION AS PER PLANS & SPECIFICATIONS ☐ YES ☐ NO DESCRIBE BELOW	
EXCLUSIONS AND QUALIFICATIONS	
ADDENDA ACKNOWLEDGEMENT **TOTAL ADJUSTMENTS**	
ADJUSTED TOTAL BID	

Figure 7.4

Means® Forms

CONDENSED ESTIMATE SUMMARY

SHEET NO. 1/1

PROJECT Gibson Office Bldg., #321

ESTIMATE NO.

LOCATION San Diego

TOTAL AREA/VOLUME

DATE 1-27-90

ARCHITECT One Line Design

COST PER S.F./C.F.

NO. OF STORIES

Bid Section 03 Concrete

EXTENSIONS BY:

CHECKED BY:

	ABC Concrete	Krackup & Brache	Miracle Works, Inc.	Bildenhope Co.	Budget
Contact	K. Kong	Pete Sake	Jack n Jill	Juan Motime	—
Phone #	555-1928	555-2189	555-3214	555-4912	—
Base Bid	229010	256550	244022	241663	266100
Does bid include: Conc., Forms, Reinf.?	✓	✓	✓	✓	✓
A bolts - Material	N/A	(✓) (1200)	N/A	N/A	(✓) (900)
Labor	✓	✓	✓	✓	✓
Finish - Cure - Protect	✓	✓	✓	✓	✓
Special Seal @ lab.	✓	✓	✓	✓	✓
Joints	✓	✓	✓	✓	✓
Driveway approach Apron	(0 8000)	✓	✓	✓	✓
Topping at Mezzanine	(0 6600)	✓	✓	✓	✓
Precast Floor/Deck	N/A	N/A	N/A	N/A	N/A
Flagpole Bases	(0 600)	✓	✓	✓	(0 600)
Sidewalks	✓	✓	✓	✓	✓
Ramps	✓	✓	✓		
Curbs	✓	✓	✓		(3200) in "02"
All taxes, fees	✓	✓	✓	✓	✓
Per Plans/Specs	✓	✓	✓	✓	✓
Permits	✓	✓	✓	✓	✓
Totals	249610	255350	244022		269000
Alternates	—	—	—		
$ after negotiations	246000	—	239950	(237000)	✓
Will meet schedule	yes	—	yes "will try"	yes ready to go	

Figure 7.5

Confirm Bid or Low Bid

The bid confirmation letter may not be used often, but it is simple to write and makes for a very inexpensive "insurance policy." In some cases, the owner may take four or five months to actually award the contract. The original subcontractor bids should be re-examined for validity.

It is easier to get the subcontractor/vendor to agree to hold a price firm for five months just after the bid has been submitted, than it is to have them confirm and hold a price they quoted five months ago when the job has finally been awarded. The subcontractor may have to stipulate that the price is good only for 30 days due to material price fluctuations.

Take the time to buy "insurance" by sending this "Confirmation of Bid" letter. Since this letter goes out prior to the project award, the price will probably stay firm for several months. Keep a log of confirmed bids or copies of the letter with the bid proposal. Also, keep a log or checklist of who has responded. If someone responds with a "bid good for 30-days" time stipulation, look for another bidder or make sure that the owner is aware that the original bid is good only for 30 days.

Chapter Eight
Pre-Contract Communications

Pre-contract communications occur in the "gray area" of time–that period between the presentation of a bid by the subcontractor, general contractor, or facility manager to the owner or user, and the time the contract is executed completely, i.e., signed by all parties.

A great deal of communication is necessary when the original documents, plans, and specifications (prepared for bidding purposes) are modified in order to:

- Clarify bid documents
- Maintain the project budget
- Change materials
- Incorporate changes required by the Building Department
- Include the final changes requested by the owner/architect

The documents issued for bid often do not incorporate all the final changes as requested by the owner. The owner is naturally anxious to find out how much the project is going to cost. Rather than wait for the final completed drawings, the bid package is released one revision shy of the finished product. Many times there are scope of work changes with a major impact included in the "final" set.

Usually the cause of such modifications to the original construction documents is the prospect of the project costs exceeding the budget. This means that to reduce the cost, there must be a change in the scope of work from the original plan. The architect or engineer may be reluctant to modify their original designs. Meanwhile, the owner, even though he realizes that there must be some cost reductions, still expects the product to be completed as originally envisioned. It is, therefore, imperative to document both the potential cost savings and reductions, as well as the effect of all changes upon the completed project .

Lack of Documentation

Failing to document decisions during this "gray period" of time can cause as many problems as all the scope of work changes that occur later in the construction project. During this period, which may last a few weeks or several months, a number of conversations and decisions take place concerning the project design and price.

Many telephone calls are made between the involved parties (owner, contractor, architect/engineer, and subcontractors), who are exchanging ideas on how to reduce the project cost. Sometimes these communications may "shortcut" the regular procedures (e.g., the owner or architect may be in direct contact with the subcontractor, rather than going through the general contractor). Not following proper procedures is risky, and should be avoided.

The following types of activities are likely to be involved during this pre-contract period:

- Redesigning the work and rebidding parts of the project. This means that all subcontractors not affected by the redesign must maintain their prices while the redesign and bidding is accomplished.
- Value engineering (i.e., substituting lower-cost materials for those specified in the original documents)
- Reducing the scope of work
- Negotiating the price
- Reviewing (reducing) the General Conditions
- Rebidding certain portions of the work
- Rebidding the total project
- Phasing the work (e.g., instead of constructing all of the building and the tenant work, delaying the tenant work until tenants are signed up.) This may not be the most cost-effective overall, but it does help on a job with a budget deficit. Naturally, this will involve rebidding a portion or all of the work at a later date.

Up to the time the bids are in and the "gray period" starts, the project is usually documented and recorded in a satisfactory manner. During the "gray period," however, the documentation of decisions, the change in the scope of work, as well as the price of the work, can get out of control.

A Typical Example

The following is a typical example of what can happen during this time.

Background: *The project is hard bid, and the project cost is over-budget by $20,000. The general contractor has been informed that he is the low bidder, and if the project can be reduced by $25,000, the job is his. The following telephone call might occur between the general contractor and the low-bidding roofer.*

General Contractor: *"Art, we need to cut some money out of the ABC project. Got any ideas?"*

Subcontractor: *"Yeah, the specs call for 4 layers of fifteen-pound felt; we could get by with just 3 layers and save about $1,200."*

General Contractor: *"Sounds like a good idea. I'll let you know."*

The following telephone call might follow between the contractor and the owner.

General Contractor: *"I wanted to review our value engineering items to see if they are acceptable to you. We can reduce the layers of felt on the roof from 4 to 3. This still results in a Class IV roof, as specified, but it will reduce the cost by $1,200."*

Owner: *"Sounds OK to me as long as I get a roof that doesn't leak. How is it coming on the rest of the cost savings?"*

General Contractor: *"Now that I can include the roofing deduct, we look very good. I think that we should have it within your budget by Friday."*

So far so good. Remember, however, that in this situation, this type of conversation will probably occur with most of the trades involved in the project. This means that there could be as many as 50 to 100 value engineering telephone calls between all parties involved up to this point. When reviewing potential cost savings ideas with the major trades, such as the HVAC and electrical bidders, there may be 20 to 30 items involved that might be factors in cost savings. It is immediately apparent that the documentation of the cost savings proposals must be done with care and diligence. The telephone conversations in the previous example are all too common. The general contractor may achieve the cost reductions of $25,000 and, without being specific, submit a summary list of trades to be reduced such as:

Gibson Office Building
Value Engineering Suggestions

Description:	**Orig. Bid**	**Revised Bid**	**Savings**
Concrete	$65,212	$61,178	$4,034
Roofing	14,165	12,965	1,200
Structural Steel	139,554	135,368	4,186
Glass and Glazing	45,649	44,280	1,369
Marble and Tile Flooring	33,910	26,973	6,937
Partitions	66,750	62,101	4,649
Wall Finishes	19,564	15,150	4,414
Total Savings			**$26,789**

For the moment all parties are happy.

- The owner is happy because the contract price now is within his budget, and meets his original criteria.
- The architect is happy because he does not have to redesign the project to come in under-budget.
- The general contractor is happy that he has the job.
- The subcontractor is happy because he, too, has the job.
- The employees of the above parties are happy because they are assured of having a job for the next few months.

Potential problems may become apparent whenever any of the following occurs. (Remember: multiply this by 20 to 30 separate trades where cost reductions were taken.)

- The subcontractor does not remember the telephone conversation regarding the $1,200 savings.
- When the work is to be done in the field and the roofer only puts down three layers of felt, the roof inspector or architect, not having knowledge of the "value engineering" plan, rejects the work.
- The owner does not remember that the value engineering was to reduce the layers of felt from four to three. Since it was not spelled out or described in the "Value Engineering Suggestions," and the specifications have not changed, the owner, on the advice of his architect, demands four layers of felt.

Documenting Activities

Using the same scenario as our previous example, the next example shows what could, or should, have been done to alleviate some of the problems when documenting activities between bidding and contract signing.

General Contractor (in telephone call to low roofing bidder): *"Art, we need to cut some money out of the ABC project. Got any ideas?"*

(Comment: Send a Transmittal to all apparent low bidders and ask for cost savings suggestions, mentioning that the project may have to be shelved if the costs cannot be reduced. It might be possible to save considerably more than the $25,000 target contract reduction. Ask that all cost savings suggestions be in writing, and describe the scope changes required. Follow up with a telephone call for any areas that can be pursued. Writing a standard type Transmittal for 20-30 subcontract bidders may be easier than playing "telephone tag" with them.)

Subcontractor: *"Yeah, the specs call for 4 layers of 15-pound felt, but we could get by with just 3 layers and save about $1,200."*

General Contractor: *"Sounds like a good idea, I'll let you know."*

(Comment: General contractor should get the proposal for cost savings in writing, with a description of any change in the scope of work for each sub-trade and how long the reduction remains valid.)

General Contractor *(in telephone call to the owner): "I just wanted to review our value engineering items to see if they are acceptable to you. We reduce the layers of felt on the roof from 4 layers to 3, while still providing a Class IV roof as specified. This will reduce the cost by $1,200."*

Owner: *"Sounds OK to me as long as I can get a roof that doesn't leak and I can still get my 15-year warranty. How is it coming on the rest of the cost savings?"*

General Contractor: *"Now that I can include the roofing deduct, it looks very good. I think that we should have it within your budget by Friday. I'll send you a confirmation of our value engineering by then."*

(Comment: Following the initial agreement, a confirming Transmittal should be sent to the owner, with a copy to the architect describing the change in scope of work and the reduction in dollars.)

Please note that it is a good idea to send a Transmittal to the owner requesting that all of the cost savings scope-of-work reductions be included in an up-to-date set of construction documents from the architect's office. A response to such a request is rarely carried out, but the fact that the letter was sent means that when there is confusion later in the job, the contractor has some defense. This documentation is also good construction practice. If the owner does not want to spend the extra money (on revised plans), then no one can spend it for him. Nevertheless, it is better to be protected by an "I told you so" than to have either the owner or the architect pointing a finger at the contractor for not "coordinating" the work.

When submitting the Value Engineering Suggestions or Cost Savings Suggestions, be sure to include in the spreadsheet exactly what and how these cost savings were achieved. The example situation would require an attachment to the cost savings proposals spelling out that the roofer will only install 3 layers of 15-pound felt, rather than 4 layers of 15-pound felt.

Cost Savings as Part of the Contract

When writing the contract between the owner and the general contractor, include the cost savings proposal as a part of the contractual documents. Back-up documentation can be included as an attachment to the contract. The back-up should spell out clearly, and in detail, any deviations from the original construction documents in order to reduce the price. If the contract has already been written, then incorporate the cost savings proposal in a *Request for Change Quote to Owner* (see Chapter 22), and include the back-up documentation as well. Document what was given up in order to achieve the cost savings.

When writing the contract to the roofing subcontractor, note the original bid price and the deduction of one layer of felt with the revised total. In this way, there will be no question about the intent or what is covered (or not covered) in the agreement.

Use of the Transmittal in the Pre-Contract Period

An efficient correspondence system should help keep the owner and the architect fully aware of all changes under consideration. The use of the Transmittal (see Chapter 3) is the ideal means for communicating changes. Transmittals are short and to the point; copies can be sent to a number of different people; and they give the owner or architect the opportunity to say, "No, I do not want you to investigate the possibility of using masonite siding instead of redwood."

It is also important to keep the owner and architect aware of any change in the completion time when submitting cost-saving proposals. If a substitute material takes six weeks longer to procure than the material specified, make sure this information is included in the back-up for the cost savings proposal.

Keep All Issues of Construction Documents

When traveling through the pre-contract "gray area," it is extremely important to date-stamp all copies of the drawings issued by the architect, and to keep at least one copy of each edition until the time the actual contract is signed, incorporating the construction documents. It is preferable to keep all of the various issues of construction documents until the project is completed and final payment has been received.

The "gray area," or time between completion of bidding and signing of the contract, can be dangerous if the project is not well documented. Decisions made at this time may not be as well documented as in the other phases of the project due to efforts to hasten the start of construction. To avoid this error, keep all who are involved with the job informed of any changes, using an efficient "correspondence system;" retain all versions of the contract documents; and make sure that all current project documentation is assembled in one workable location.

Chapter Nine
Contracts

The contract is the basis for the project's construction, profits and, possibly, litigation. It is the foundation of the project, and it will, to some degree, dictate the orderliness of the construction process. The contract is, therefore, one of the most important pieces of paperwork in the project. There is an old saying, "good fences make good neighbors," a concept that also applies to the Construction Contract. A good contract will help make a good project by protecting all parties and, at the same time, ensuring that all uphold their respective obligations.

A "good" contract means one that does not favor one group over another, but treats all fairly and covers all issues. Thought must be given to as many conceivable occurrences as possible, and a fair resolution chosen in the event that such situations arise. The contract acts as a "bible" for that particular project, spelling out the codes or rules governing the project.

There are many types of contracts. However, this book is primarily concerned with the following general categories.

- The Contract between the Owner and Architect
- The Contract between the Owner and the General Contractor
- The "contractual agreement" between the user and the facility manager
- The Contract between the General Contractor and the Subcontractors
- The Contract between the Facility Management Department and a General Contractor, or directly with the Subcontractors

Pre-printed Contract and General Conditions Forms

Use one of the standard pre-printed forms that are available through the American Institute of Architects (AIA), the Associated General Contractors (AGC), or the National Association of Home Builders (NAHB). These forms may be modified simply by writing in additional information and/or crossing out the paragraphs and sections that do not apply. A major advantage of these pre-printed forms is a proven track record....they have been tested in court. Refer to the bottom of AIA Contract 101, where a note states that this document has been copyrighted in 1915, 1918, 1937, 1951, 1958, 1961, 1963, 1967, 1974, and 1977. Clearly, it has stood the test of time.

Many contractors feel they can write a contract better than a lawyer. They should keep in mind what their own reaction would be if a lawyer tried to build a house instead of utilizing the experience of a good contractor. If litigation results, remember that it will be played out in the lawyer's court.

Types of Contracts

Contract for a Stipulated Sum

The *Standard Form of Agreement between Owner and Contractor for a Stipulated Sum* (1977 Edition, AIA Document A101, by the American Institute of Architects and approved and endorsed by the Associated General Contractors of America) includes seven major Articles, as well as the information about the contracting parties and effective dates. A copy of this four-page form is shown in the Appendix. Review this document closely.

This type of contract basically states that the work, as described by the construction documents, will be completed for a fixed amount of money. Any change or addition will be considered as extra to the contract. This is the most common type of contract issued.

Contract for the Cost of the Work Plus a Fee

If a firm is involved in negotiated work or "cost of the work plus a fee," refer to *The Standard Form of Agreement between Owner and Contractor, where the basis for payment is the Cost for the Work Plus a Fee.* The 1978 Edition (AIA Document A111, approved and endorsed by the AGC) includes 16 major Articles, as well as information about the contracting parties and effective dates.

This type of contract states that the work, as described by the construction documents, will be completed for the actual costs of the project plus a predetermined amount as a fee. (Sometimes a percentage of the cost is used in lieu of a fee). This contract form is used when the scope size or quality of a project is not specifically known and the owner needs "protection" against large change orders. An advantage of this contract is that cost savings can be directly realized by an owner. The contract types include: guaranteed maximum–not to exceed, unit cost, and design/build.

Note: Many books focus only on contracts. Choose one and make it available to construction personnel as a reference text. Obtaining a good working knowledge of the "ins and outs" of construction law is a complex, if not downright difficult, business. The following book is quite thorough, although it covers only the AIA forms: *Handbook of Modern Construction Law,* by J.D. Lamber and L. White, Prentice Hall, 1982.

Contract Forms by the AIA

The American Institute of Architects has produced forms for almost every conceivable construction situation. The combined cost of research review, discussion, litigation review, and publishing required to produce one's own contract forms is tremendous. It never pays to reinvent the wheel, so choose existing forms suited to each project and leave the contract writing to the lawyers, experts, and organizations with money to spend.

The following is partial list of documents available through the local chapters of the American Institute of Architects at a nominal charge:

- A101-*Standard Form of Agreement between Owner and Contractor, Where the basis of payment is a Stipulated Sum* (1977)
- A111-*Standard Form of Agreement between Owner and Contractor, Where the basis of payment is the Cost of the Work plus a Fee* (1980)
- A201-*General Conditions of the Contract for Construction* (1976)
- A210/CM-*General Conditions of the Contract for Construction (Construction Management Edition*–1980)
- A401-*Subcontract, Standard Form of Agreement between Contractor and Subcontractor* (1978)
- A501-*Bidding Procedure and Contract Awards* (1969)
- B141-*Standard Form of Agreement between Owner and Architect (Construction Management Edition*, 1960)
- B801-*Standard Form of Agreement between Owner and Construction Manager* (1980)

Copies of these documents can be obtained by contacting the American Institute of Architects at 1735 New York Avenue, Washington, D.C. 20006.

Guide for Bidding Procedures

The Form A501-*Bidding Procedure and Contract Awards*, is not used often. No signatures are required, but they can be included as a part of the contract. The document refers to the responsibilities of the owner, the architect, the contractor, and subcontractors, as well as the time of completion, contracts, and subcontracts. This form may not be required for every contract, but it is useful to identify the various responsibilities, and where they are spelled out, in AIA form 201–*General Conditions*.

Manual of Contract Documents

The Associated General Contractors publishes a *Manual of Contract Documents* which includes everything from the General Conditions to the Equipment Rental Agreement, as well as several forms of construction agreements. The manual also includes AIA documents. It probably contains more contract forms than the average contractor will ever use, but it is one of those items that everyone should have. The list of documents included in the Manual is shown in Figure 9.1. Several standard AGC contracts have been reproduced in the Appendix, courtesy of the Associated General Contractors.

Summary

A good contract is the foundation for a good job. There are many types of contracts for construction, including the *Contract for a Stipulated Sum* and the *Contract for the Cost of the Work Plus a Fee.* Both the Associated General Contractors and the American Institute of Architects publish a collection of sample contracts. Use these contracts to take advantage of the accumulated knowledge and experience of construction professionals. Many informative books are also available on the subject of construction contracts. Contractors should build, and let attorneys practice law.

AGC Contract Documents	
200	Standard General Conditions of the Construction Contract (EJCDC 1910-8)
201	Contract Documents for Construction of Federally Assisted Water & Sewer Projects
202	Standard Form of Agreement Between Owner and Contractor on the Basis of a Stipulated Price (EJCDC 1910-8-A-1)
203	Standard Form of Agreement Between Owner and Contractor on the Basis of Cost-Plus (EJCDC 1910-8-A-1)
204	Change Order (EJCDC 1910-8-B)
205	Certificate of Substantial Completion (EJCDC 1910-8-D)
206	Application for Payment (EJCDC 1910-8-E)
207	Work Directive Change (EJCDC 1910-8-F)
209	Guide to the Preparation of Supplementary Conditions (EJCDC 1910-17)
210	Suggested Bid Form and Commentary for Use (EJCDC 1910-18)
211	Engineer's Letter to Owner Requesting Instructions re: Bonds and Insurance During Construction (EJCDC 1910-20)
212	Owner's Instructions to Engineer re: Bonds and Insurance During Construction (EJCDC 1910-21)
213	Notice of Award (EJCDC 1910-22)
214	Notice to Proceed (EJCDC 1910-23)
230	EPA Standard Proposal Form
300	Standard Form of Agreement Between Owner and Contractor (AIA A101)
301	General Conditions of the Contract for Construction (AIA A201)
305	Abbreviated Form of Agreement Between Owner and Contractor (AIA A107)
310	Standard Form of Agreement Between Owner and Contractor – Cost Plus Fee (AIA A111)
317	Abbreviated Form of Agreement Between Owner and Contractor (AIA A117)
320	Contractor's Qualification Statement (AIA A305)
327	Guide to Supplemental Conditions (AIA A511)
330	Change Order (AIA G701)
335	Application and Certificate for Payment and Continuation Sheet (AIA G702 & G703)
340	Certificate of Substantial Completion (AIA G704)
400	Preliminary Design-Build Agreement
410	Standard Form of Design-Build Agreement and General Conditions Between Owner and Contractor
415	Standard Form of Design-Build Agreement and General Conditions Between Owner and Contractor (Where the basis of compensation is a lump sum)
420	Standard Form of Agreement Between Contractor and Architect
430	Conditions Between Contractor and Subcontractor for Design-Build
440	Change Order/Contractor Fee Adjustment
450	Standard Design-Build Subcontract Agreement with Subcontractor not Providing Design
450-1	Standard Design Build Subcontract Agreement with Subcontractor Providing Design
500	Standard Form of Agreement Between Owner and Construction Manager
501	Amendment to Owner-Construction Manager Contract
510	Standard Form of Agreement Between Owner and Construction Manager (Owner Awards all Trade Contracts)
520	General Conditions for Trade Contractors Under Construction Management Agreements
525	Change Order/Construction Manager Fee Adjustment
600	Standard Subcontract Agreement for Building Construction
603	Short Form Subcontract
605	Standard Subbid Proposal
610	Subcontractor's Application for Payment
625	AGC Certificate of Substantial Completion
645	Standard Form of Negotiated Agreement Between Owner and Contractor
907	AGC Equipment Rental Agreement

Figure 9.1

Chapter Ten
Subcontracts

Like the contracts between the owner and architect, and the owner and contractor, subcontracts also need careful attention. Normally, a great deal of time and effort must be expended by both the architect and owner preparing a good set of construction documents that will result in an architecturally pleasing and functional project. It remains for the general contractor, the facility manager, and the subcontractors to bring all of the pieces together to make the project happen within budget and with a high level of craftsmanship.

The architect's plans do not isolate and individually identify the work of each trade. The specifications list the materials and methods for installing materials and, to some degree, a level of acceptable workmanship. There is, however, a great deal of interpretation in such phrases as "acceptable in the industry." Whose industry? "Acceptable," not "above average?" These are tough questions that must be answered for every project.

It is the responsibility of the general contractor and the facility manager to accurately define the scope of work for each subcontract. Good project management is the act of "defining the scope of work and reaching a consensus with the subcontractor on that definition." Well-written subcontracts are understandable and enforceable. They serve as a basis for a well-run project, in terms of time and quality.

Standard Subcontract Agreements

As recommended in Chapter 9, "Contracts," a standard subcontract form should be used. The Associated General Contractors (AGC) form 600, *Standard Subcontract Agreement for Building Construction*, and the shorter form 603, *Short Form Subcontract*, have generally been quite acceptable to both the general contractor and the subcontractor. (See the Appendix for copies of both documents.)

AGC Standard Subcontract Form 600, Long Form

The Standard Subcontract is intended to take into consideration the conditions that exist on construction sites today. The subcontract includes 16 articles covering such provisions as scope of work, schedule of work, contract price, payment, contractor's and subcontractor's obligations, indemnification, insurance, arbitration, and contract interpretation. One section allows for the inclusion of provisions specific to the subcontract, such as the scope of work, common temporary services, insurance coverage amounts, and special provisions. Appropriate labor relations provisions also may be tailored specifically and inserted into the subcontract.

AGC Subcontract Form 603, Short Form

This convenient form is designed specifically for use on subcontract work of limited dollar value to be completed within a relatively short time frame. The form has instructions for all fill-in provisions, and has nine general articles:

- Contract Payment
- Scope of Work
- Schedule of Work
- Changes
- Failure of Performance
- Insurance
- Indemnification
- Warranty
- Special Provisions

Pre-printed Subcontract Forms

Pre-printed subcontract forms may be modified by writing in the information that applies to a particular project, and then crossing out the paragraphs that do not apply. Again, the advantage of using the pre-printed forms available from AGC or AIA is that they have a proven track record. These documents have been to court. Some may disagree with certain statements in particular documents, but should recognize that the documents have been "tested."

The other big advantage of using pre-printed forms is that people in the industry have become accustomed to dealing with them and accept them, even with some of the built-in biases of a few of the documents. The pre-printed form can be adapted for particular uses. The alternative, a completely original and typewritten contract, forces the participants to have an attorney review the contract for each and every project to ensure complete and proper coverage. The possibility of change creates doubt, doubt creates fear, and fear leads to higher prices. No one wants higher prices than those offered by the competition.

Preferences for one form of document over another are usually based on which one each individual is comfortable with and used to. In addition to the AIA and AGC standard contracts, the National Association of Home Builders (NAHB) and the Building Industry Association (BIA) publish many standard forms and contracts. These forms are generally geared to smaller contracts and are quite adequate for use by general contractors, facility managers, and owner/builder projects. Choose a pre-printed and "tested" contract form and stick with it. This applies to contracts with owners/users as well as subcontractors.

Subcontract Scope of Work

The subcontract agreement is not complete without a clear, understandable description of the scope of work. This is an essential ingredient in the contract and is imperative to good project management.

The Three Basic Forms

There are three basic forms used to document the scope of work in the subcontract agreement. The choice of which to use depends on the type of project, the trade involved, and the degree of completeness of the construction documents. It is quite possible to use all three types on the same job, depending on the circumstances of the work to be accomplished, the trade involved, and the particular subcontractor. Each of the three versions is described below.

- **The Short Version:** The "brief as possible" method should be used only where the construction documents are complete, accurate, and coordinated. Readable plans and clear specifications must be available. The following is an example of a short version description.

  "All Sheet Metal work is included in the plans and specifications for the referenced project, but is not limited to the work described in Specification Section 0880."
- **The Medium Version:** Generally, this version is used when the construction documents are less than complete. The scope of work should be as detailed as possible, using the construction documents as a foundation, and including the subcontractor proposal as well as any additional items that need to be added for clarity and enforcement. How much is enough? There may never be enough detail, and gathering it all might take too much time. The amount of detail to include must be a judgment call on the part of the subcontract writer, based on the clarity of the scope of work, and the reputation of the subcontractor–including his/her financial health.
- **The Long Version:** The scope of work description in the long version contract should be very detailed. Projects that involve developer/builders usually require this type of description. In the case of a tract subdivision, for example, specifications may not exist, and the plans may not indicate material quality or level of workmanship. In such cases, the writer must "write the specifications" completely for the trade involved.

Drafting the Subcontract

Chapter 9, "Contracts," recommends the *Handbook of Modern Construction Law*, by Jeremiah Lambert and Lawrence White, both of the firm Peabody, Rivlin, Lambert and Meyers. The following excerpt from this volume highlights important points to consider in the preparation of subcontracts.

"A starting point may be the form of subcontracts promulgated by the AGC in 1966 and the American Institute of Architects (AIA Document A401) in 1972, both widely recognized as neutral forms, the terms of which favor neither the general contractor nor the subcontractor.

"A subcontractor should be familiar with the terms of the AGC and AIA subcontract forms, and, if circumstances permit, should not hesitate to insist on the substitution of a neutral clause from these forms for an inequitable provision suggested by the general contractor.

"Sometimes a subcontractor may wish to substitute another form for the one the general contractor wishes to use. This will normally be rejected as the general contractor will wish to keep all the forms alike so he knows exactly where to find specific clauses in all subcontracts.

"The AGC subcontract form is more readily accepted by general contractors than the AIA subcontract form because the AGC's form is sponsored and approved by the general contractor's own national association. In any event, either the AIA or the AGC standard subcontract form contains terms that protect both the general contractor and the subcontractor, and both forms are certainly more

neutral in respect to the subcontractors than the loaded subcontract form devised by the general contractors.

"Many subcontractors include stipulations in their bids that the subcontractor's bid be conditioned upon use of an AIA or AGC standard subcontract form between the parties if their bid is accepted. If the subcontractor's bid is accepted, the use of the stipulated standard form of subcontract may then be legally required to be used by the general contractor. However, even this requirement can be negotiated before acceptance."

Key Items to be Included

In addition to the normal items to include in the subcontract agreement (such as Contract Payment, Scope of Work, Schedule of Work, Changes, Failure of Performance, Insurance, Indemnification, Warranty, and Special Provisions), two key paragraphs or items also should be reviewed and included. These are:

- **CPM Clause:** A clause in the subcontract agreement referring to the Critical Path Method schedule is essential. The following paragraph is an example of such a clause.

  "Subcontractor agrees to follow the Critical Path Method schedule as prepared by the General Contractor. Subcontractor will, during the course of construction, inform the General Contractor if, for any reason, including delivery of materials, strikes, or interferences in the performance of the work described, the subcontractor will be unable to perform the work in the time stipulated as shown in the CPM Schedule."

Refer to Chapter 14, "Schedules," for additional information about documentation of the construction schedule.

- **Cleanup:** A provision for the method of cleanup should be included in the subcontract agreement. The important thing is to include cleanup in the subcontract agreement from the very beginning. In this way, there is an opportunity for reasonable discussion before going straight to a heated debate. In most cases, when the cleanup issue turns into a problem, it turns into a backcharge by the general contractor against the subcontractor's retainage.

Tracking Subcontract Agreements and Purchase Orders

Contract agreements sent to subcontractors, consultants, and vendors should be tracked and followed up if the contract is not returned in a timely manner. Changes to the subcontract also must be initialed, signed, and returned. The general contractor may run into serious problems later if either the original contract or the changes to the contract are not signed and returned. An individual or company should not be allowed to work on a job or an assignment if they have not returned the contract as issued. The Transmittal (see Chapter 3) can be used as a cover letter for this purpose, with a due date for return of the subcontract (or a change to a subcontract). A subsidiary list or log should also be kept. This will make compliance easy to track. Be sure that all of the subcontracts and changes are signed and returned. The Transmittal can be used for additional identification and, most importantly, makes it easier to account for the location and status of the Transmittal.

Company policy may be to issue contracts to subcontractors and vendors in sequence, assigning numbers to the contracts in the order written and not necessarily in the job order. Whoever is responsible for writing the contracts should have a log to record the date, contract number, job number, addressee, and return date.

It is important to the job records to know who has not signed and returned their contract. The Transmittal letter used as a cover letter to track subcontracts would look like Figure 10.1 and include the pertinent information about the contract. Again, a due date should be stated on the Transmittal. This procedure allows "tracking" of the Transmittal and helps to ensure that a signed contract is returned by the subcontractor or vendor.

Subcontracts and Insurance Certificate Log

This log can be created before actual work begins, but after the general contract is signed. It will list all the subcontractors and/or vendors (only major suppliers who need to be bound by a contract) and their status (which contract documents have been sent to them). It is very important that this log be checked before the individual subcontractors proceed with the work on site. This ensures that they are bound by a contract and have submitted proper insurance documents. Figures 10.2a and 10.2b are manual and computer-generated versions of these types of logs.

DELTA CONSTRUCTION COMPANY

6065 Mission Gorge Road, #193
San Diego, CA 92120
(619) 582-5829

*** TRANSMITTAL ***

Date: 2/07/90

Job Number: 321-SC
Transmittal Number: 3

TO: APPLIED ROOFING COMPANY
7654 DAMP STREET
LAKESIDE, CA 92345

ATTENTION: G. SMELL

JOB NUMBER: 321-SC GIBSON OFFICE BLDG / SUBCONTRACTS & INS

FROM: J. EDWARD GRIMES

SUBJECT: TRANSMITTAL NUMBER: 3
SUBCONTRACT

We are transmitting to you the following which is ENCLOSED/ATTACHED for your RESPONSE REQUIRED by: 2/21/90.

Please sign and return two copies of the subcontract in the amount of $50,000. The subcontract must be executed and returned prior to your starting work on the project.

Please contact me as soon as possible if you need any further information or if you have any questions.

Sincerely,

J. Edward Grimes

dp: 2/07/90 9:51 **PaperWorks, a Construction Software Program**

Figure 10.1

Delta Construction Co., Job #321 - Gibson Office Building
SUBCONTRACTOR/INSURANCE LOG

#	TO	SUBJECT	DATE SENT	DATE DUE	DATE REC'D	COMMENTS
1	Lot's of Rebar	Subcontract	2/7	2/21		
2	Lot's of Rebar	Insurance Cert.	2/7	2/21		
3	Applied Roofing	Subcontract	2/7	2/21		
4	Applied Roofing	Insurance Cert.	2/7	2/21		
5	Cool Air Cond.	Subcontract	2/7	2/21		
6	Cool Air Cond.	Insurance Cert.	2/7	2/21		

Figure 10.2a

```
                         DELTA CONSTRUCTION COMPANY
================================================PaperWorks Version 1.23F==
                               TRANSMITTAL LOG                        .TR1
            Job: 321-SC    GIBSON OFFICE BLDG / SUBCONTRACTS & INS.
Date: 02/28/90                                                   Page:    1
==========================================================================
Transmittal    Name of Addressee        Date       Date      Date         Status
 Number                                 Sent       Due       Received
--------------------------------------------------------------------------
    1                                 02/07/90   02/21/90
  subj: SUBCONTRACT
        Please sign and return two copies of the subcontract in the
        amount of $98,000.  The subcontract agreement must be executed
        prior to your starting work on the project.
  LOT'S OF REBAR                  for: RESPONSE REQ         02/18/90 Returned
--------------------------------------------------------------------------
    2                                 02/07/90   02/21/90
  subj: INSURANCE CERTIFICATES
        Please have your insurance carrier forward to us copies of your
        certificates for Public Liability and Public Damage as well as
        a certificate of your Workman's Compensation Coverage.
  LOT'S OF REBAR                  for: RESPONSE REQ         02/22/90 Returned
--------------------------------------------------------------------------
    3                                 02/07/90   02/21/90
  subj: SUBCONTRACT
        Please sign and return two copies of the subcontract in the
        amount of $50,000.  The subcontract must be executed and returned
        prior to your starting work on the project.
  APPLIED ROOFING COMPANY         for: RESPONSE REQ         NOT RET. Printed
--------------------------------------------------------------------------
    4                                 02/07/90   02/21/90
  subj: INSURANCE CERTIFICATES
        Please instruct you insurance carrier to provide us with
        insurance certificates as required by the Specifications.
        Name us as additionally insured.
  APPLIED ROOFING COMPANY         for: RESPONSE REQ         NOT RET. Printed
--------------------------------------------------------------------------
    5                                 02/07/90   02/21/90
  subj: SUBCONTRACT
        Please sign and return two copies of the subcontract in the
        amount of $325,000.  The subcontract agreement must be executed
        and returned prior to your starting work on the project.
  COOL AIR CONDITIONING           for: RESPONSE REQ         02/18/90 Returned
--------------------------------------------------------------------------
    6                                 02/07/90   02/21/90
  subj: INSURANCE REQUIREMENTS
        Please instruct you insurance carrier to provide us with the
        insurance certificates as required by the specifications.
        Please name us as additionally insured.
  COOL AIR CONDITIONING           for: RESPONSE REQ         NOT RET. Printed
--------------------------------------------------------------------------
```

Figure 10.2b

Chapter Eleven
Submittal Requests

There are three separate, but connected, parts of the Submittal process–Submittal Request (SR), Submittal to Architect (SA), and Returned Submittal from Architect (RS). A fourth part might be the contract documents and specifications, which call out the required materials, equipment, and finishes. In this chapter, we will concentrate on Submittal Requests. SA's and RS's will be covered in Chapters 12 and 13.

It is not *necessary* to utilize the Submittal Request format letters, but it is a good practice. Using this organized method can help to avoid a lot of headaches later in the project.

The Concept

The concept behind the Submittal Request procedure is to have the manager review the Contract Specifications. This is done to ensure that both general contractor and the subcontractors submit and use approved materials and products on the project. If a Submittal varies from the specification, or the item is never submitted, it is an early warning of a problem. By the same token, if there are substitutions, they should be submitted and addressed. The Submittal Request process is a good way of getting to know the job–especially when the firm has a separate estimating department. By completing the Submittal Request process, the manager gets to know what he and the subcontractors are obligated to provide and perform.

By reviewing the specifications of the trades and writing down the items needed, one gains deeper insight into the problems that might arise involving such items as shop drawings, delivery time, and curing time. It is strongly advised that the specifications be reviewed in detail prior to preparing the CPM construction schedule. Some of the detail items in the specifications should be noted as they will affect the schedule. (Marble, for example, must usually have samples approved, and then be quarried. It must then be cut, fabricated, finished, and shipped from Italy to its U.S. destination.)

The fact is, specifications can make for very dry reading. However, those who *do* read them succeed much more often than those who do not. The following is an example of the kinds of problems that can occur when specifications are not reviewed carefully, and Submittal data is not expedited.

A project with an underground garage had specifications calling for a particular type of waterproofing. The Submittal data arrived late, after the garage walls were up. Only then did it become apparent that the waterproofing product required a 21-day curing time between each of three coats, and another 21 days curing time before backfilling. The general contractor, of course, tried to get the product changed, but was unable to do so in time. The net result of this late Submittal was a construction time delay of over two months.

If the product data in this example had been submitted promptly and reviewed by the general contractor, there would have been time to select a different waterproofing product, or the general contractor may have elected to build the job differently. On this particular project, the job site overhead alone was about $20,000 a month, not to mention that the personnel occupied on this job could have been elsewhere making money. The total loss to the contractor was somewhere in the range of $40,000. Considering these figures, it would have been better to offer a credit to the owner to change the material and get on with the project. As it was, the originally specified material had already been ordered, and it was too late to change.

Expediting Submittals

One of the keys to successful job management is getting tasks accomplished on time in the overall scheme of construction. By assigning due dates for Submittals, the manager can follow up on the subcontractors and get them "trained" to respond by a specified date. However, it is up to the contractor to set the appropriate due dates. In the following example, the specifications require (and the manager must obtain) these items:

- Material list, catalog cuts of pipe specs, type of connectors, etc.
- Plumbing fixtures, catalog cuts of fixtures, and maybe a physical sample of a faucet to verify finish
- As-built drawings
- Warranty
- Insurance certificates (before any subcontractors arrive on the site to begin work)

Assume (for the purpose of the example) that this is a fast track job, and the site is ready to build on. Today is June 15. The job will start July 1 and should be complete by February 15. An individual Submittal Request letter is sent, with assigned due dates for the items as follows:

Underground material list:	July 1
Above-ground material list:	July 13
Sample of faucet to verify finish:	July 24
Plumbing fixtures–catalog cuts and description:	July 15
As-builts and Warranty:	Feb. 1

Why go through the extra work of sending five separate Submittal Request letters, when you could send one long one? Remember, paper is one of the cheapest things we build with. It can be a lot easier to follow up on the items that have not been submitted if each is a separate piece of paper with a unique identification number, such as Job 1101, SR #31. Another reason to send separate Submittals is that some of the information is more difficult for the subcontractor to assemble, and would, therefore, hold up the responses to the other included items. (Note that if this was not a fast track job, the general contractor might ask for the material lists all at one time, rather than asking for five separate Submittals.)

In the case of the sample brass faucet, the general contractor should call the subcontractor and ask to have a sample delivered as soon as possible. He should also send the written Submittal Request. Keep in mind the fact that anytime you request a sample, the time required to make a decision is multiplied by the number of people who have to review the sample.

Remember, the plumber cannot order the plumbing fixtures without knowing the approved finish for the fixtures. Sometimes the selection of particular finishes can totally disrupt construction progress because the supplier may say: "Yes, brushed brass is a standard finish. We made a bunch of them last month, but those are all gone. We are not scheduled to make brushed brass finish again until December 20, so you *should* be able to have them by February 1." Such predictions should not be relied upon too heavily. Using a computerized system, the long-term Submittal items, such as as-builts and warranties, will not show up on the "delinquent due" log report until late in the job, when they are needed. However, if the "delinquent" log is used like an account receivable list, there is no need to wait until these items are late and already required on the job. There is some advance warning.

The sample computer-generated Submittal Request Letter shown in Figure 11.1a has all the information and tracking data that is normally needed. A sample handwritten Transmittal letter used as a Submittal Request is shown in Figure 11.1b.

DELTA CONSTRUCTION COMPANY

6065 Mission Gorge Road, #193
San Diego, CA 92120
(619) 582-5829

*** SUBMITTAL REQUEST ***

Date: 2/05/90

Job Number: 321
Submittal Request Number: 3

TO: COOL AIR CONDITIONING
7776 MISSION GORGE PLACE
SAN DIEGO, CA 92344

ATTENTION: LARRY ICEBERG

JOB NUMBER: 321 GIBSON OFFICE BUILDING

FROM: J. EDWARD GRIMES

SUBJECT: SUBMITTAL REQUEST NUMBER: 3
HVAC SUBMITTALS AS REQUIRED BY CONTRACT

Please send us the following information NO LATER THAN 2/19/90 and refer to Submittal Request Number: 3 when submitted, as well as the applicable Specification Line number in box below.

1 SHEET METAL DUCTWORK FOR GARAGE EXHAUST

2 HVAC DUCT DRAWINGS FOR FLOORS 2 THROUGH 6

3 HVAC DUCTWORK DRAWINGS FOR 1ST FLOOR AND LOBBY

4 VAV BOXES, MANUF LIT AND CATALOG CUTS

5 COOLING TOWER SPECIFICATIONS

Comments:

THE COOLING TOWER IS A LONG LEAD ITEM, LETS'S GET IT SUBMITTED!

dp: 2/05/90 10:11 **PaperWorks, a Construction Software Program**

Figure 11.1a

Means Forms

LETTER OF TRANSMITTAL

Delta Construction Company
6678 Mission Road
San Diego, CA 92120
(619) 581-6315

FROM: J. Edward Grimes
Project Manager

TO: Cool Air-Conditioning, Inc.
7776 Mission Gorge Pl.
San Diego, Ca. 92344

DATE 2/5/90
PROJECT #321 Gibson Office Bldg.
LOCATION
ATTENTION Larry Iceberg
RE:
Submittal Request

SR 321.03

Gentlemen:

WE ARE SENDING YOU ☑ HERewith ☐ DELIVERED BY HAND ☐ UNDER SEPARATE COVER

VIA ______ THE FOLLOWING ITEMS:

☐ PLANS ☐ PRINTS ☐ SHOP DRAWINGS ☐ SAMPLES ☐ SPECIFICATIONS

☐ ESTIMATES ☐ COPY OF LETTER ☑ submittal request

COPIES	DATE OR NO.	DESCRIPTION
	1.	Sheet metal ductwork for garage exhaust
	2.	HVAC duct drawings for floors 2 through 6
	3.	HVAC ductwork drawings for 1st floor and lobby
	4.	VAV boxes, Manuf lit and catalog cuts
	5.	Cooling tower Specifications

THESE ARE TRANSMITTED AS INDICATED BELOW

☐ FOR YOUR USE ☐ APPROVED AS NOTED ☐ RETURN ______ CORRECTED PRINTS

☐ FOR APPROVAL ☐ APPROVED FOR CONSTRUCTION ☐ SUBMIT ______ COPIES FOR ______

☐ AS REQUESTED ☐ RETURNED FOR CORRECTIONS ☐ RESUBMIT ______ COPIES FOR ______

☐ FOR REVIEW AND COMMENT ☐ RETURNED AFTER LOAN TO US ☐ FOR BIDS DUE ______

☐ ______

REMARKS: Please send us the above listed information No Later Than 2-19-90 and refer to Submittal Request Number: 3 when submitted, as well as the applicable Spec. line number.

The cooling tower is a long lead item, let's get it submitted!

IF ENCLOSURES ARE NOT AS INDICATED, PLEASE NOTIFY US AT ONCE.

SIGNED: J Edward Grimes

Figure 11.1b

Follow-up Logs and Reports

The follow-up and tracking of Submittal Requests is extremely important. Not only does it get the job started on a positive and professional note, but it can also "train" subcontractors to take your requests and deadlines seriously. The subcontractor's attitude, as established during the Submittal process, will carry forward into the pricing of scope changes and other responses that become necessary during the course of the job.

Following up to see that the subcontractor does submit what is required is made easier utilizing logs. A dedicated computer program can generate more selective logs than those that are usually kept by hand. Figure 11.2a is an example of a manually kept log (adapted from Means Job Progress Report Form). Figure 11.2b is an example of a Submittal Request Log. Figure 11.2c is a report that lists only when Submittals are due. Figure 11.2d is a report that lists only those companies that are "past due." A manager can give such a list to someone in the office and say, "Call all these people and find out why they haven't submitted, and when they are going to submit the required information." The next step, of course, is to follow up with the Submittal Request letters for documentation.

Means® Forms

JOB Submittal Log

Delta Construction Company
6678 Mission Road
San Diego, CA 92120
(619) 581-6315

SHEET ____ OF ____

PROJECT Gibson Office Building

JOB NO. 321

LOCATION San Diego

PREPARED BY J. Edward Grimes

YEARS

Contractor	Subject	Milestone Dates																							
		Requested	Due from Sub.	Rec'd from Sub.	Sent to Arch.	Due from Arch.	Rec'd from Arch.	Status	Returned to sub.	Due from Sub.	Rec'd from Sub.	Sent to Arch.	Due from Arch.	Rec'd from Arch.	Status	Returned to sub.	Due from Sub.	rec'd from Sub.	Sent to Arch.	Due from Arch.	rec'd from Arch.	Status	Returned to Sub		
4 Strong Steel	Shop Dwgs.	2/5	2/19																						
	OSHA Er. Proc.	2/5	2/19																						
5 Cool Air Cond.	As builts	2/5	9/1																						
	Warranty	2/5	9/1																						
6 Applied Roofing	Mat'ls	2/5	2/19																						
7 Applied Roofing	Warranty/Bond	2/5	9/1																						
8 Sparks Elec.	Fixture E-1	2/5	2/27																						
	E-2	2/5	2/27																						
	E-3	2/5	2/27																						
9 Sparks Elec.	Switchgear	2/5	2/19																						
10 Clear Glass	Glass Sample	2/13	2/27																						
	Alum. Sample	2/13	2/27																						
	Detail Dwgs.	2/13	2/27																						

Figure 11.2a

```
                              DELTA CONSTRUCTION COMPANY
=====================================================PaperWorks Version 1.23F==
                          *** SUBMITTAL REQUEST LOG ***                    .SR1
        Job #: 321        GIBSON OFFICE BUILDING
Date: 03/30/90                                                        Page:   1
================================================================================
SR #    Name of SUB/VENDOR                              SENT TO         DATE DUE
            Subject Summary                             SUB/VEN         SUB/VEN
--------------------------------------------------------------------------------
   1        BENT SHEET METAL                            02/05/90        02/19/90
1.1     ROOF HATCH
                                                        DATE REFERENCED: 02/20/90
                                                REFERENCED BY SUBMITTAL:
1.2     SKYLIGHT
1.3     SHEET METAL SHAPES SHOP DRAWINGS
                                                        DATE REFERENCED: 02/20/90
                                                REFERENCED BY SUBMITTAL:
1.4     MATERIAL LIST
                                                        DATE REFERENCED: 02/20/90
                                                REFERENCED BY SUBMITTAL:
  COMMENTS: SUBMIT AS SOON AS POSSIBLE AS ARCHITECT IS GOING ON VACATION
            IN TWO WEEKS.
--------------------------------------------------------------------------------
   2        LOT'S OF REBAR                              02/05/90        02/19/90
2.1     REBAR SHOP DRAWINGS FOR FOOTING TYPE F1
                                                        DATE REFERENCED: 02/20/90
                                                REFERENCED BY SUBMITTAL:
2.2     REBAR SHOP DRAWINGS FOR FOOTING TYPE F2
                                                        DATE REFERENCED: 02/20/90
                                                REFERENCED BY SUBMITTAL:
2.3     REBAR SHOP DRAWINGS FOR FOOTING TYPE F3
                                                        DATE REFERENCED: 02/20/90
                                                REFERENCED BY SUBMITTAL:
  COMMENTS: SUBMIT AS SOON AS POSSIBLE.  WE HOPE TO POUR THE FIRST FOOTINGS
            BY FEB. 20
--------------------------------------------------------------------------------
   3        COOL AIR CONDITIONING                       02/05/90        02/19/90
3.1     SHEET METAL DUCTWORK FOR GARAGE EXHAUST
                                                        DATE REFERENCED: 02/27/90
                                                REFERENCED BY SUBMITTAL:
3.2     HVAC DUCT DRAWINGS FOR FLOORS 2 THROUGH 6
                                                        DATE REFERENCED: 02/27/90
                                                REFERENCED BY SUBMITTAL:
3.3     HVAC DUCTWORK DRAWINGS FOR 1ST FLOOR AND LOBBY
                                                        DATE REFERENCED: 02/27/90
                                                REFERENCED BY SUBMITTAL:
3.4     VAV BOXES, MANUF LIT AND CATALOG CUTS
                                                        DATE REFERENCED: 02/27/90
                                                REFERENCED BY SUBMITTAL:
3.5     COOLING TOWER SPECIFICATIONS
                                                        DATE REFERENCED: 02/27/90
                                                REFERENCED BY SUBMITTAL:
  COMMENTS: THE COOLING TOWER IS A LONG LEAD ITEM, LET'S GET IT SUBMITTED!
--------------------------------------------------------------------------------
```

Figure 11.2b

DELTA CONSTRUCTION COMPANY
PaperWorks Version 1.23F
*** SUBMITTAL REQUEST DUE LOG *** .SR3
Job #: 321 GIBSON OFFICE BUILDING
Date: 03/30/90 Page: 1

SR #	NAME OF SUB/VENDOR summary	SENT TO SUB/VEN	DATE DUE
1	BENT SHEET METAL	02/05/90	02/19/90
1.2	SKYLIGHT		
	COMMENTS: SUBMIT AS SOON AS POSSIBLE AS ARCHITECT IS GOING ON VACATION IN TWO WEEKS.		
4	STRONG STRUCTURAL STEEL	02/05/90	02/19/90
4.2	OSHA APPROVED ERECTION PROCEDURE		
	COMMENTS: GET THE ANCHOR BOLT INFO IN SOON AS WE'LL BE POURING FOOTINGS ABOUT 2/20		
5	COOL AIR CONDITIONING	02/05/90	09/01/90
5.1	AS BUILT DRAWINGS		
5.2	WARRANTY FOR ALL HVAC EQUIPMENT		
	COMMENTS:		
7	APPLIED ROOFING COMPANY	02/05/90	09/01/90
7.1	20 YEAR BOND FROM MANUFACTURER AND INSTALLER		
7.2	ONE YEAR LABOR WARRANTY FROM APPLIED ROOFING COMPANY		
	COMMENTS:		
8	SPARKS ELECTRIC COMPANY	02/05/90	02/27/90
8.1	LIGHT FIXTURES TYPE E1		
8.2	LIGHT FIXTURES TYPE E2		
8.3	LIGHT FIXTURES TYPE E3		
8.4	LIGHT FIXTURES TYPE E4		
8.5	LIGHT FIXTURES TYPE E5		
	COMMENTS: FIXTURE E5 GOES IN THE LOBBY AND THE ARCHITECT WANTS TO REVIEW ALL THE DATA AS SOON AS POSSIBLE (POSSIBLE CHANGE). HURRY.		
9	SPARKS ELECTRIC COMPANY	02/05/90	02/19/90
9.1	ELECTRICAL SWITCH GEAR.		
	COMMENTS:		

Figure 11.2c

DELTA CONSTRUCTION COMPANY

PaperWorks Version 1.23F

*** SUBMITTAL REQUEST DELINQUENT LOG *** .SR5

Job #: 321

Date: 03/30/90 Page: 1

SR #	NAME OF SUB/VENDOR summary	SENT TO SUB/VEN	DATE DUE
1	BENT SHEET METAL	02/05/90	02/19/90
1.2	SKYLIGHT		
COMMENTS:	SUBMIT AS SOON AS POSSIBLE AS ARCHITECT IS GOING ON VACATION IN TWO WEEKS.		
4	STRONG STRUCTURAL STEEL	02/05/90	02/19/90
4.2	OSHA APPROVED ERECTION PROCEDURE		
COMMENTS:	GET THE ANCHOR BOLT INFO IN SOON AS WE'LL BE POURING FOOTINGS ABOUT 2/20		
7	APPLIED ROOFING COMPANY	02/05/90	09/01/90
7.1	20 YEAR BOND FROM MANUFACTURER AND INSTALLER		
7.2	ONE YEAR LABOR WARRANTY FROM APPLIED ROOFING COMPANY		
COMMENTS:			
8	SPARKS ELECTRIC COMPANY	02/05/90	02/27/90
8.1	LIGHT FIXTURES TYPE E1		
8.2	LIGHT FIXTURES TYPE E2		
8.3	LIGHT FIXTURES TYPE E3		
8.4	LIGHT FIXTURES TYPE E4		
8.5	LIGHT FIXTURES TYPE E5		
COMMENTS:	FIXTURE E5 GOES IN THE LOBBY AND THE ARCHITECT WANTS TO REVIEW ALL THE DATA AS SOON AS POSSIBLE (POSSIBLE CHANGE). HURRY.		
9	SPARKS ELECTRIC COMPANY	02/05/90	02/19/90
9.1	ELECTRICAL SWITCH GEAR.		
COMMENTS:			

Figure 11.2d

Chapter Twelve
Submittals to Architect

Purpose

In Chapter 11, "Submittal Requests," we strongly recommended utilizing the Request for Submittal procedure. Starting a job with a detailed Submittal Request program, along with a well-thought-out CPM schedule, creates a solid foundation for the remainder of the job. In tennis or golf, you must meet the ball when you swing, but you need a proper follow-through to make a good shot. The Submittal to Architect/Engineer, Job Minutes, and the day-to-day construction correspondence are the follow-through that make it all happen.

The Submittal (from the contractor) to the Architect/Engineer (SA) is the correspondence vehicle used to transmit or submit the information required by the specifications. Included in this "package" should be the Submittals received from subcontractors and vendors, as well as the Submittal that the contractor is required to make. This Submittal normally is made to the architect, but in some cases, also may be sent to the owner for review. This step is a formalization and documentation of the items that are to be used to construct the project.

The manager needs the architect/engineer's prompt completion of the Submittal to Architect so that he has approved documentation of the materials and/or equipment to be installed (by the contractor or subcontractor) in the project. By completing the submittal process, the owner, architect, general contractor, and subcontractor all have a record of the items to be included in the project.

Upon completion of the submittal process, the general contractor's representative in the field can knowledgeably "inspect" to ensure that the materials and equipment used in the project are approved items. Often, a less than desirable item is included in the project and then the cost of removal, either in time or cash, is so great that the owner must settle for something less than what he/she planned and paid for.

The submittal process (Submittal Request, Submittal to Architect/Engineer, and Returned Submittal) is completed when an approved submittal is returned to the contractor by the architect/engineer.

The General Contractor as Broker

More and more general contractors are becoming "brokers," or "construction managers," who oversee the job, but do not perform any of the construction work with their own forces. The contractor (whether general contractor or construction manager) must make sure he knows (and has approval from the architect/owner) what materials and finishes are to be included in the project. Without a "scorecard," it is difficult for the on-site people to know if the materials and equipment being installed are correct.

Architects often provide generic specifications that are so general one could install "Brand X" for almost all of the materials on the job. If the contractor installs Brand X, and the owner says he wanted Brand Y later, it could be difficult proving the two brands are the same. The Submittal process should leave no unanswered questions about the materials to be used. It is much easier to get these items resolved early, while there is still some time, rather than waiting until the day before the project is to be occupied, when tempers are hot and there is no money left. The owner gets angry, the architect blames the general contractor, the general contractor blames the subcontractors, and then the owner explodes, saying he will not have red oak in the building when he wanted white oak (as called for by the original specifications).

It is especially important to keep the KISS (Keep It Simple, Silly) principle in mind when submitting items to the architect. Do not be afraid to send many individual Submittals. If they are simple, they get processed more quickly and easily. They are also less likely to be rejected than complicated, multi-subject Submittals. Never, never include several unrelated items as a single Submittal to Architect letter with one identification number. Combining items not only delays the response, but can be confusing, particularly if the architect must send one of these items to an electrical and/or mechanical engineer for additional approval or review.

Again, the key to the Submittal process is to keep it simple so that as few people as possible have (or think they have) the obligation to review it. Every time one additional person must review an item, a minimum of three-to-five extra days are added to the review cycle. On a large job, it may be possible to make an arrangement with the principal architect whereby someone in his or her office is designated responsible for processing the paperwork. Submittals may be sent to that designated party. However, meeting minutes or pricing documents should be delivered directly to the responsible architect.

We mentioned in the previous chapter that Submittal Requests are a good means of getting the subcontractors "trained" to respond promptly. Now, we get a chance to train the architect. We will train the owner in the pricing area. As we deal with the issue of training others, we must remember to train ourselves as well. If we do not learn to follow through, then we are not going to succeed in getting the desired response from others. It is a two-way street. Contractor, architect, facilities manager: train thyself.

Part of keeping Submittals simple and straightforward is knowing what to do if there is a problem with one Submittal item. The first step is to make sure you send that item as a separate Submittal. For example, the HVAC subcontractor sends the general contractor a Submittal for materials (e.g., duct types, supply and return registers, and the cooling tower) in one Submittal document. The subcontractor tries to substitute this particular type of cooling tower for the original as specified and his bid is based on acceptance of this substitution. However, it may not be a simple matter to get the mechanical engineer to accept this cooling tower as an equivalent.

At this point, a Submittal should be sent for all of the HVAC material items that should be approved without any problem. The cooling tower should be a separate Submittal. The contractor should note on the Submittal that he would like to meet the mechanical engineer, the subcontractor, and the cooling tower manufacturer to discuss the substitution.

Even if the contractor cannot "sell" the engineer on the substitute cooling tower, at least he has not held up the Submittal for the other HVAC items. Therefore, the subcontractor can get most of the "nuts and bolts" lined up and get on with the job. Remember: our prime objective is to complete the job within the time and cost framework originally estimated, not to get into a "pushing match" with anyone. The goal is to keep the job moving; that is the way we all– owner, architect, general contractor, subcontractor, and manufacturer–will make as much money as possible.

If a Submittal is totally out of line, incomplete, or inaccurate, then it should be returned to the subcontractor. The contractor should call and ask that the subcontractor redo the Submittal, and should then follow through and make sure they do. As we have said before, this is the time when precedents are established and attitudes formed. If the contractor is lax on Submittals, then the subcontractor will just keep providing partial information and make the job more difficult.

In the pre-construction meeting, while everyone is still friendly, the turn-around time for Submittals (in the architect's office) should be agreed upon. Once the Submittal has been sent to the architect, the project manager should make sure that the architect abides by the timetable. If the architect's office cannot comply, the architect should be contacted and asked if he wants to change the turn-around time. He should also be asked to call the owner to explain why he cannot return the Submittal in a timely manner. Before this happens, make sure that you have your act together. Figure 12.1a is an example of a computer-generated letter of transmittal of Submittals to the Architect. Figure 12.1b is an example of a manual letter of transmittal of Submittals to the Architect.

DELTA CONSTRUCTION COMPANY

6065 Mission Gorge Road, #193
San Diego, CA 92120
(619) 582-5829

*** SUBMITTAL ***

Date: 2/27/90

Job Number: 321
Submittal Number: 7

TO: ONE LINE DESIGN
2001 DREAM STREET
SAN DIEGO, CA 92100

ATTENTION: ROD WRIGHT

JOB NUMBER: 321 GIBSON OFFICE BUILDING

FROM: J. EDWARD GRIMES

SUBJECT: SUBMITTAL NUMBER: 7
HVAC SHOP DRAWINGS AND VAV BOX SUBMITTAL

We are sending the following submittal information ENCLOSED/ATTACHED received from COOL AIR CONDITIONING.
Please review and return this information to us by 3/13/90.

1 Ductwork shop drawings for floor 2 - 6, Shts AC 1 to 6

2 Ductwork shop drawings for first floor lobby AC 7 & 8

3 VAV mixing box submittal. Xtracool #56-98

Comments:

Please review the VAV box, as it is a substitution, as soon as possible.

dp: 2/27/90 10:13 **PaperWorks, a Construction Software Program**

Figure 12.1a

Means Forms

LETTER OF TRANSMITTAL

Delta Construction Company
6678 Mission Road
San Diego, CA 92120
(619) 581-6315

FROM: J. Edward Grimes
Project Manager

DATE 2/27/90
PROJECT Gibson Office Bldg. #321
LOCATION
ATTENTION Rod Wright
RE: Sub to Arch #7

SA 321.7

TO: One Line Design
2001 Dream Street
San Diego, Ca. 92100

Gentlemen:
WE ARE SENDING YOU ☒ HEREWITH ☒ DELIVERED BY HAND ☐ UNDER SEPARATE COVER
VIA ______ THE FOLLOWING ITEMS:
☐ PLANS ☐ PRINTS ☒ SHOP DRAWINGS ☐ SAMPLES ☐ SPECIFICATIONS
☐ ESTIMATES ☐ COPY OF LETTER ☒ details

COPIES	DATE OR NO.	DESCRIPTION
		1 Ductwork shop drawings for floor 2-6, Shts Ac 1 to 6
		2 Ductwork shop drawings for first floor lobby Ac 7 & 8
		3 VAV mixing box submittal. Xtracool #56-98

THESE ARE TRANSMITTED AS INDICATED BELOW
☐ FOR YOUR USE ☐ APPROVED AS NOTED ☐ RETURN ______ CORRECTED PRINTS
☐ FOR APPROVAL ☐ APPROVED FOR CONSTRUCTION ☐ SUBMIT ______ COPIES FOR ______
☐ AS REQUESTED ☐ RETURNED FOR CORRECTIONS ☐ RESUBMIT ______ COPIES FOR ______
☐ FOR REVIEW AND COMMENT ☐ RETURNED AFTER LOAN TO US ☒ ~~FOR BIDS~~ DUE 3/13/90
☐ ______

REMARKS: Please review the VAV box, as it is a substitution, as soon as possible.

IF ENCLOSURES ARE NOT AS INDICATED, PLEASE NOTIFY US AT ONCE.

SIGNED: J Edward Grimes

Figure 12.1b

Follow-up Logs and Reports

The follow-up and tracking of Submittals to the Architect are extremely important. Not only does this get the job started on a positive and professional note, but it can also "train" subcontractors to take your requests and deadlines seriously. The subcontractor's attitude, as established during the Submittal process, will carry forward into the pricing of scope changes and other responses that become necessary during the course of the job.

Following up to see that the subcontractor does submit what is required is made easier by utilizing logs. It may be easier to follow up using a dedicated computer program, as the logs that it generates can be more selective than those that are usually kept by hand. Figure 12.2a is an example of a manually kept log. Figure 12.2b is an example of a Submittal Request Log. Figure 12.2c is a report that lists only when Submittals are due, and Figure 12.2d is a report that lists only those companies that are "past due." A manager can give such a list to someone in the office and say, "Call all these people and find out why they haven't submitted, and when they are going to submit the required information." The next step, of course, is to follow up with the Submittal Request letters for documentation.

Means® Forms

JOB Submittal Log

Delta Construction Company
6678 Mission Road
San Diego, CA 92120
(619) 581-6315

SHEET ______ OF ______

PROJECT Gibson Office Building

JOB NO. 321

LOCATION San Diego

PREPARED BY J. Edward Grimes

Contractor	Subject	Milestone Dates: Requested	Due from Sub.	Rec'd from Sub	Sent to Arch.	Due from Arch.	Rec'd from Arch.	Status	Returned to Sub.	Due from Sub.	Rec'd from Sub	Sent to Arch.	Due from Arch.	Rec'd from Arch.	Status	Returned to Sub.	Due from Sub.	Rec'd from Sub	Sent to Arch.	Due from Arch.	Rec'd from Arch.	Status	Returned to sub.
1 Lot's of Rebar	Footing F-1	1/31	2/20	2/20	2/20	3/6	3/5	A	3/5														
	F-2	1/31	2/20	2/20	2/20	3/6	3/5	A	3/5														
	F-3	1/31	2/20	2/20	2/20	3/6	3/5	R	3/5	3/19													
2 Bent Sht MtL	Roof Hatch	1/31	2/27	2/20	2/20	3/6 (circled)																	
	Sht MTL 1-5	1/31	2/27	2/20	2/20	3/6 (circled)																	
	Mat'L List	1/31	2/27	2/20	2/20	3/6 (circled)																	
3 Strong Steel	Shop Dwgs 1-27	1/31	2/22	2/19	2/20	3/6	3/1	↓															
	1-13, 15-17, 19-27							A	3/1														
	14, 18							R	3/1	3/19													
4 Applied Roofing	U/G Waterproof	2/13	2/28	2/27	2/27	3/13	3/4	R	3/4	3/19													
	Procedures	2/13	2/28	2/27	2/27	3/13	3/4	R	3/4	3/19													
	Appl. Cert.	2/13	2/28	2/27	2/27	3/13	3/4	R	3/4	3/19													
5 Cool Air Cond.	GE 1 & 2	1/31	2/27	2/26	2/27	3/13 (circled)																	

↓ = See below circled = late A = approved R = rejected

Figure 12.2a

DELTA CONSTRUCTION COMPANY

PaperWorks Version 1.23F

SUBMITTAL LOG .SA1

Job 321 GIBSON OFFICE BUILDING

Date: 03/25/90 Page: 1

Submittal Number	Name of Sub/Ven Summary	Sent To Arch	Date Due Back	Date Ret To Sub	Status
1	LOT'S OF REBAR	02/20/90	03/06/90	03/05/90	
1.1	Shop Drawing for Footing type F-1				APPROVED
1.2	Shop Drawing for Footing type F-2				APPROVED
1.3	Shop Drawing for Footing type F-3				REV/RESU
	Date Due Resubmittal from SUB: 03/19/90				
	comment: We will be pouring these footings very soon. Please give special consideration to the shop drawing review.				
2	BENT SHEET METAL	02/20/90	03/06/90		
2.1	Roof Hatch catalog cuts and specifications				NOT RET.
2.2	Sheet Metal shop drawings SM-1 to SM-5				NOT RET.
2.3	Material List				NOT RET.
	comment:				
3	STRONG STRUCTURAL STEEL	02/20/90	03/06/90	03/01/90	
3.1	Structural Steel Shop Drawings Sheet SS-1 to SS-27				APPROVED
3.2	Sheets SS-1 to SS-13				APPROVED
3.3	Sheets SS-14				REJECT
	Date Due Resubmittal from SUB: 03/19/90				
3.4	Sheets SS 18				REJECT
	Date Due Resubmittal from SUB: 03/19/90				
3.5	Sheets SS 15, 16, 17 and 19-27				APPROVED
	comment:				
4	APPLIED ROOFING COMPANY	02/27/90	03/13/90	03/04/90	
4.1	Underground waterproofing system specifications				REJECT
	Date Due Resubmittal from SUB: 03/19/90				
4.2	Manuf. recommended procedures				REJECT
	Date Due Resubmittal from SUB: 03/19/90				
4.3	Manufacturer's application inspection certificate				REJECT
	Date Due Resubmittal from SUB: 03/19/90				
	comment: We will have to schedule the Manuf. rep to be here when the work is being done and then he will fill out the certificate.				
5	COOL AIR CONDITIONING	02/27/90	03/13/90		
5.1	Sheet metal shop drawings for garage exhaust GE 1 & 2				NOT RET.
	comment:				
6	COOL AIR CONDITIONING	02/27/90	03/13/90		
6.1	Cooling tower specifications and details				NOT RET.
	comment:				
7	COOL AIR CONDITIONING	02/27/90	03/13/90		
7.1	Ductwork shop drawings for floor 2 - 6, Shts AC 1 to 6				NOT RET.
7.2	Ductwork shop drawings for first floor lobby AC 7 & 8				NOT RET.
7.3	VAV mixing box submittal. Xtracool #56-98				NOT RET.
	comment: Please review the VAV, box as it is a substitution, as soon as possible.				

Figure 12.2b

```
                         DELTA CONSTRUCTION COMPANY
=================================================PaperWorks Version 1.23F==
                         SUBMITTAL DUE FROM SUBS                       .SA3
              Job 321     GIBSON OFFICE BUILDING
Date: 03/25/90                                                   Page:    1
===========================================================================
Submittal        Name of Sub/Ven             Sent To     Date due
 Number            Summary                    Arch       from Arc      Status
---------------------------------------------------------------------------
    2 BENT SHEET METAL                       02/20/90    03/06/90
2.1   Roof Hatch catalog cuts and specifications                     Not Ret.
2.2   Sheet Metal shop drawings SM-1 to SM-5                         Not Ret.
2.3   Material List                                                  Not Ret.
   summary: SHEET METAL SHOP DRAWINGS AND SUBMITTALS
---------------------------------------------------------------------------

    5 COOL AIR CONDITIONING                  02/27/90    03/13/90
5.1   Sheet metal shop drawings for garage exhaust  GE 1 & 2         Not Ret.
   summary: SHEET METAL SHOP DRAWINGS AND SUBMITTALS
---------------------------------------------------------------------------

    6 COOL AIR CONDITIONING                  02/27/90    03/13/90
6.1   Cooling tower specifications and details                       Not Ret.
   summary: COOLING TOWER SUBMITTAL
---------------------------------------------------------------------------

    7 COOL AIR CONDITIONING                  02/27/90    03/13/90
7.1   Ductwork shop drawings for floor 2 - 6, Shts AC 1 to 6         Not Ret.
7.2   Ductwork shop drawings for first floor lobby AC 7 & 8          Not Ret.
7.3   VAV mixing box submittal.  Xtracool #56-98                     Not Ret.
   summary: HVAC SHOP DRAWINGS AND VAV BOX SUBMITTAL
---------------------------------------------------------------------------

   11 CLEAR GLASS COMPANY                    03/05/90    03/19/90
11.1  Sample of mirror and frame for use in restrooms                Not Ret.
   summary: MIRROR SUBMITTAL
---------------------------------------------------------------------------

   13 CONTINENTIAL HARDWARE                  03/05/90    03/19/90
13.1  Sample of proposed substitute door closure                     Not Ret.
13.2  Specification of proposed substitute door closure              Not Ret.
   summary: DOOR CLOSURES
---------------------------------------------------------------------------

   15 GREEN LANDSCAPING                      03/12/90    03/26/90
15.1  Plant list                                                     Not Ret.
15.2  Photos of 36" box trees                                        Not Ret.
15.3  Material List                                                  Not Ret.
   summary: LANDSCAPING SUBMITTAL
---------------------------------------------------------------------------

   16 UP AND DOWN ELEVATOR CO                03/12/90    03/26/90
16.1  Elevator shop drawings                                         Not Ret.
16.2  Equipment List                                                 Not Ret.
16.3  Cab interior details                                           Not Ret.
   summary: ELEVATOR SUBMITTAL
```

Figure 12.2c

DELTA CONSTRUCTION COMPANY

PaperWorks Version 1.23F

SUBMITTAL DELINQUENT LOG--AS OF 03/25/90 .SA5

Job 321 GIBSON OFFICE BUILDING

Date: 03/25/90 Page: 1

Submittal Number	Name of Sub/Ven Summary	Sent To Arch	Date due from Arc	Status
2	BENT SHEET METAL	02/20/90	03/06/90	
2.1	Roof Hatch catalog cuts and specifications			Not Ret.
2.2	Sheet Metal shop drawings SM-1 to SM-5			Not Ret.
2.3	Material List			Not Ret.
comment:				
5	COOL AIR CONDITIONING	02/27/90	03/13/90	
5.1	Sheet metal shop drawings for garage exhaust GE 1 & 2			Not Ret.
comment:				
6	COOL AIR CONDITIONING	02/27/90	03/13/90	
6.1	Cooling tower specifications and details			Not Ret.
comment:				
7	COOL AIR CONDITIONING	02/27/90	03/13/90	
7.1	Ductwork shop drawings for floor 2 - 6, Shts AC 1 to 6			Not Ret.
7.2	Ductwork shop drawings for first floor lobby AC 7 & 8			Not Ret.
7.3	VAV mixing box submittal. Xtracool #56-98			Not Ret.
comment:	Please review the VAV box, as it is a substitution, as soon as possible.			
11	CLEAR GLASS COMPANY	03/05/90	03/19/90	
11.1	Sample of mirror and frame for use in restrooms			Not Ret.
comment:				
13	CONTINENTIAL HARDWARE	03/05/90	03/19/90	
13.1	Sample of proposed substitute door closure			Not Ret.
13.2	Specification of proposed substitute door closure			Not Ret.
comment:	These closures can be painted without affecting the UL approval.			

Figure 12.2d

Chapter Thirteen

Returned Submittals

Returned Submittals are the third part of the Submittal process, following Submittal Requests (SR) and Submittal to Architect (SA). Processing the Returned Submittal is like rounding third base in the paperwork game, at which point it is time to concentrate on getting to home plate (getting the project built).

An approved Submittal means that the following has happened:

- The subcontractor has made the Submittal to the general contractor.
- The actual Submittal has been made to the architect.
- The architect has approved the Submittal and all items have been approved as meeting the specifications in the contract documents or are acceptable to the architect/owner.

A Returned Submittal to the general contractor contains information originally submitted from a subcontractor and now returned to that subcontractor or vendor. The architect usually uses a common stamp, with all or some of the following terms listed, and boxes beside each entry, to be checked. After the appropriate box is checked, the architect initials the notation. In this way, it is possible to get approval even if the Submittal is not perfect.

__ Approved

__ Approved, as noted

__ Approved as noted; revise and re-submit for the record

__ Revise and resubmit

__ Rejected

__ Note: ______________________________

The approval/rejection notes listed above are fairly standard, but should be reviewed nevertheless in the pre-construction meeting. For instance, some architects/engineers like to have the entire document resubmitted, even if only one page requires revision. This helps to ensure that the one revised page remains connected with the original Submittal.

There are actually two Returned Submittals for each Submittal– one from the architect to the general contractor and one from the general contractor to the subcontractor.

Returned Submittal Procedure

The Returned Submittal presents no problem if it is approved. The general contractor returns the approved (or approved as noted) Submittal to the subcontractor or vendor, keeps a file copy for his own records, and sends a copy to the job superintendent. Sending a copy to the superintendent is highly recommended. Many managers erroneously put the "approved" copy in the file cabinet and do not inform their field personnel about what has been approved or not approved.

It may be just as important to know what was *not* approved as it is to know what was approved. A copy of the Returned Submittal should, therefore, be sent to the superintendent. A copy should also be sent to the subcontractor. If the Submittal is approved, then include all submitted items with the copy of the Returned Submittal, including: shop drawings, catalog cuts, product descriptions, etc. Do not send the enclosures to the field when the Submittal is made to the architect, or if the Returned Submittal is rejected. It is better for field personnel to have *only* the approved plans, shop drawings, catalog cuts, etc. Having unapproved documents can only create confusion. However, it is a good idea to let field personnel know that an item has been rejected or requires resubmittal.

The Source of Problems

Potential problems in record keeping and tracking arise when the architect stamps the Returned Submittal with one of the following:

- Resubmit for record
- Revise and resubmit
- Rejected, revise, and resubmit

This is where a good communication and tracking system can really pay off. The purpose of many of the letters and logs that we use in connection with Submittal Requests and Submittals to Architect is to make sure we do not let the non-submitted and rejected items "fall through the cracks." Failing to reprocess and resubmit a rejected Submittal is as bad as not making the Submittal in the first place. In fact, it can be worse. Having seen the Submittal sent the first time, field personnel may assume everything has been worked out.

Returned Submittal Letters

Figure 13.1a shows a computer-dedicated program letter and Figure 13.1b is a manual transmittal letter. Two of the advantages of using a dedicated program (such as the one used to prepare this sample Transmittal, the Papertrak system) are as follows. First, it automatically fills in most of the standard information, so that logs are always up-to-date. Second, it guides the person writing the letter in choosing which information should be included.

Every item submitted to the architect should be listed and tracked. The following tools can be very helpful in keeping up-to-date records.

DELTA CONSTRUCTION COMPANY

6065 Mission Gorge Road, #193
San Diego, CA 92120
(619) 582-5829

*** RETURNED SUBMITTAL ***

Date: 03/05/90

Job Number: 321
Submittal Number: 3-2

TO: STRONG STRUCTURAL STEEL
4567 IRON STREET
SAN DIEGO, CA 92312

ATTENTION: ROBERT STRONG

JOB NUMBER: 321 GIBSON OFFICE BUILDING

FROM: J. EDWARD GRIMES

SUBJECT: SUBMITTAL NUMBER: 3-2
STRUCTURAL STEEL SHOP DRAWINGS

We are sending the following submittal information enclosed or attached.

The submittal was received from ONE LINE DESIGN.
The submittal review is indicated following the submittal item.
Resubmittal is required. Please resubmit by 03/19/90 referenced by number.

1 Structural Steel Shop Drawings Sheet SS-1 to SS-27
ACTION: APPROVED AS SUBMITTED.

2 Sheets SS-1 to SS-13
ACTION: APPROVED AS SUBMITTED.

3 Sheets SS-14
ACTION: REJECT & RESUBMIT. REVISE & RESUBMIT BY: 03/19/90

4 Sheets SS 18
ACTION: REJECT & RESUBMIT. REVISE & RESUBMIT BY: 03/19/90

5 Sheets SS 15, 16, 17 and 19-27
ACTION: APPROVED AS SUBMITTED.

dp: 3/05/90 10:14 **PaperWorks**, a Construction Software Program

Figure 13.1a

Means Forms

LETTER OF TRANSMITTAL

Delta Construction Company
6678 Mission Road
San Diego, CA 92120
(619) 581-6315

FROM: J. Edward Grimes
Project Manager

TO: Strong Structural Steel
4567 Iron Street
San Diego, Ca. 92312

DATE 3/5/90
PROJECT Gibson Office Bldg. #321
LOCATION San Diego, Ca.
ATTENTION Bob Strong
RE: Submittal #3-2
Structural Steel Shop Dwg's.

RS 321.03-2

Gentlemen:

WE ARE SENDING YOU ☒ HEREWITH ☐ DELIVERED BY HAND ☐ UNDER SEPARATE COVER

VIA ______ THE FOLLOWING ITEMS:

☐ PLANS ☐ PRINTS ☒ SHOP DRAWINGS ☐ SAMPLES ☐ SPECIFICATIONS

☐ ESTIMATES ☐ COPY OF LETTER ☐ ______

COPIES	DATE OR NO.	DESCRIPTION
3	2/21/90	Structural Steel Dwg sheets SS-1 thru SS-27

THESE ARE TRANSMITTED AS INDICATED BELOW

☐ FOR YOUR USE ☒ APPROVED AS NOTED ☐ RETURN ______ CORRECTED PRINTS

☐ FOR APPROVAL ☐ APPROVED FOR CONSTRUCTION ☐ SUBMIT ______ COPIES FOR ______

☐ AS REQUESTED ☐ RETURNED FOR CORRECTIONS ☒ RESUBMIT 4 COPIES FOR 3/19/90

☐ FOR REVIEW AND COMMENT ☐ RETURNED AFTER LOAN TO US ☐ FOR BIDS DUE ______

☐ ______

REMARKS: Drawings SS-1 thru SS-13, SS-15 thru SS-17, and SS-19 thru SS-27 are all approved as submitted

Revise and resubmit drawings SS-14 and SS-18 by 3/19/90

IF ENCLOSURES ARE NOT AS INDICATED, PLEASE NOTIFY US AT ONCE.

SIGNED: J Edward Grimes

Figure 13.1b

Returned Submittal Follow-up and Tracking

This section could be entitled "Penny-Wise, Pound-Foolish," or "How to Shoot Yourself in the Foot." What follows are a couple of horror stories that cost someone a lot of potential project profit:

> *"The Case of the Missed Dimension"*
> *Due to a change in the north architectural elevations caused by interference of the structural steel columns, the window sizes in a 20-story building had to be altered. The window shop drawings were submitted prior to resolution of the conflict, and the architect properly noted on the returned shop drawings that the north elevation had to be corrected and resubmitted. Due to poor communication and follow-up by the window subcontractor (and general contractor), the window shop drawings were not revised. Consequently, the aluminum and glass for that elevation were manufactured incorrectly.*

It may have been the window subcontractor's fault, but this error affected everyone by delaying job progress by almost three months. The continuing job site overhead ate away at the profit of both the general contractor and the other subcontractors who could not complete their work. Everyone suffered– even those subcontractors who had already finished their work, who had their retention held. The atmosphere on the job changed from cooperative to finger-pointing and laying of blame. The inefficiencies and hidden costs of that sort of job atmosphere are difficult to absorb and even more difficult to track. More money is made by everyone in a cooperative job atmosphere than in a confrontational one.

In the end, it makes little difference whose fault it was. What matters is who had the most to lose and their contractual liabilities.

The **window subcontractor** lost a great deal. Not only did he have to provide all new aluminum and glass for the north elevation, but he could have been sued by the general contractor, the owner, and maybe even some of the other subcontractors. All of these dire consequences can be traced back to improper handling and follow-up of the Returned Submittals.

The **general contractor** also had a lot to lose, not only in terms of extended job overhead, but also in the possibility of a liquidated damages suit from the owner. The general contractor is also responsible to the owner for producing a finished product on time and for a stipulated sum. While it was the subcontractor's fault, it was the general contractor's obligation to track that rejected Submittal. The AIA *Standard Contract* makes the statements shown in Figure 13.2 in regard to the obligations of the general contractor.

The **architect** has a fiduciary duty to see that the shop drawings are resubmitted. Most architects/engineers rely on the language in the original construction documents which state that the contractor shall comply with construction documents regarding Submittals. The architect/engineer should follow up and see that the correct Submittals are made by the general contractor. The following excerpt from the AIA contract between an owner and architect (AIA Form B141 - 1977) provides some indication of what is expected from the architect/engineer:

> ***Paragraph 1.5.13***: *The Architect shall review and approve or take other appropriate action upon the contractor's Submittals, such as shop drawings, product data, and samples, but only for conformance with the design concept of the work and with the information given in the contract documents. Such action shall be taken with reasonable promptness so as to cause no delay.*

In this AIA contract, the architect is responsible for reviewing the shop drawings, but he is not responsible for ensuring that rejected Submittals are resubmitted as required.

The **owner** is usually the most angry of all in this situation, and the one subject to the most hidden costs. The owner in this example lost potential revenue. In fact, if he had had a Triple A firm lease in hand for the new building space, he probably would be able to sue the contractor and collect. He also thought he had hired the best architect and the best general contractor available to get the building built on time and within budget–and look what happened.

One can just imagine the atmosphere around the job site in this situation, not to mention the mood in the owner's check-writing department. Not good! The whole mess could have been avoided by proper tracking of that one rejected Submittal by either or both the architect/engineer or general contractor/facility manager.

Obligations of the General Contractor

A201-1976 General Conditions of the Contract for Construction Between an Owner and Contractor.

Article 4.3 Supervision and Construction Procedures

Article 4.3.1 The Contractor shall supervise and direct the Work, using his best skill and attention. He shall be solely responsible for all construction means, methods, techniques, sequences and procedures and for coordinating all portions of the Work under the Contract.

Article 4.3.2 The Contractor shall be responsible to the Owner for the acts and omissions of his employees, Subcontractors, and their agents and employees, and other persons performing any of the Work under a contract with the Contractor.

Article 4.3.3 The Contractor shall not be relieved from his obligations to perform the Work in accordance with the Contract Documents either by the activities or duties of the Architect in his administration of the Contract, or by inspections, tests, or approvals required or performed under Paragraph 7.7 by persons other than the Contractor.

Figure 13.2

"The Case of the Unknown Detail"
The structural steel drawings are one of the first Submittals made. (The steel layout details–especially the anchor bolt lay-out–influence the foundation layout.) There are 28 sheets of structural steel drawings submitted. On Sheet No. 19 there is a missing detail for the supports of the window washing davits. The architect returns sheets 1-18, and 20-28 "Approved as Noted." He returns Sheet No. 19 as "Revise and Resubmit." The manager is to obtain this information on the type and size of davit supports for the structural steel fabricator to design into the structure. There is no question that the steel subcontractor must provide the supports for the davits; the problem is that he does not know the extent of support to be provided.

The Returned Submittal was forwarded to the steel subcontractor, and everyone went their merry way. The rejected Submittal was not resubmitted and no one noticed. It was at the very beginning of the job and since everyone was up to page 18 of the logs, they simply overlooked the fact that Sheet S-19 had not been resubmitted. The steel subcontractor was waiting for information from the general contractor. The field personnel were not aware that they were missing davit supports, and went on building the structure, including pouring the roof deck.

Fourteen stories later, the structural steel workers had to beef up all the steel at the roof perimeter. The concrete deck had to be chipped out and drilled at great and unnecessary cost. The steel work delayed the installation of the curtain wall, and what followed was a classic example of the domino theory. The job lasted six weeks longer than necessary, and the problems were aggravated by finger-pointing and bad feelings.

The out-of-pocket costs for the project delay totalled about $25,000. Add the potential profit (opportunity lost because the same team was not working on another project) for a total loss of $65,000 to $105,000. Even that figure does not include the lost revenue to the owner. With that kind of money at stake, it is short-sighted not to invest in the relatively minor cost of additional clerical help, a computer, and/or a construction correspondence program.

The moral of these examples: get organized and establish an in-house system to track Returned Submittals, whether you are the general contractor, the architect, or the facilities manager. Everyone suffers if a rejected Submittal gets lost in the shuffle. In this day and age, with the processing tools available, no one can afford to operate haphazardly without risking costly penalties.

Follow-up Logs and Reports

The follow-up and tracking of Returned Submittals is extremely important. Not only does it keep the job progressing on a positive and professional note, but it can also "train" subcontractors and architects to take your requests and deadlines seriously. The subcontractor's and architect's attitude, as established during the Submittal process, will carry forward into the pricing of scope changes and other responses that become necessary during the course of the job.

Following up to see that the architect and subcontractor return what is required is made easier utilizing logs. A dedicated computer program can be helpful, as the logs that it generates can be more selective than those that are usually kept by hand. Figure 13.3a is an example of a manually kept Submittal log. Figure 13.3b is a Returned Submittal Log; Figure 13.3c is a report that lists the dates and status of Submittals returned to subs; and Figure 13.3d is a report that lists when re-submittals are "due." Figure 13.3e is a listing of subcontractors who are delinquent in resubmitting. A manager can give such a list to someone in the office and say,

"Call all these people and find out why they haven't submitted, and when they are going to submit the required information." The next step, of course, is to follow up with the Submittal Request letters for documentation.

Means® Forms

JOB Submittal Log

SHEET 1 OF 7

PROJECT Gibson Office Building

JOB NO. 321

LOCATION San Diego

PREPARED BY J. Edward Grimes

Contractor	Subject	Milestone Dates: Requested	Due from Sub.	Rec'd from Sub.	Sent to Arch.	Due from Arch.	Rec'd from Arch.	Status	Returned to sub.	Due from Sub.	Rec'd from Sub.	Sent to Arch.	Due from Arch.	Rec'd from Arch.	Status	Returned to sub.	Due from Sub.	Rec'd from Sub.	Sent to Arch.	Due from Arch.	Rec'd from Arch.	Status	Returned to sub.
1 Lot's of Rebar	Shop Dwgs.																						
	Ftg F-1	2/8	2/22	2/20	2/20	3/6	3/5	✓	3/5														
	F-2	2/8	2/22	2/20	2/20	3/6	3/5	✓	3/5														
	F-3	2/8	2/22	2/20	2/20	3/6	3/5	R	3/5	3/19													
2 Bent Sht Mtl.	Roof Hatch	2/2	2/19	2/20	2/20	(3/6)																	
	Shop Dwgs.	2/2	2/19	2/20	2/20	(3/6)																	
	Mat'l. List	2/2	2/19	2/20	2/20	(3/6)																	
3 Strong Steel	SS-1 – 27	2/8	2/19	2/19	2/20	3/6	3/1	↓															
	SS-1 → 13						3/1	✓															
	SS15, 16, 17						3/1	✓															
	SS 19 – 27						3/1	✓															
	SS14, SS18							R	3/1	3/19													
4 Applied Roof	S/G Waterproof	2/13	2/27	2/27	2/27	3/13	3/4	R	3/4	3/19													
	MFG Procedures	2/13	2/27	2/27	2/27	3/13	3/4	R	3/4	3/19													
	Apl. Certificate	2/13	2/27	2/27	2/27	3/13	3/4	R	3/4	3/19													

✓ = Approved R = Rejected; Revise & Resubmit ↓ = See Following Listings

Figure 13.3a

```
                     DELTA CONSTRUCTION COMPANY
==========================================PaperWorks Version 1.23F==
                            SUBMITTAL LOG                       .SA1
         Job 321      GIBSON OFFICE BUILDING
Date: 03/25/90                                             Page:   1
=====================================================================
Submittal      Name of Sub/Ven             Sent To  Date Due   Date Ret
 Number           Summary                   Arch      Back      To Sub  Status
---------------------------------------------------------------------
    1  LOT'S OF REBAR                     02/20/90  03/06/90  03/05/90
1.1    Shop Drawing for Footing type F-1                            APPROVED
1.2    Shop Drawing for Footing type F-2                            APPROVED
1.3    Shop Drawing for Footing type F-3                            REV/RESU
                            Date Due Resubmittal from SUB: 03/19/90
 comment: We will be pouring these footings very soon.  Please give special
          consideration to the shop drawing review.
---------------------------------------------------------------------
    2  BENT SHEET METAL                   02/20/90  03/06/90
2.1    Roof Hatch catalog cuts and specifications                   NOT RET.
2.2    Sheet Metal shop drawings SM-1 to SM-5                       NOT RET.
2.3    Material List                                                NOT RET.
 comment:

---------------------------------------------------------------------
    3  STRONG STRUCTURAL STEEL            02/20/90  03/06/90  03/01/90
3.1    Structural Steel Shop Drawings Sheet SS-1 to SS-27           APPROVED
3.2    Sheets SS-1 to SS-13                                         APPROVED
3.3    Sheets SS-14                                                 REJECT
                            Date Due Resubmittal from SUB: 03/19/90
3.4    Sheets SS 18                                                 REJECT
                            Date Due Resubmittal from SUB: 03/19/90
3.5    Sheets SS 15, 16, 17 and 19-27                               APPROVED
 comment:

---------------------------------------------------------------------
    4  APPLIED ROOFING COMPANY            02/27/90  03/13/90  03/04/90
4.1    Underground waterproofing system specifications              REJECT
                            Date Due Resubmittal from SUB: 03/19/90
4.2    Manuf. recommended procedures                                REJECT
                            Date Due Resubmittal from SUB: 03/19/90
4.3    Manufacturer's application inspection certificate            REJECT
                            Date Due Resubmittal from SUB: 03/19/90
 comment: We will have to schedule the Manuf. rep to be here when the
          work is being done and then he will fill out the certificate.
---------------------------------------------------------------------
    5  COOL AIR CONDITIONING              02/27/90  03/13/90
5.1    Sheet metal shop drawings for garage exhaust  GE 1 & 2       NOT RET.
 comment:

---------------------------------------------------------------------
    6  COOL AIR CONDITIONING              02/27/90  03/13/90
6.1    Cooling tower specifications and details                     NOT RET.
 comment:

---------------------------------------------------------------------
    7  COOL AIR CONDITIONING              02/27/90  03/13/90
7.1    Ductwork shop drawings for floor 2 - 6, Shts AC 1 to 6       NOT RET.
7.2    Ductwork shop drawings for first floor lobby AC 7 & 8        NOT RET.
7.3    VAV mixing box submittal.  Xtracool #56-98                   NOT RET.
 comment: Please review the VAV, box as it is a substitution, as soon as
          possible.
---------------------------------------------------------------------
```

Figure 13.3b

```
                       DELTA CONSTRUCTION COMPANY
=================================================PaperWorks Version 1.23F==
                *** SUBMITTALS RETURNED TO SUBS LOG ***                .RS1
             Job 321      GIBSON OFFICE BUILDING
Date: 03/25/90                                                    Page:   1
===========================================================================
Submittal      Name of Sub/Ven               Date Ret    Date Due
 Number           Summary                     To Sub   Resubmittal      Status
---------------------------------------------------------------------------
    1  LOT'S OF REBAR                        03/07/90   03/19/90
1.1    Shop Drawing for Footing type F-1                              Approved
1.2    Shop Drawing for Footing type F-2                              Approved
1.3    Shop Drawing for Footing type F-3                              Rev/Resu
                                        Date Due Resubmit: 03/19/90
---------------------------------------------------------------------------

    3  STRONG STRUCTURAL STEEL               03/05/90   03/19/90
3.1    Structural Steel Shop Drawings Sheet SS-1 to SS-27             Approved
3.2    Sheets SS-1 to SS-13                                           Approved
3.3    Sheets SS-14                                                   Reject
                                        Date Due Resubmit: 03/19/90
3.4    Sheets SS 18                                                   Reject
                                        Date Due Resubmit: 03/19/90
3.5    Sheets SS 15, 16, 17 and 19-27                                 Approved
---------------------------------------------------------------------------

    4  APPLIED ROOFING COMPANY               03/05/90   03/19/90
4.1    Underground waterproofing system specifications                Reject
                                        Date Due Resubmit: 03/19/90
4.2    Manuf. recommended procedures                                  Reject
                                        Date Due Resubmit: 03/19/90
4.3    Manufacturers application inspection certificate               Reject
                                        Date Due Resubmit: 03/19/90
---------------------------------------------------------------------------

    8  QUICK PLUMBING COMPANY                03/12/90   03/26/90
8.1    Plumbing Fixtures, Items 1 through 15                          Approved
8.2    Item 7                                                         Reject
                                        Date Due Resubmit: 03/26/90
8.3    Item 11                                                        Reject
                                        Date Due Resubmit: 03/26/90
---------------------------------------------------------------------------

    9  SPARKS ELECTRIC COMPANY               03/12/90
9.1    General Electric Switchgear information and submittal          Approved
---------------------------------------------------------------------------

   10  JONES CEILING COMPANY                 03/12/90
10.1   Sample and specifications of 2 x 2 reveal ceiling tile         Approved
10.2   Sample and specifications of 2 x 4 fissured tile               Approved
---------------------------------------------------------------------------

   12  CONTINENTAL HARDWARE                 03/21/90   03/26/90
12.1   Finish Hardware Groups                                         Approved
12.2   Sample of lockset with brass finish                            Reject
                                        Date Due Resubmit: 03/26/90
---------------------------------------------------------------------------
```

Figure 13.3c

```
                         DELTA CONSTRUCTION COMPANY
==================================================PaperWorks Version 1.23F==
                 ***  RE-SUBMITTALS DUE FROM SUBS ***                    .RS3
          Job: 321       GIBSON OFFICE BUILDING
Date: 03/25/90                                                     Page:   1
=============================================================================
Submittal      Name of Sub/Ven             Date Returned   Date Due
 Number           Summary                     to Sub     Resubmittal     Status
-----------------------------------------------------------------------------
  1  LOT'S OF REBAR                        03/07/90       03/19/90
1.3  Shop Drawing for Footing type F-3                                 Rev/Resub
 summary: REBAR SHOP DRAWINGS
-----------------------------------------------------------------------------

  3  STRONG STRUCTURAL STEEL               03/05/90       03/19/90
3.3  Sheets SS-14                                                      Reject
3.4  Sheets SS 18                                                      Reject
 summary: STRUCTURAL STEEL SHOP DRAWINGS
-----------------------------------------------------------------------------

  4  APPLIED ROOFING COMPANY               03/05/90       03/19/90
4.1  Underground waterproofing system specifications                   Reject
4.2  Manuf. recommended procedures                                     Reject
4.3  Manufacturers application inspection certificate                  Reject
 summary: WATERPROOFING SUBMITTALS
-----------------------------------------------------------------------------

  8  QUICK PLUMBING COMPANY                03/12/90       03/26/90
8.2  Item 7                                                            Reject
8.3  Item 11                                                           Reject
 summary: PLUMBING FIXTURE SUBMITTAL
-----------------------------------------------------------------------------

 12  CONTINENTAL HARDWARE                 03/21/90       03/26/90
12.2 Sample of lockset with brass finish                               Reject
 summary: FINISH HARDWARE SUBMITTAL
-----------------------------------------------------------------------------
```

Figure 13.3d

```
                              DELTA CONSTRUCTION COMPANY
=====================================================PaperWorks Version 1.23F==
             *** PAST DUE RE-SUBMITTALS DUE FROM SUBS as of 03/25/90 ***   .RS5
             Job: 321          GIBSON OFFICE BUILDING
Date: 03/25/90                                                         Page:  1
===============================================================================
SR #                NAME OF SUB/VENDOR                 Date Ret    Date Due
                       summary                          to Sub    Resubmittal   Status
--------------------------------------------------------------------------------
__
     1    LOT'S OF REBAR                              03/07/90     03/19/90
1.3    Shop Drawing for Footing type F-3                                    Rev/Resub
  Summary:    REBAR SHOP DRAWINGS
-------------------------------------------------------------------------------

     3    STRONG STRUCTURAL STEEL                     03/05/90     03/19/90
3.3    Sheets SS-14                                                         Rejected
3.4    Sheets SS 18                                                         Rejected
  Summary:    STRUCTURAL STEEL SHOP DRAWINGS
-------------------------------------------------------------------------------

     4    APPLIED ROOFING COMPANY                     03/05/90     03/19/90
4.1    Underground waterproofing system specifications                      Rejected
4.2    Manuf. recommended procedures                                        Rejected
4.3    Manufacturers application inspection certificate                     Rejected
  Summary:    WATERPROOFING SUBMITTALS
-------------------------------------------------------------------------------
```

Figure 13.3e

Chapter Fourteen
Construction Schedule

Construction scheduling is a fundamental management tool for effective communication. The schedule should communicate to all concerned the dates when project activities should be started and completed. The schedule must be clear in order to be understood by everyone concerned with the project: owner, architect, engineer, facility manager, facility departments, subcontractors, vendors, and the field personnel of all companies. Just as important as clarity is proper distribution of the schedule to all of the aforementioned parties.

The schedule shows everyone's activities and the impact of these activities on everyone else. The schedule should also show how an activity not completed per the schedule may delay an entire sequence of activities, as well as the entire project.

The Owner and General Contractor

The construction schedule, or completion schedule, is of the greatest concern to the owner (or user) and the general contractor or facility manager. These parties stand to lose the most if the project is not completed on time. By the same token, they both tend to gain significantly if the project is finished ahead of schedule.

Given advance notice derived from a good schedule, the owner should be able to formulate plans with reasonable certainty that he will be able to occupy and use the project as soon as it is ready. Usually, the owner is very happy if the job is completed on or ahead of schedule, and very unhappy when its completion lags significantly behind the schedule. The owner must be kept informed of the anticipated date of completion. (See Chapter 17, "Meetings and Minutes.")

Stipulated Time of Completion Versus Developed Schedule

The time stipulated in the contract documents quite often turns out to be insufficient. It is what the owner considers to be a "target" date, or "Contract Date of Completion." However, a well thought-out, printed, and distributed schedule, prepared with input from the field personnel, subcontractor, and vendors, normally presents a *realistic and obtainable* completion date.

Usually, a carefully developed schedule presented at the beginning of the project will be accepted by the owner, even if its completion date differs from that stated in the Contract Documents. Too many organizations and individuals responsible for developing the construction schedule "force" the schedule to "fit" what the owner wants, even if it is not realistic. Then, at the end of the job, "all hell breaks loose." This is not good communication, nor does it help anyone involved in the project.

Scheduling as a Tool

Scheduling is often misinterpreted as the key to successful total project management. Some scheduling enthusiasts suggest that a good schedule, well followed-up, will resolve all worries for the project. Not so! It is true that effective scheduling is one of the most important parts of project management, but it is simply one of the many tools an effective manager uses.

Producing a construction schedule is like buying a shovel. The shovel is not purchased just "to have and to hold." What you really want is a hole; the shovel is just a tool you use to get what you want. By the same token, producing a construction schedule is not the goal; completing the project is the goal, and the schedule is just a tool to help get there on time.

Why Schedule?

Why schedule? The answer is simple: experience shows that jobs with a well thought-out, well-documented schedule are more likely to be completed on time. Travelers with a good map have a better chance of reaching their destination on time.

A good schedule is also an aid with which to foresee problems and conflicts. Organizations that produce a good schedule and use it to communicate with the owner, subcontractors, and suppliers most often follow that schedule pretty closely. The schedule is in so many people's hands that the firm starts to lose credibility if the schedule is not followed.

It is vital to the progress of the job to:

- Develop and complete a well thought-out schedule.
- Maintain the overall schedule by whatever means are available.
- Update the schedule to reflect changes in scope, lack of decisions, etc.
- Keep all parties informed of changes in the schedule, and give them an opportunity to adjust, modifying the work schedule or occupancy timetable as required.
- Document the dates the schedules are sent to the owner, architect, subcontractor, and vendor.

Developing the Schedule

The general contractor is usually responsible for creating the overall schedule. The two most time-consuming portions of a job for the general contractor are at the beginning and the end of projects, when one job is finishing while another starts. At this point, a conflict often develops. Most of the time, the schedule suffers. It is easy to blame someone else for not submitting the required information.

A schedule that is not communicated to all parties until halfway through the job is of very little help. At this point, decisions have been made already, without benefit of the overview, insight, and work dependencies that a schedule gives to the owner, user, general contractor, facility manager, subcontractor, and vendor.

If an outside scheduling service is needed in order to get the schedule published within a few weeks of contract award, **use one!** It will be money well spent. In fact, with the sophistication of many of the existing computerized scheduling packages, it makes sense to have the work done outside by someone who does that kind of work every day. It is, however, essential that the contractor, subcontractor, architect, engineer, and owner provide input to the professional scheduler. Otherwise, the schedule will not accurately reflect the organization responsible for construction completion. The project manager should have the primary responsibility for the construction schedule, but that does not mean actually drawing it up or inputting it to the computer.

Note: The information going into the schedule must be reliable. If faulty data (i.e., one week to form and pour 4,000 l.f. of foundation and wall in January) is input, then the schedule will also be nonsense (the "garbage in, garbage out" syndrome).

Maintaining the Overall Schedule

Maintaining the schedule means following up and solving problems, rather than letting them persist. If the original logic is faulty, or an unexpected material delivery problem occurs, figure out a way to get around the immediate problem. This may involve issuing a new schedule highlighting any change, or it may require the substitution (with approval) of a different material.

Updating the Schedule

Nothing defeats the original scheduling effort faster than an outdated schedule that does not reflect the job as it is *today*, or that shows a completion date which is no longer valid. Even as changes are made, the work force and the people charged with completing the project will respect a schedule that is updated and includes scope of work changes, weather delays, etc., in a realistic, responsible, and timely manner.

Keeping all Parties Informed

Changes in the schedule, whether caused by changes in the scope of work or by substantial delays, should be highlighted on the schedule. Give all of the participants an opportunity to adjust to the changed conditions, so they can modify their work schedule or occupancy timetables as required.

Sometimes, changes in the scope of work are resisted in an attempt to complete the project as planned. The sentiment is that the owner should have known what he or she wanted to build when the project started. However, always remember: it is the owner's right to make changes in the scope of the work; it is his or her building. The owner can make any changes required to use the completed building for its designed purpose. One project manager's experience shows how easily this concept can get lost in the process of building:

The project manager on a 320,000 S.F. hospital project was aghast at all of the changes in scope and their costs (complete wings were gutted and rebuilt), and made her thoughts known. The owner's representative explained that only 1.5% of the hospital income dollar was going into the original construction cost, while over 50% went toward staff direct labor costs. Consequently, floor layout improvements and the addition of other potential labor-saving devices was much more cost-effective in the long run than worrying too much about the dollar amount of construction changes.

Types of Schedules

There are two basic scheduling modes: long- and short-term scheduling. The long term scheduling activity generally covers the total time period of the project. The short-term schedule is for one, two, or possibly four weeks in advance. Short-term schedules usually identify the activities by subcontractor and/or areas of work.

The following types of schedules are normally used in the construction industry:

- Bar Chart
  - Long-term (See Figure 14.1)
  - Short-term (See Figure 14.2)
- Critical Path Method (arrow)
  - Long-term (See Figure 14.3)
  - Short-term (See Figure 14.4)
- Precedence Schedule and PERT (Program Evaluation Review Technique)
  - Long-term (See Figure 14.5)
- Text Listing
  - Short-term (See Figure 14.6)

The following are the advantages and disadvantages of each of the four scheduling types.

Bar Chart (Gantt)

Advantages

- It is good for presentation purposes.
- It is simple.
- It can be understood by most people.
- Field superintendents like them because they are simple to read.
- It is useful for short-term, two-week schedules.

Disadvantages

- It is too easy to back the schedule up from the intended completion date and "prove" (erroneously) that the project can be completed with the time constraints.
- It becomes unclear when large numbers of activities are involved.
- It does not show the effect of changes in scope of work, material, or weather delays.
- It is not useful for dependency planning.
- Field superintendents like them because they are so general that they (superintendents) cannot be pinned down.
- They are difficult to update.
- It can be difficult to pinpoint project status.
- Intuition may be misapplied.

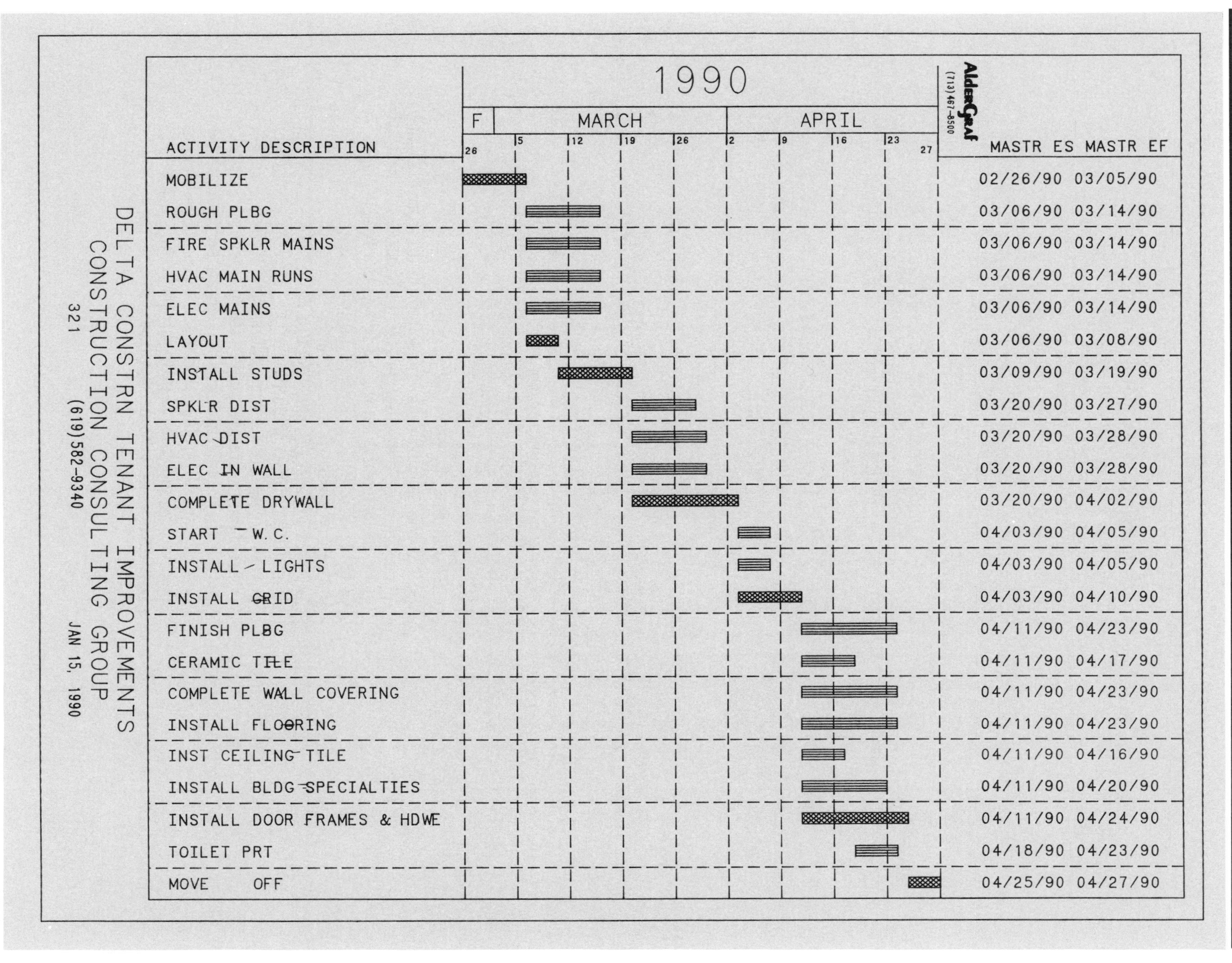
1990
F
MARCH
APRIL
26
5
12
19
26
2
9
16
23
27
AlderGraf
(713) 467-8500
ACTIVITY DESCRIPTION
MASTR ES MASTR EF
MOBILIZE 02/26/90 03/05/90
ROUGH PLBG 03/06/90 03/14/90
FIRE SPKLR MAINS 03/06/90 03/14/90
HVAC MAIN RUNS 03/06/90 03/14/90
ELEC MAINS 03/06/90 03/14/90
LAYOUT 03/06/90 03/08/90
INSTALL STUDS 03/09/90 03/19/90
SPKLR DIST 03/20/90 03/27/90
HVAC DIST 03/20/90 03/28/90
ELEC IN WALL 03/20/90 03/28/90
COMPLETE DRYWALL 03/20/90 04/02/90
START W. C. 04/03/90 04/05/90
INSTALL LIGHTS 04/03/90 04/05/90
INSTALL GRID 04/03/90 04/10/90
FINISH PLBG 04/11/90 04/23/90
CERAMIC TILE 04/11/90 04/17/90
COMPLETE WALL COVERING 04/11/90 04/23/90
INSTALL FLOORING 04/11/90 04/23/90
INST CEILING TILE 04/11/90 04/16/90
INSTALL BLDG SPECIALTIES 04/11/90 04/20/90
INSTALL DOOR FRAMES & HDWE 04/11/90 04/24/90
TOILET PRT 04/18/90 04/23/90
MOVE OFF 04/25/90 04/27/90
DELTA CONSTRN TENANT IMPROVEMENTS
CONSTRUCTION CONSULTING GROUP
321 (619)582-9340 JAN 15, 1990

Figure 14.1

Means® Forms

PROJECT SCHEDULE

PROJECT Gibson Office Bldg. #321

DATE 4-6-90

BY Bill Hardy

CALENDAR PERIOD Days

NO.	DESCRIPTION	APRIL 9	10	11	12	13	16	17	18	19	20	23	24	25	26	27	COMMENTS
	Drywall 2nd floor																
	3rd floor																
	Pull Elec. Homerun																
	2nd floor																
	Tape 1st floor																
	Tape 2nd floor																
	Elec. Conduit 3rd floor																Need inspection
	4th floor																
	HVAC 3rd floor																
	4th floor																
	Fire Sprinkler																
	3rd floor																
	4th floor																
	5th floor																

Figure 14.2

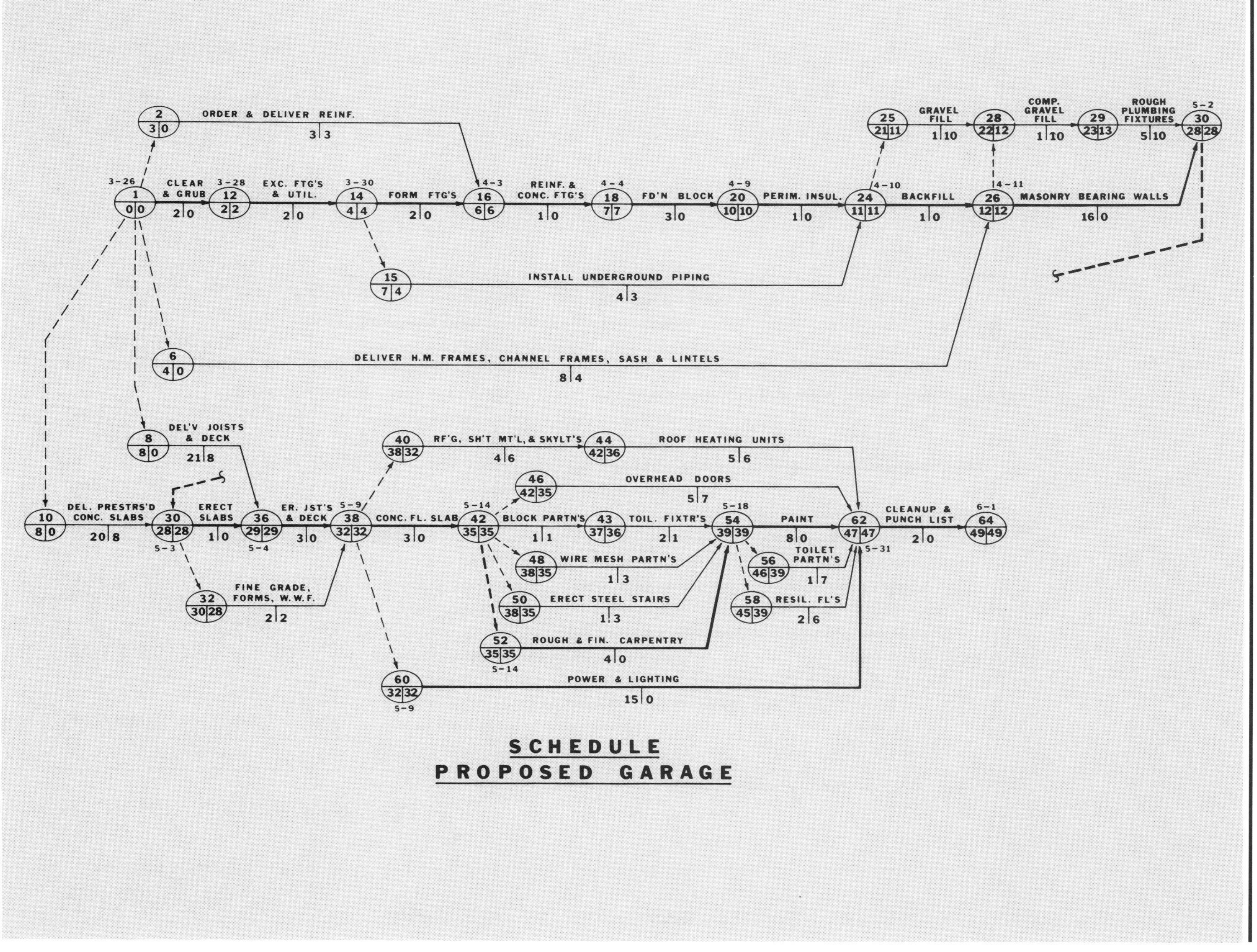

SCHEDULE
PROPOSED GARAGE
ORDER & DELIVER REINF.
CLEAR & GRUB
EXC. FTG'S & UTIL.
FORM FTG'S
REINF. & CONC. FTG'S
FD'N BLOCK
PERIM. INSUL.
BACKFILL
MASONRY BEARING WALLS
GRAVEL FILL
COMP. GRAVEL FILL
ROUGH PLUMBING FIXTURES
INSTALL UNDERGROUND PIPING
DELIVER H.M. FRAMES, CHANNEL FRAMES, SASH & LINTELS
DEL'V JOISTS & DECK
DEL. PRESTRS'D CONC. SLABS
ERECT SLABS
ER. JST'S & DECK
FINE GRADE, FORMS, W.W.F.
CONC. FL. SLAB
RF'G, SH'T MT'L, & SKYLT'S
ROOF HEATING UNITS
OVERHEAD DOORS
BLOCK PARTN'S
TOIL. FIXTR'S
WIRE MESH PARTN'S
ERECT STEEL STAIRS
ROUGH & FIN. CARPENTRY
POWER & LIGHTING
PAINT
TOILET PARTN'S
RESIL. FL'S
CLEANUP & PUNCH LIST

Figure 14.3

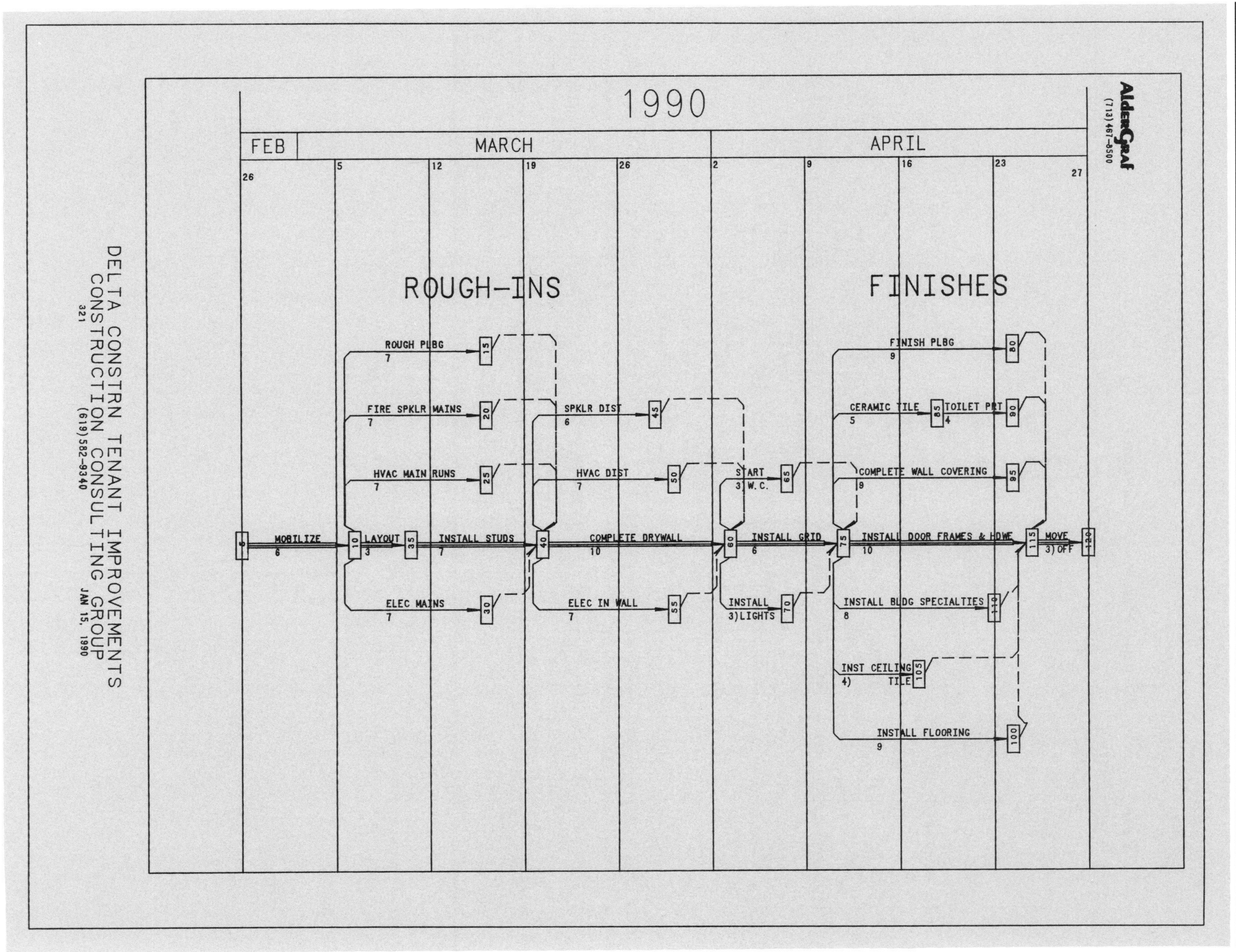

AlderGraf
(713) 467-8500
1990
FEB
MARCH
APRIL
26
5
12
19
26
2
9
16
23
27
ROUGH-INS
FINISHES
MOBILIZE 6
LAYOUT 3
INSTALL STUDS 7
COMPLETE DRYWALL 10
INSTALL GRID 6
INSTALL DOOR FRAMES & HDWE 10
MOVE 3) OFF
ROUGH PLBG 7
FIRE SPKLR MAINS 7
HVAC MAIN RUNS 7
ELEC MAINS 7
SPKLR DIST 6
HVAC DIST 7
ELEC IN WALL 7
START 3) W.C.
INSTALL 3) LIGHTS
FINISH PLBG 9
CERAMIC TILE 5
TOILET PRT 4
COMPLETE WALL COVERING 9
INSTALL BLDG SPECIALTIES 8
INST CEILING 4) TILE
INSTALL FLOORING 9
DELTA CONSTRN TENANT IMPROVEMENTS
CONSTRUCTION CONSULTING GROUP
321
(619) 582-9340
JAN 15, 1990

Figure 14.4

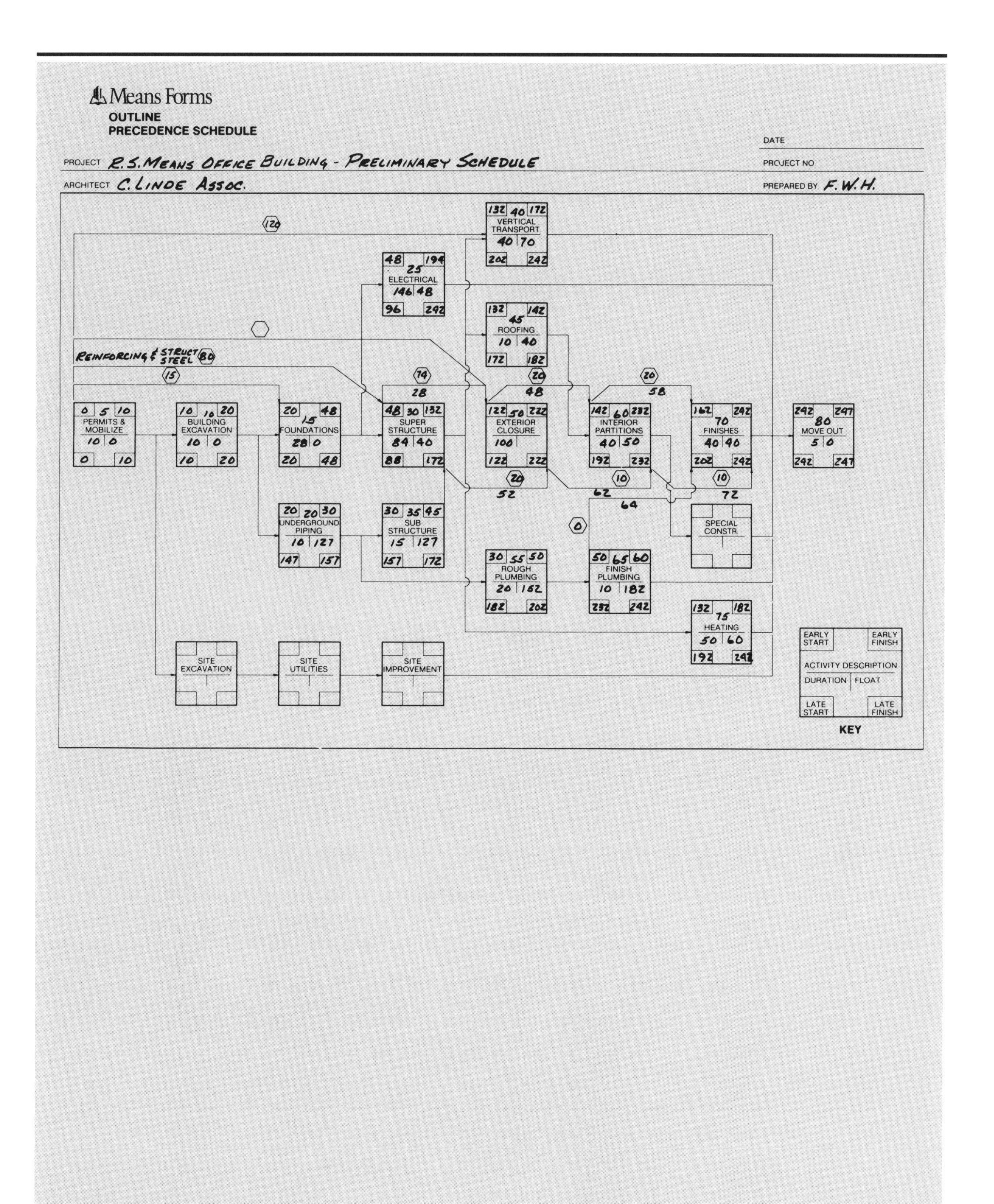

Means Forms
OUTLINE
PRECEDENCE SCHEDULE
DATE
PROJECT R.S. MEANS OFFICE BUILDING - PRELIMINARY SCHEDULE
PROJECT NO.
ARCHITECT C. LINDE ASSOC.
PREPARED BY F.W.H.
VERTICAL TRANSPORT
ELECTRICAL
ROOFING
REINFORCING & STRUCT STEEL
PERMITS & MOBILIZE
BUILDING EXCAVATION
FOUNDATIONS
SUPER STRUCTURE
EXTERIOR CLOSURE
INTERIOR PARTITIONS
FINISHES
MOVE OUT
UNDERGROUND PIPING
SUB STRUCTURE
ROUGH PLUMBING
FINISH PLUMBING
SPECIAL CONSTR.
HEATING
SITE EXCAVATION
SITE UTILITIES
SITE IMPROVEMENT
EARLY START
EARLY FINISH
ACTIVITY DESCRIPTION
DURATION
FLOAT
LATE START
LATE FINISH
KEY

Figure 14.5

```
PROJECT:  Interior Tenant Improvement in Existing Buildings

Week One
            Layout
            Install sprinkler main pipe runs
            Install HVAC main runs
            Install electric main runs

Week Two
            Complete installation studs/drywall one side
            Complete installation sprinkler main pipe runs
            Complete installation HVAC main runs
            Complete installation electric main runs

Weeks Three and Four
            Install fire sprinkler distribution
            Install HVAC distribution
            Install electric in walls and ceilings
            Complete hanging drywall
Week Five
            Install lights
            Install grid
            Start wall covering

Week Six
            Install flooring
            Install door frames, hardware
            Install cabinet work
            Continue wall covering
            Install ceiling tile
            Install ceramic tile
            Install specialties

Week Seven
            Complete flooring
            Complete door frames, hardware
            Complete cabinet work
            Complete wall covering
            Install plumbing fixtures
            Install toilet partition and accessories

            Punch List
            Move off
```

Figure 14.6

Critical Path Method (Arrow Diagram)

Advantages:

- It shows critical task and delivery items.
- It is activity-oriented.
- It shows a logical work flow.
- It provides for "number-crunching"–an analytical method for reviewing overall impact.
- It provides an effective method for focusing effort.
- It provides a framework for identifying schedule and resource conflicts before the start of work.
- It is dependent upon an analytical method and does not require as much intuitive experience.

Disadvantages:

- It requires a level of detail that is difficult to generate accurately and manage effectively.
- It can be abused, thereby resulting in a false schedule that no one understands.
- If drawn by hand, the schedule must be totally redrawn in case of a significant change in the scope of work, major delays, or faulty logic. (Computer-generated CPM schedules overcome these disadvantages.)

Precedence or PERT Schedules

Advantages:

- It is event-oriented.
- It is best suited for projects that have never been built before and have many unknown activity durations.
- It provides many of the same advantages of CPM.
- It reflects task dependencies that are not tied to absolute completion of the previous task.
- It provides a method for predicting the most probable completion date, using optimistic, pessimistic, and realistic task durations.

Disadvantages:

- It is usually not time-scaled.
- It can be even more confusing than CPM to users.
- It demands an even more detailed approach than CPM.

Text Listing

Advantages:

- It can be more descriptive of the scope of work.
- It is useful in minutes and reports.
- It is good for short schedules.
- It is easy to produce.
- It can be most easily adapted to the background of the scheduler.
- It is easily understood by the layman.

Disadvantages:

- It is difficult to show dependencies.
- It does not work for long schedules.
- It lacks discipline and, therefore, dependability.
- There are no checks and balances for impact.
- It is not adequate for planning purposes.
- It requires more experience and building "intuition."

Recommended Use of Schedules

The Critical Path Method works best for the long-term schedule, while the manual bar chart is most effective for the two- or four-week short-term schedule. Using the Critical Path Method, it is easier to organize long-term schedules (with a larger amount of detail) into a usable form. The precedence type schedule can also be quite satisfactory for this purpose–if it is time-scaled. This makes it easier for the participants (and laymen) to understand the tasks and their relationships to time and to one another.

Note: Many computerized scheduling systems employ precedence style diagramming with CPM analysis–though most lack the probability element offered by a true PERT system. Manual bar charts, on the other hand, are easy and quick to produce, but are very intuitive (i.e., one must have experience to create an effective bar chart), and they show less detail.

If the project is long-term, there should be three levels of schedules:

- Total project:
  Critical Path Method, if reliable information is available
- Short-range (2-4 months)
  Critical Path Method
  Possible bar chart
- Short-term (2-4 weeks)
  Bar Chart
  Text Listing

If the short-term and long-term schedules become totally divergent, it is time to rethink, redraw, and republish the long-term project schedule. Most computer CPM programs have the capability to print "windows" of two weeks to three months, which can be used for short-range or intermediate schedules. Computerized CPM information can also be "translated" to produce bar charts. However, in many cases, because of the way the initial information is entered in detailed CPM segments, a bar chart translation can become unwieldy.

Construction scheduling is part of the communication process. To be effective, it must be received, understood, documented, and agreed upon. To tell the owner of the project that the job will be completed on or about December 1st, without showing a schedule, is not proper documentation. For that matter, neither is it good business. It is imperative that a written schedule be produced and transmitted to all parties involved so that:

- The owner can plan occupancy and deliveries (such as furniture).
- The owner can predict cash flow.
- The general contractor and subcontractors can schedule manpower, deliveries, etc.
- The subcontractors will know when and what is expected of them.
- The effect of scope of work changes, or adverse weather conditions can be visually presented and readily understood as they relate to the original construction and completion schedule.

All of these reasons represent tools that must be used by the participants planning the project. Not finishing a project by the stipulated or agreed upon time is one of the most frequent causes of litigation. The *most* frequent causes for not completing a project "as scheduled" are:

- The project was not well thought out and scheduled in the first place.
- The stipulated completion date in the construction documents was "wished for," rather than "planned for."
- The impact of changes in the scope of work was not fully considered.
- The effects of adverse weather conditions were not fully considered. In many cases, weather affects a project to a much greater extent than just the days that it actually rains or snows. The resulting mud and frozen dirt both impact construction completion and, therefore, costs.
- Owners and participants are not informed of revised anticipated completion dates.
- The individual responsible for the schedule did not reach a consensus with all of the participants that the target dates were achievable. The participants include the owner, architect, contractor, subcontractors, and vendors.

Managers should use the communication process to include all project participants in the schedule preparation. If the schedule is deemed unattainable by those who have to work within it (but who were not consulted in its preparation), then you will have taken a big step toward creating a negative job-site attitude. The last item, agreement and consensus to a completion schedule, *must not* be left out of the process.

Each subcontractor on the project must have a *final* deadline for completing various phases of their work. No single function or activity should be longer than ten working days. If so, break it up into smaller, more specific "pieces." For example:

Install hangers	4 days
Install VAV boxes	15 days
Install ductwork	23 days
Test system	3 days

Promises from subcontractors, such as "I'll be done in time!", are of little help, and generally represent wishful thinking.

The best and, in fact, only way to give everyone confidence in a final completion date is to meet each and every intermediate milestone date consistently. This is an absolute. If the schedule is not met, then the schedule is not relevant, and will not be relied upon. Once the schedule is not relied upon and followed, there is a great potential for chaos.

A good schedule must include many specific *intermediate* milestones that can be used to identify problems early on. The key to correcting the problem is everyone's concurrence, *at the beginning of the project*, that missing intermediate milestone dates means that the project will be late. This approach leads everyone to focus on fixing the problem instead of ignoring its existence with statements like: "It's too early in the job to have an impact," or "Don't worry about it, we'll catch up." No one should be allowed to explain away missed dates with such statements, because that attitude may pervade the job and poison the project.

Communications

The owner and subcontractors can be kept informed of job progress through the Job Minutes (see Chapter 17) and through requests for extension of contract time for weather or for changes in the scope of work dictated by the owner.

The Transmittal format can be used as a cover letter to send the construction schedule to the owner, architect, vendor, lender, or other contractors. By using the format in Chapter 3, it is easy to document who received the construction schedule, and when, where, and how they got it.

The letter with the construction and completion schedule shown in Figure 14.7 should be sent to the owner, user, architect, and engineer. The message here is: if the owner/user changes his or her mind, it will affect the completion of the project.

The letter shown in Figure 14.8 should be sent to the subcontractors and vendors. This letter puts the subcontractor on notice that he must inform the scheduler if the work cannot be completed or if difficulties arise. Without such a letter, a subcontractor might claim that he never agreed to the schedule, or never even saw it. This kind of communication documentation will save many arguments during the course of construction.

Some subcontractors do not look at the schedule until it is time for them to begin work on the job. This is why it is especially important to document the fact that they have been sent the schedule, along with a request that they notify the manager if they cannot complete their work in the stipulated time. It is wise to plan to have the subcontractor's field manager visit the site two weeks before their scheduled start, and then again a few days before their scheduled start, to ensure that they can start on schedule, "claim" their lay-down area, and get their materials delivered to the site. This step can help to head off many subsequent problems.

When owners, departments, divisions, architects, engineers, contractors, subcontractors, and vendors are, or may be, affecting the construction or design schedule, they should be notified using the Transmittal format described in Chapter 3. It provides documentation of when it was sent, to whom, and the subject matter.

March 15, 1990

Mr. Rod Wright
One Line Design
1000 Dream Street
San Diego, CA 90000

Dear Mr. Wright:

Reference: Job #321 - Gibson Office Building
Subject: CPM Schedule

Enclosed is our Critical Path Method Schedule for the Gibson Office Building, our Job #321.

The various items of work shown on the enclosed are not commitments, but are estimates based only on conditions and other representations in the contract documents, weather conditions, materials shortages, and labor conditions.

Because many items of work are dependent upon or preceded by other items of work, any changes in these conditions, including changes in scope requested by the Owner or Architect, or obstacles to our starting and proceeding with the activities listed on the enclosed schedule, will necessarily affect our schedule.

This schedule is preliminary and is currently being reviewed by our subcontractors; it is, therefore, subject to revision.

Sincerely,

J. Edward Grimes
Project Manager

Enclosure

Figure 14.7

To: All Subcontractors
From: J. Edward Grimes, Project Manager

Reference: Job #321 - Gibson Office Building
Subject: CPM Schedule
Date: March 15, 1990

Enclosed is the Critical Path Method schedule for the Gibson Office Building project.

I have reviewed this schedule with you. Please review it again carefully and advise us if you anticipate any delivery or manpower problems.

This is a preliminary schedule and we will assume that you can complete your work in the time framework as shown. Please let me know no later than March 1, 1990 if you have any problems at all complying with this schedule.

If, during your review of the schedule or during the course of construction, you foresee any problems with your capability to perform your portion of the work, such as delivery of material, strikes, or any other item that might interfere with your completion of your portion of the contracted work as scheduled, please contact me as soon as possible.

J. Edward Grimes
Project Manager

Figure 14.8

Summary

The project schedule is an important communication tool for the construction industry. Remember, though, that no tool will do any real good unless it is used properly. Proper project scheduling should include the following items.

- Input from *all* parties to the project: contractor, subcontractors, vendors, architect, engineers, and owner.
- Attention to long lead time items.
- A well thought-out plan of attack.
- Realistic and obtainable durations and completion dates.
- A scheduling format that is proper for the situation and user. (You may develop the best schedule ever assembled, but if the field superintendent and/or subcontractors cannot understand it, then it is a wasted effort.)
- A consistent effort to inform all parties to the project of any changes in the schedule (whether due to scope changes, lack of direction and decisions, weather, or any other reason) by issuing updates and/or addenda to the schedule.
- A concerned management team who will actively strive to maintain the schedule by all means available.

The final item in the previous list is especially important. Even if you get the proper input, identify all of the long lead time items, plan out the project, assign all scheduled items reasonable completion times, and develop a really good schedule, it is all for naught if the project leaders do not keep after the subcontractors and vendors to maintain the schedule.

Remember, the schedule is a very effective weapon in your arsenal in the battle for success and profit. Use it right and you may not have to fight as hard to attain victory.

Chapter Fifteen
Special Use Letters

Concept

One of the most important concepts in construction correspondence is that of giving others the opportunity to say no, as in: "No, do not do what you told me you were going to do."

This "opportunity to decline" is extremely important between **all** contractually-bound parties. Accountability within these relationships is typically as follows.

Party	Accountable to
General Contractor	Owner
Architect	Owner
Engineer	Owner
Consultant Disciplines	Architect, Engineer, or Owner (depending on who enlisted and pays for their services)
Facility Manager	User Department
Subcontractor	General Contractor

The issues covered in this chapter may occur infrequently, but the way in which they are handled is of crucial importance. They must be addressed precisely and accurately and must be dealt with as soon as possible.

Failure to Respond to Special Letters

Once formally contacted through the appropriate channels with a special letter or Transmittal, the addressee may choose not to respond. This failure to respond may be interpreted as implicit authorization to proceed with the plan as explained in the special letter. The recipient (individual or company) has an obligation to issue instructions to stop work on tasks for which they are not willing to pay or do not want performed. The courts have been quite specific in ruling that an owner cannot knowingly benefit from the actions of another party, if that owner has knowledge that the items are not, or will not be, covered in the contractual agreement. In other words, an individual should only get what he or she pays for.

Most of the letters shown in this chapter are "notification" letters, in which notice is given of an action about to be performed. These letters are not "asking" for anything, but are, instead, notifying the person(s) paying for the work that a task is either under way or about to be started. This form of letter does not ask a question. It states a fact, thereby making it the *recipient's* responsibility to respond.

Follow-up

Unlike other correspondence, the follow-up for special letters is minimal. The most important requirement is to make sure the correct parties receive the letter. After that, a lack of response usually indicates authorization to proceed with the action described in the special letter. Consequently, there is little to track.

A Word of Caution

Do not utilize special use letters without first considering alternative action, and do not "trap the owner in a corner," if it can be avoided. Also, carefully choose what is to be included in this type of correspondence. Do not include superfluous and extraneous questions in special use letters, as they will simply cloud the issue the letter intends to address. Once the true intent of a special use letter is violated, others may also forsake the rules, and the value of this tool is lost.

Types of Special Use Letters

The first seven special use letters, listed below and shown in Figures 15.1 through 15.7, illustrate the need for effective communications in construction. These special letters directly relate to getting the work done, whether or not a written or verbal contract exists. These seven letters, with some text variations for special conditions, can be used by the *architect, engineer, facility manager, general contractor, and the subcontractor.*

- Extra work and extension of time (Figure 15.1)
- Directive to accelerate work (Figure 15.2)
- Order not to start work (Figure 15.3)
- Order to stop work (Figure 15.4)
- Notice to the owner that the work will be stopped (Figure 15.5)
- Notice of a deferred submittal of a cost proposal (Figure 15.6)
- Notice of reserving impact costs (Figure 15.7)

The next twelve letters, listed below and shown in Figures 15.8 through 15.19, apply chiefly to the general contractor/owner, and the subcontractor/general contractor. Items such as "Letter confirming bid" and "Defective work by a prior contractor" may be of special interest to the facility manager.

- Differing site conditions (Figure 15.8)
- Notice of interference (Figure 15.9)
- Defective work by prior contractor (Figure 15.10)
- Extension of contract time due to weather (Figure 15.11)
- Extension of time for causes beyond your control (Figure 15.12)
- Questioning the price limits of change orders (Figure 15.13)
- Confirm bid or low bid (Figure 15.14)
- Notice to proceed with work (Figure 15.15)
- Letter of intent to award a contract (Figure 15.16)
- Notification of termination of contract with subcontractors (Figure 15.17)
- Stop work order (Figure 15.18)

The Construction, or Work, Schedule letter (in reference to the project completion) could also be included in this list. However, that particular letter is discussed as a separate topic in Chapter 14, "Construction Schedules."

Extra Work and Extension of Time

In theory, a general contractor should not have to deal with requests for extra work and time extensions because the plans and the construction documents should reflect accurately what is to be built. The fact is, requirements for extra work and appropriate extensions of construction time are part of every job. This is the case because a construction job is never static; it is always changing. Consequently, changes in the scope of work and completion time are inevitable. They should be looked upon as a challenge, not an attempt to cause the contractor problems.

The authority to commit to additional costs must be assigned and understood by all early in the project, preferably at the pre-construction meeting. If such authority has not been established, then a special use letter with the beginning paragraph as shown in Figure 15.1 ("Extra work time extension" letter) is entirely appropriate. The axiom "No trust without testing" certainly applies to this situation.

When an architect issues direction for work, he or she should be willing to provide a corresponding Field Order, or whatever change of scope of work authorization is appropriate for the project. When such documentation is provided, it is assumed that the architect feels that these instructions will not cost any money. However, the general contractor, subcontractors, and field personnel should not alter their work without written instructions to do so. The architect has the right of review on cost changes, and can argue the point of the costs at that time.

Traditionally, architects and engineers do not look favorably on Price Requests and Field Orders. On the other hand, the issuance of a Price Request can, and does, provide the architect with a vehicle and reason to charge the owner extra for work *beyond the scope of his original agreement and contract with the owner.*

The letter shown in Figure 15.1 is a bit long for the Transmittal format. In this case, the Transmittal format (shown in Chapter 3) can be used as a cover letter to track the document. An answer is not required from the owner, but the letter is one more occasion on which to give the owner an opportunity to say "no" before trouble starts. This letter assumes that the answer is "yes," until the addressee says "no."

June 18, 1990

Mr. Rod Wright
One Line Design
1000 Dream Street
San Diego, CA 90000

Reference: Job 321 - Gibson Office Building
Subject: Directions to Remove Block Wall

Dear Mr. Wright:

This letter will serve to confirm that your Mr. Jones directed us to remove the block wall between rooms 101 and 102. This work was not a part of the scope of work contract.

We are performing this extra work with the understanding that we will be paid all additional costs incurred and will receive a commensurate extension of contract time.

(If this letter were sent to the owner, the following sentence might be included: We understand that you have authorized your architect to direct us to perform this extra work.)

In order not to delay the work and, at the same time to expedite completion of the project, we understand that you want us to proceed. We are performing this extra work at the present time and will be issuing a Request for Change when we have compiled the costs involved.

Sincerely,

J. Edward Grimes
Project Manager

Figure 15.1

Directive to Accelerate Work

Like the "Extra work" letter, this directive is intended to confirm and notify. It also gives the responsible party an opportunity to say "no" to acceleration of the work–a condition that will add costs which they are expected to pay.

The Transmittal can be used as a cover letter for the directive to accelerate work. The Transmittal documents indicate when, where, to whom, and the subject matter. The Transmittal can also be logged, so that the pertinent information can be retrieved easily. See Figure 15.2 for a sample "Directive to accelerate work."

Order Not to Start Work

This type of letter is intended to confirm instructions received *not* to start a project or a particular portion of the project. Such an order could have a severe impact on a firm's operations and its ability to meet other commitments. The owner should pay for any such order that negatively affects the firm performing the work. See Figure 15.3 for an example.

Order to Stop Work

This letter is similar to, but more straightforward than, a letter indicating that work must stop unilaterally due to a lack of information. In this case, direction to stop is given *after* work begins. See Figure 15.4 for an example.

It is imperative that a letter similar to this be sent to both the owner and the architect/engineer. This type of letter serves as a notification that additional costs will/may be incurred. It establishes the fact that the project will cost more and makes everyone aware of the ramifications. Such a letter sometimes forces a decision for an alternate plan, rather than stopping the work altogether.

Few people understand the impact that stopping work can have on a project. The losses can be huge and unrecoverable for everyone–the owner, architect, engineer, consultants, general contractor, and subcontractors. In many cases, the workers in the field also lose because they are not able to find other work immediately.

If the work stoppage only lasts for one or two days, the project will probably not be affected too severely. However, once a work stoppage begins, the time slips from one day to ten days all too easily. That is why it is so important to notify the person paying the bills that the work stoppage is costing money.

October 19, 1990

Mr. Rod Wright
One Line Design
1000 Dream Street
San Diego, CA 90000

Reference: Job #321 - Gibson Office Work
Subject: Direction to Accelerate Work

Dear Mr. Wright:

This letter will serve to confirm the directive given by your Mr. Ralph today that we accelerate our work on this project by putting on more men and working overtime.

We are not behind schedule on this project and thus this directive to accelerate constitutes a directive for us to perform extra work for which we are entitled to be compensated for all the additional costs incurred in the acceleration, including any loss of efficiency that will result.

We are commencing acceleration of the work as directed on October 18, 1990 with the understanding and agreement that we will be paid these additional costs.

Sincerely,

J. Edward Grimes
Project Manager

Figure 15.2

January 17, 1990

Mr. Rod Wright
One Line Design
1000 Dream Street
San Diego, CA 90000

Reference: Job #321 - Gibson Office Building
Subject: Order Not to Start Work

Dear Mr. Wright:

This letter is to confirm our telephone conversation today in which you directed us not to start work at the job site, since our work areas will not be made available until a later date.

You further advised us that we cannot start the work until you notify us.

We bid this project on the basis of your having the work on the project in such a condition that we could start our work and continue uninterrupted until its conclusion, utilizing one crew. We have scheduled our entire operation to have this work completed by April 27, 1990.

Please be advised that this stoppage will require us to reschedule our crew: store our materials at the job site and then rehandle them; and to perform our work in periods of higher wage rates, out of the normal sequence for such work, and at a time of adverse weather which otherwise would not have been encountered.

All of the above will entail additional labor at additional cost to us and, as we advised you on the telephone today, all of these additional costs will be on your account. We expect prompt payment for these additional costs, as well as a commensurate extension of contract time.

Sincerely,

J. Edward Grimes
Project Manager

Figure 15.3

Delta
Construction
Company

6678 Mission Road
San Diego, CA 92120
(619) 581-6315

January 30, 1990

Mr. Rod Wright
One Line Design
1000 Dream Street
San Diego, CA 90000

Reference: Job #321 - Gibson Office Building
Subject: Confirmation of Work Stoppage Order

Dear Mr. Wright:

This letter confirms our telephone conversation today in which you directed us to stop work on the job site, for reasons for which we are not responsible.

You further advised us that we cannot recommence work until you notify us.

We bid this project on the basis of your having the work on the project ready for us to begin and continue uninterrupted until its conclusion, utilizing one crew. We have scheduled our entire operation to have this work completed by August 15, 1990.

This stoppage will require us to reschedule our crew, store our materials at the job site and rehandle it; and then perform our work in periods of higher wage rates, out of the normal sequence of doing such work and at a time of adverse weather which otherwise would not have been encountered.

All of the above will entail additional labor cost to us and, as we advised you on the telephone today, these additional costs will have to be passed on to you. We expect prompt payment for these additional costs, as well as a commensurate extension of contract time.

(Optional Paragraph: We are continuing to work on other portions of the contract, although our efficiency is severely impacted.)

Sincerely,

J. Edward Grimes
Project Manager

Figure 15.4

Notice to Owner That Work Will Stop Unless....

This is a tough letter to write. However, at times, it is necessary, especially when the project is going nowhere, when instructions conflict, and the situation must be "brought to a head." This type of letter (See Figure 15.5) is more likely to be necessary in highly structured owner-architect-contractor relationships, such as public contract work, where it takes a long time to get Price Requests and Change Orders approved and implemented, and Price Requests wind up affecting other Price Requests that have not yet been approved.

Again, the key is to "give notice" and (implicitly) require a response from the addressee. When no response is forthcoming, that lack of response may serve to establish everyone's position itself. If common sense says that the owner should respond, and he does not, then find out what happened. Never knowingly waste the owner's money nor allow work to progress without proper information or authorization. Juries cannot be counted on to deliver a consistent verdict when "intent" is in question. On the other hand, the costs of work stoppage must also be taken into consideration.

Do not send this letter arbitrarily, or without thorough consideration of the reason, content, and consequences. Make sure of the facts and then try all, repeat *all*, other avenues and carefully document the effort. This letter should be a last ditch measure.

Deferred Submittal of a Cost Proposal

In many cases, owners, the owner's representatives, or architects will send a request for a cost proposal for additional work, stating that any costs must be returned within the time stipulated, or they will assume that the change will cost nothing. An example is shown in Figure 15.6.

The responding letter may state that an accurate price cannot be generated in the time stipulated, but that a cost proposal will be submitted when all of the facts are accumulated, thereby deferring the submittal of the actual cost proposal.

Many owners consider most cost proposals to be too high. They may be right, but until the "real" costs are established, taking guesses at the estimate will only cause harm. People tend to remember the first estimate they hear. If that first off-the-cuff estimate was very wrong, resentment, suspicion, and conflict may result from hazarding a guess too early.

June 16, 1990

Mr. Rod Wright
One Line Design
1000 Dream Street
San Diego, CA 90000

Reference: Job #321 - Gibson Office Building
Subject: Notice of Work Stoppage

Dear Mr. Wright:

We regret to inform you that we will be forced to stop work on Building "B" of referenced project on Friday, June 10, 1990 at 3:00 PM for the following reason(s):

We have not received authorization to proceed with PR #15, which affects long lead items as well as the staffing of our site personnel.

The work specified in PR #15 is so all-encompassing that we cannot work around it without the expenditure of a great deal of money, which was not included in the price quoted to you for PR #15 nearly four months ago, on March 2, 1990.

We have been informed that you intend to proceed with the scope of work outlined in PR #15, but we have not received specific authorization for the expenditures of money on your behalf. We cannot in good faith and under the terms of our contract "act in the best interests of the owner" and continue to construct work that will have to be torn apart when PR #15 is fully executed into our contract.

This stoppage will require us to reschedule our crew; store our materials at the job site; rehandle them; and perform our work in periods of higher wage rates out of the normal sequence of doing such work, and a time of adverse weather which otherwise would not have been encountered. All of the above will entail additional labor and additional cost to us and, as we advised you on the telephone today, these additional costs will be passed on to you, as they are not covered by our contract. We also expect and are entitled to receive a commensurate extension of contract time.

We are continuing to work on other portions of the contract, although our efficiency is severely impacted.

Sincerely,

J. Edward Grimes
Project Manager

Figure 15.5

May 18, 1990

Mr. Rod Wright
One Line Design
1000 Dream Street
San Diego, CA 90000

Reference: Job #321 - Gibson Office Building
Subject: Field Order #15

Dear Mr. Wright:

We received the directive (either a Field Order, Price Request, or Clarification) on May 15, 1990. This work is being implemented at the present time. Because of the complexity of this change, we are unable to submit a proposal for an equitable adjustment in price at this time.

This directive has caused and will cause us to incur additional costs. Therefore, we intend to alert you to a claim for an increase in our contract costs, with a commensurate extension of contract time.

We anticipate submitting a response to the directive, FO #15, on or about July 15, 1990.

Sincerely,

J. Edward Grimes
Project Manager

Figure 15.6

Reserving Impact Costs

A special letter to reserve for unknown impact costs may be used in the following situation. A Major Price Request is issued and the response includes a significant contingency for potential unanticipated work. The price might be $150,000, with a contingency of $45,000. The owner/architect does not allow the contingency and issues a Change Order for $105,000, with instructions to proceed with the work described. The contractor might respond with a letter similar to that shown in Figure 15.7, reserving the right to claim impact costs at a later date, when many unknown ramifications creep into the "scope of work."

Sometimes it is difficult to see all the items that a change may affect. For example:

> *The owner requests that all doors be changed from 6'-8" to 8'-0". The response must cover not only the additional cost of the doors, but also additional painting, different door frames, and maybe even the cost for different finish hardware. Futhermore, to install an 8' door, an allowance of about 2-1/2" must be included for the frame. There are two options. The ceiling must either be raised or the doors cut down. These two options may cost more than the cost cited in the original proposal, especially since this new cost may not be discovered until the doors and frames are ready to be installed.*

Consider including a statement about reserving cost impact in each quote to the owner/architect.

Differing Site Conditions

Unknown conditions can make the difference between beggars and millionaires. If a contractor on the project runs into unforeseen obstacles, such as a differing site condition, the owner must be notified:

> *"promptly, and before such conditions are disturbed," according to government contract (23A), III-Article 4.*

> *Notice must be given "within 20 days of the first observance of the differing site conditions," according to AIA Document A-201, II-Article 12.2.1.*

In both cases (the government contract and the AIA document), claims for differing site conditions will be rejected nevertheless if the stipulations of the contract are not followed. Notification should be given using the Transmittal cover letter so the information is logged and recorded, and a record of the information exists in a form other than just the "hard" copy (see Figure 15.8). If the letters notifying the owner are lost, and the copies are misplaced or misfiled, it will be difficult (without a Transmittal or a log of correspondence) to produce documentation to show when (and if) the letter was sent. In reality, just where do you file, "unforeseen obstacles," general correspondence, Price Requests, and Requests for Change? How do you keep track of where the document is located if and when you get an answer? That tracking ability may have significant impact on potential claims against the project. With the use of the Transmittal format as described in Chapter 3, that information will always be available. Providing the owner/architect/engineer with a copy of the logs will ensure that they, too, have a record that the documentation was issued. Consider sending this Transmittal and letter via certified mail, and/or hand delivery.

April 30, 1990

Mr. Rod Wright
One Line Design
1000 Dream Street
San Diego, CA 90000

Reference: Job #321 - Gibson Office Building
Subject: Signed CO #18 and Potential Impact Costs

Dear Mr. Wright:

Enclosed is a signed copy of Change Order #18 which compensates us for the direct costs incurred in performing the extra scope of work referred to in Price Request #24 and shown in Change Order #18.

Please be aware that, in the performance of this extra work, we may incur additional unknown costs and expenses in performing the balance of the contract work which we otherwise would not have incurred.

We understand that this change order has been issued so that we will be reimbursed for our direct costs. We are thereby executing and returning to you the Change Order with the understanding that it is limited to that purpose, and with the further understanding that we have reserved and preserved our claim for the impact costs which we may incur, and that our right to proceed with our impact claim does not in any way adversely affect the execution of this Change Order.

Sincerely,

J. Edward Grimes
Project Manager

Figure 15.7

February 14, 1990

Mr. Rod Wright
One Line Design
1000 Dream Street
San Diego, CA 90000

Reference: Job #321 - Gibson Office Building
Subject: Differing Site Conditions

Dear Mr. Wright:

This letter confirms our conversation of (date), in which we advised you that we encountered rock during our excavation today on the east side of the project. We have had to cease our operations in this area, while awaiting directions from you on how to proceed.

This contract was bid on the basis that the underground work would involve excavating only soil. Therefore, we are entitled to be reimbursed by you for all additional costs incurred as a result of these conditions. We are also entitled to an extension of time, commensurate with these delays which have already begun.

Please advise us immediately since this condition is delaying completion of our work.

Sincerely,

J. Edward Grimes
Project Manager

Figure 15.8

Notice of Interference

This letter (see Figure 15.9 for an example) is very useful for remodeling and renovation work. It is appreciated by those who have torn out a wall only to discover plumbing that cannot be moved. The question is: Who pays for rebuilding the wall (or doing whatever must be done to resolve the problem)?

This is the occasion for a "Notice of interference" letter. If severe ramifications and downstream impact costs become involved, a record exists showing that the proper parties were notified.

Defective Work by a Prior Contractor

An example of this letter is shown in Figure 15.10. A typical occasion for sending this letter would be as follows. A contract is let for the wallpapering part of a project. The wallpaper contractor arrives on the job and discovers that the drywall has not been completed in a condition adequate to apply wallpaper.

Other such instances can arise when owners or facility departments issue separate contracts for work that depends on other contractors, or on work accomplished at some time in the past.

A note of caution: Double check the nature of the defect to make sure the complaint is truly valid. Check the schedules and plans to ensure that the prior contractor did not leave the situation incomplete or incorrect for a valid reason. Never use this letter to make excuses. This tool is weakened if it is misused.

August 8, 1990

Mr. Rod Wright
One Line Design
1000 Dream Street
San Diego, CA 90000

Reference: Job #321 - Gibson Office Building
Subject: Notice of Interference

Dear Mr. Wright:

According to the contract documents and information provided us for this project, no interference with our work was shown or detailed in any way.

We have encountered substantial interference in the areas of the wall B-6 on the 3rd floor. This letter confirms that we are incurring additional costs as a result of this interference and are entitled to reimbursement of all costs and expenses so incurred.

Furthermore, any delays resulting from this additional work will entitle us to a commensurate extension of contract time, as well as any proportional overhead and profit costs.

Sincerely,

J. Edward Grimes
Project Manager

Figure 15.9

July 3, 1990

Mr. Rod Wright
One Line Design
1000 Dream Street
San Diego, CA 90000

Reference: Job #321 - Gibson Office Building
Subject: Work of prior contractor is defective

Dear Mr. Wright:

Our pipe on this project is to be hung from hangers which are attached to steel embedded in the concrete slab by others. We came to the job site for inspection and discovered that in some areas no receptacle pieces are embedded, and in others, the pieces are embedded in the wrong place.

We have pulled our forces off the job while we wait for the preparatory work to be accomplished or corrected, until we receive other directions from you.

We are entitled to reimbursement for the costs incurred in moving on and off the job site and, if these defects are not remedied immediately, we are entitled to further reimbursement for additional costs for increased labor rates and for performing this work under conditions and sequences different than those on which our bid was based. We are also entitled to receive a commensurate extension of contract time.

Sincerely,

J. Edward Grimes
Project Manager

Figure 15.10

Extension of Contract Time Due to Weather

An "Extension of time due to weather conditions" letter (see Figure 15.11) may be awarded in the following situation. A storm front passes through and leaves water-laden footings, caved-in trenches, wet drywall, lost time, lost wages, and lost profits in its wake. Under most contracts, an "Extension of contract time due to weather" will be allowed.

AIA A-201 (1976) III, Articles 8.3.1 and 8.3.2 state that a claim must be filed within 20 days of the commencement of the delay. This means that if it rains for a month, you are required to notify the architect that you are delayed before you know when the rain is going to end.

In the pre-construction meeting, or at least in the job site minutes, the project manager should make it a point to mention how lost job time will be monitored. Letters sent once a month as an update on this situation will usually suffice. This is an easy way to do it: At the first of every month, the "rain letter" is sent out–even if there is no lost time–usually summarizing the total lost days. If your claim is challenged, you may want to include any weather information from the newspapers or from local airports to bolster your claim.

The critical point: The extent of rain damage to the project will be unknown until somewhere downstream (no pun intended). For example, the steel fabrication shop, even though it is away from the construction site, may be under water for five weeks, and this could have a severe effect on the construction schedule. This is the reason for the "catch-all" statement at the end of the sample letter, which states: "We will advise you of the impact on the schedule." This statement is intended to cover the unknowns. This letter can utilize the Transmittal format.

Extension of Time for Causes Beyond Your Control

This letter (See Figure 15.12) is a straightforward notification of fact. The owner may already be aware of this fact (because it was on the front page of the newspaper, for example), but he or she (as well as all parties stated in the contract, i.e., the architect, the owner, and sometimes the lending agency) must be formally notified. Never assume that everyone knows that an event affects the project.

This letter can use the Transmittal format, or the Transmittal can be used as a cover letter.

January 19, 1990

Mr. Rod Wright
One Line Design
1000 Dream Street
San Diego, CA 90000

Reference: Job #321 - Gibson Office Building
Subject: Weather Delays

Dear Mr. Wright:

This letter confirms that we have encountered unusual weather conditions which have delayed our work. We are entitled to a commensurate extension of contract time which may be in excess of the following actual rain or muddy days.

Rain: Jan 15, Jan 16, Jan 17, Mud: Jan 18

We will advise you regarding the extension of contract time and impact on completion time when we are fully able to determine the extent of the delay.

Sincerely,

J. Edward Grimes
Project Manager

Figure 15.11

May 1, 1990

Mr. Rod Wright
One Line Design
1000 Dream Street
San Diego, CA 90000

Reference: Job #321 - Gibson Office Building
Subject: Extension of time for causes beyond our control

Dear Mr. Wright:

This letter confirms that we have encountered a delay in transporting the material to the job site due to a strike by teamsters, which is beyond our control.

The strike will delay our completion of the work. We are entitled to a commensurate extension of contract time.

As soon as we know the extent of the delay, its effect on the completion of the project, and whether or not any extra costs are incurred, we will advise you.

Sincerely,

J. Edward Grimes
Project Manager

Figure 15.12

Questioning the Price Limits of Change Orders

Use a Transmittal and a letter (as shown in Figure 15.13) when receiving a directive with broad ramifications to change the scope of work on a project. Receipt of a "low" *not-to-exceed* dollar allowance usually falls into this category.

It is easy for a third party to say: "That shouldn't cost more than $10,000." The catch is *they* do not have to do the work for that price. The other guy's work can always be done in less time and at less cost than the owner thinks. The fact is that he is able to remain in that trade or business because he can make a profit, and he is the one responsible for completing the work.

This type of letter can be sent by the architect, engineer, contractor, or anyone who receives a not-to-exceed directive. Facility managers can utilize this letter to cover potential increases over the sum "*dictated*" by company management.

Confirm Bid or Low Bid

This letter (see Figure 15.14) is not often used, but it should be! It is simple to prepare, and acts as a very inexpensive "insurance policy." For example, if the owner/agency takes four or five months to actually award the contract, the original bid and subcontractor bids should be–and need to be–re-examined for validity. It is easier to get the subcontractor/vendor to agree–just after the bid is submitted–to hold his price for five months, than it is to have him confirm and hold his price when you call five months later and tell him that the job has finally been awarded. The subcontractor is liable to say that his price was good for only 30 days, and that all his costs have gone way up. The "Confirmation of bid" letter is the ideal tool for this purpose.

Ask a subcontractor, prior to getting the job, about pouring the slab. The answer will probably be, "It'll take two weeks." Ask the same question after the job is awarded and the answer may be "I didn't figure on all the cold joints, this is going to take six weeks at least." The problem is that the bid contained the "two week" number.

Take the time to "buy insurance" by sending this letter. Since it goes out before the project is awarded, the price indicated will probably remain firm for several months. Keep a special log or copies of the letter with the bid proposal. Also, keep a checklist of who has responded. If someone responds, but includes a 30-day time limit, look for another bidder, or tell the owner that the original bid is good for 30 days only, if that trade is critical to your bid.

July 20, 1990

Mr. Rod Wright
One Line Design
1000 Dream Street
San Diego, CA 90000

Reference: Job #321 - Gibson Office Building
Subject: Ceiling price on FO #15

Dear Mr. Wright:

We are in receipt of Field Order (or Change Order) #15 which directs us to implement a change at a price not to exceed $10,000.00. While this might be a reasonable estimate of the cost of the scope of work at this time, we have not yet fully implemented the change, nor conducted a detailed analysis of the cost impact of the directive.

In view of these circumstances, we cannot and do not agree to be bound to the not-to-exceed amount stated in the Field Order (or Change Order).

We are proceeding to implement the directive Field Order with the understanding that our right to receive an equitable adjustment is not limited to such ceiling amount as stated in Field Order (or Change Order) #15.

Sincerely,

J. Edward Grimes
Project Manager

Figure 15.13

December 19, 1989

Bildenhope Co.
1475 Outada Way
San Diego, CA 90002
Attn: Juan Motyme

Subject: Confirmation of Telephone Bid

Dear Mr. Motyme:

We received your telephone bid on December 18, 1989 at 1:45 PM PDT. Please return one copy of this confirming letter to us signed by an officer of the firm. Your bid may or may not be selected, and this letter is a confirmation of your bid only.

This will confirm the following:

Project: Gibson Office Building

Bid Amount: $237,000

The bid is based on plans and specifications prepared by:

One Line Design

dated: October 3, 1989

and addendums: 1, 2, 3, 4

Bid is valid for 90 calendar days.

Items covered: Spec Section 03: Concrete
(Scope of work Description.)

BILDENHOPE CO.

Authorized Signature

Figure 15.14

Letter/Notice to Proceed with the Work

This letter (see Figure 15.15) is not used very often. However, it is highly recommended to those interested in achieving complete documentation. Its purpose is to inform the contractor/subcontractor that they have the "green light," and can proceed with the work.

This letter is also useful if, for some reason, the contractor/subcontractor starts late, and then wants an extension of time or extra money for working overtime in order to meet the schedule. This signed letter documents the date on which the contractor/subcontractor was told to proceed, thereby invalidating claims for extensions or overtime expense. This letter should be used if you do not send out the CPM schedule (as discussed in Chapter 14).

Letter of Intent to Award a Contract

More commonly known as a "Letter of Intent," this letter (shown in Figure 15.16) is especially useful if there is little time between receipt of bids and the need to start work (as in a fast track or very tight schedule situation). This letter format can be utilized by the owner or facility manager, as well as the general contractor.

The Letter of Intent informs the addressee that he/she has been awarded the work, that you are now writing the contract (which may take one week or longer for various approvals), and that he/she is to order all materials necessary, proceed with the work, and submit all documents (submittals, shop drawings, insurance certificates, etc.) as required by the specifications.

March 15, 1990

Mr. Thomas Thum
Green Landscaping
3456 Ivy Street
San Diego, CA 92343

Reference: Job #321 - Gibson Office Building
Subject: Notice to Proceed

Dear Mr. Thum:

This is your notice to proceed immediately with the contracted work for the above-referenced project.

This notice confirms our conversation of this morning, Monday, March 15, 1990.

Sincerely,

J. Edward Grimes
Project Manager

Figure 15.15

February 9, 1990

Bildenhope Co.
1475 Outada Way
San Diego, CA 90002
Attn: Juan Motyme

Subject: Confirmation of Telephone Bid

Dear Mr. Motyme:

In confirmation of our conversation of February 8, 1990, you have been awarded the contract for the above referenced work. We are now in the process of writing your contract, which will cover the following:

Section 03: Concrete	
Base Cost:	$237,000
Alternates - None	0
Total Cost of Contract	$237,000

This is in accordance with your bid documents dated January 2, 1990 and as per plans and specifications.

Please be advised that this letter is your authorization to proceed with the work. All submittals should be forwarded to us as soon as possible, to facilitate the purchasing of long lead time items.

Note that no work can start on the site until we are in receipt of valid insurance certificates as per the specified requirements.

We look forward to working with you toward a successful completion of this project.

Sincerely,

J. Edward Grimes
Project Manager

Figure 15.16

Termination of Subcontractors

This letter (see Figure 15.17) is one that no one likes to write, but which occasionally must be written. It is not one to be taken lightly, nor issued without considering all alternatives.

If a subcontractor fails to live up to his part of the contract (i.e., does not work diligently, fails to maintain a schedule, does not have adequate personnel on the job, or does unacceptable work), and does not respond to a heart-to-heart talk, then it may be necessary to terminate this contract according to the terms of the Contract for Construction.

Note, however, that terminating a contract raises several new issues and potential pitfalls. For example, can the subcontractor be replaced? Can he be replaced in reasonable time so as not to further strain the schedule? Does the subcontractor have materials on site or on order that cannot be readily replaced or quickly procured? Can you afford to pay the subcontractor off for work done and also pay to have this work finished? These are some of the questions that must be addressed before this ultimatum is issued and carried out. If appropriate action is specified in the subcontract agreement, remember that a subcontract can (with proper justification and according to contract requirements) be cancelled at any time. However, the project will be hurt if letters of termination are given without proper forethought.

Once the decision is made to terminate a contract, the letter must be sent with proof of delivery (i.e., courier with return signature, telegram, or hand-delivered with a witness). Without proper documentation, there may be problems later on. Without a confirmed delivery, the old "I never got that letter" routine will, no doubt, come into play.

Stop Work Order

This letter (shown in Figure 15.18) is issued for a number of reasons including, but not limited to: the drawings for work being done are being revised; lack of payment from owner; or conditions beyond general contractor's or owner's control (heavy mud, bad site conditions), unsafe conditions, conflict of workers, etc.

Again, this letter should not be issued without careful thought, because when you stop the flow of work, you incur shutdown and start-up costs; workers who have been sent to other sites must "relearn" the job, which costs time and money; you may not get the workers back (they may not be able to return if they found work elsewhere); you may damage the schedule beyond the point of feasibility; and you may ruin your credibility with the subcontractor(s) as to seriousness of your intent to maintain a schedule.

This letter should not be sent unless notification of work stoppage has been sent to the owner.

March 15, 1990

Mr. Harry Leggs
Terra Firma Remova Company
88 Orderinda Court
San Diego, CA 90002

Reference: Job #321 - Gibson Office Building
Subject: Termination of Contract

Please be advised that you are in violation of your contract due to lack of work performed.

As per Section 16 of your contract with us, you have 3 days in which to remedy the situation to our satisfaction.

Please let us know your intentions as soon as possible.

Sincerely,

J. Edward Grimes
Project Manager

VIA CERTIFIED MAIL

Figure 15.17

March 15, 1990

Mr. Edward Waterhouse
Quick Plumbing Company
7654 32nd Street
San Diego, CA 92110

Reference: Job #321 - Gibson Office Building
Subject: Order to Stop Work on Waste Piping System

Dear Mr. Waterhouse:

As per our conversation of this morning, March 15, 1990, stop all work on the underground waste piping system. We have been advised by Rod Wright, the architect, that due to a revision in the process equipment, the layout is being revised.

Please continue with your work on the storm drain system and water supply system as these, I have been advised, will not be affected.

Sincerely,

J. Edward Grimes
Project Manager

Figure 15.18

Chapter Sixteen

On-Site Reports

Importance of Field Documentation

The most important place to document the physical construction facts of a project is in the field, because that is where the action takes place. This is where the plans and specifications are transformed into something physical and real. This is where the project is built, and ideally, becomes a useful part of our economic and social system.

Problems

It is crucial that on-site construction personnel record and document the project events. Such events include the decisions, lack of decisions, accidents, and required changes in the work as reflected in Field Orders, backcharges, and other developments (such as subcontractors not showing up).

Unfortunately, in most cases, the closer one gets to the actual job site construction, the poorer the documentation usually becomes. This is due to several factors, including:

- Little or no clerical help is available to act as an office "conscience," to help and prod job site personnel to perform the required documentation.
- The job site superintendent is generally interested in building construction, not keeping records.
- The immediacy of job site activities seems to preclude time to sit down at the desk and write reports.
- The multitude of problems tend to take priority over preparing reports and documentation.
- Field personnel generally do not like to commit themselves by writing things down. Most field people, like most people in general, would rather leave themselves an "out," if at all possible. They would much rather say, "Remember when I told you this wouldn't work?" rather than "In my weekly summary report, I recommended..." The problem of selective memory arises.
- There is no strong upper management emphasis or company policy for proper field reporting and documentation. Again, "People do well what the boss checks." If the boss does not check the field reporting, then the field personnel do not make it a priority.
- Job site personnel may not understand the importance of keeping proper records.
- Many decisions in the field are of the "knee jerk," or reaction

type and, therefore, are viewed in the past tense once they occur. The attitude seems to be "Why worry about something that has already happened?"

Sometimes the management of a construction-related firm (including architect, engineers, facility management departments, general contractors, subcontractors, and vendors) tend to treat some of their "prima donna" personnel in a way that rewards non-documentation. Such personnel tend to respond with statements such as, "If I do all that paperwork, then I can't make money for you on the job," or "I'll never have time to get out on the job." Naturally, if the top people react this way, this attitude will be parroted by others.

Correcting the Problems

Strong Company Policy

This section is concerned with on-site construction activities and is, therefore, of primary interest to the general contractor and the facility manager. However, the principles outlined here are the same for any firm connected with a construction project. Virtually all of the recommended solutions to the problems of on-site or field reporting and documentation can be addressed, if not solved, by adopting a company policy for the processing of paperwork. This policy should address all of the following points, and should also require consistent review by top management.

Easy Documentation Methods

When something is easy to do, it is much more likely to be accomplished on a regular basis. Make it easy for field personnel to do what is required, and let them know what is expected. This includes using checklist-type forms–preprinted ones, if possible. Use a form that can be filled in by hand rather than typed. Above all, adopt a standard operating procedure so that field personnel know what is expected.

Monitoring

The system should also make it easy to monitor compliance. If procedures are too difficult, the daily or weekly job reports may fall behind by two or three weeks, and must then be written based on memory, instead of fact.

Reports from the field must be submitted *promptly*. Almost everyone turns in a time sheet of one sort or another. It should, therefore, be a requirement that all field reports accompany the time cards. Have the payroll clerk keep track of what is or is not included. Review the log once every couple of weeks and then "counsel" those field people who are not abiding by company policies. Remember the axiom: "People do well what the boss checks."

It makes life easier if the company policy, or "standard operating procedure," includes the way the files are set up, the proper forms, and necessary clerical help, as required.

Proper Forms: Having the proper forms for on-site documentation is crucial. It is easier for everyone to fill out a form rather than develop a special report "from scratch." The use of a form also helps to ensure that all of the information is included. All forms commonly used in the field can be printed on carbonless, multi-copy stock.

Time Cards: Time cards, or time sheets, are among the most important documentation that originates at the job site. Time card information is used to:

- Pay the people involved
- Account for time spent
- Calculate future bids

Each organization, by necessity, requires its own version of the time card to identify the use of time for their own unique purposes. Figure 16.1 is a Means Weekly Time Sheet (from *Means Forms for Building Construction Professionals*) designed for a "per job basis," to list all of the people who are working on the job. This type of form is best utilized on a fairly large project with a stable work population.

For a firm with many different projects, where job site personnel are transferred back and forth from project to project, the time card shown in Figure 16.2 is more appropriate.

Extra Work Order Form: Field Orders, Directives, and Extra Work Orders are all necessary elements in the construction process. All of these devices have the same purpose: *they are orders to proceed with the work described.* Unless properly documented, written, and signed, these orders can create chaos and may be costly!

In Chapter 20, we review the steps necessary to quote to the owner the Extra Work Order, or Directive. The form in Figure 16.3 is typical of an "on-site" or "field" Extra Work Order. It is meant to be filled out in the field so that authorization is received when the work is directed. This form must be signed prior to proceeding with the work. All Extra Work Orders should be numbered consecutively, filled out, and sent to the office for processing as described in Chapter 18.

This form may be used between an owner and general contractor, a general contractor and a subcontractor, a facility management department and subcontractor, or a subcontractor and a sub-subcontractor.

Means Forms

DAILY TIME SHEET

PROJECT R.S.M. OFFICE BUILDING

FOREMAN R. MEANS

WEATHER CONDITIONS

TEMPERATURE

DATE 4-15-86

SHEET NO. 1 OF 1

NO.	NAME	DESCRIPTION OF WORK	HANG DOORS 1ST FLOOR	INSTALL HARDWARE	PANEL AT CONFERENCE ROOM	UNLOAD MATERIAL					TOTALS REG-ULAR	TOTALS OVER-TIME	RATES REG-ULAR	RATES OVER-TIME	OUTPUT
101	R. MEANS	HOURS	8								8		$21^{\underline{50}}$		.67
		UNITS	12								12				
		HOURS													
		UNITS													
102	S. MEANS	HOURS			8						8		$17^{\underline{50}}$		.03
		UNITS			240 SF						240				
		HOURS													
		UNITS													
107	B. SMITH	HOURS			8	2					8	2	$12^{\underline{20}}$	$18^{\underline{30}}$	.03
		UNITS			240 SF						240				
		HOURS													
		UNITS													
110	J. JONES	HOURS	8								8		$17^{\underline{50}}$		.80
		UNITS	10								10				
		HOURS													
		UNITS													

Figure 16.1

Means Forms

WEEKLY
TIME SHEET

EMPLOYEE	S. MEANS	EMPLOYEE NUMBER	102	WEEK ENDING	1-5-86
FOREMAN	R. MEANS	HOURLY RATE	17.50	PIECE WORK RATE	

PROJECT	NO.	DESCRIPTION OF WORK	DAY	W	TH	F	Sa	M	Tu		TOTAL		HOURS	RATE	TOTAL COST	UNIT COST
RSM	86-1	INSTALL H.M. DOORS	HOURS	8	8							REG	16	17.50	280.00	10.77
OFFICE		EA	UNITS	14	12							O.T.				
		INSTALL WOOD DOORS	HOURS			8		8				REG	16	17.50	280.00	15.55
		EA	UNITS			10		8				O.T.				
		INSTALL FINISH HDW	HOURS						8			REG	8	17.50	140.00	5.83
		SETS	UNITS						24			O.T.				
		TRIM @ CONFERENCE RM	HOURS				4					REG				
		L.F.	UNITS				200					O.T.	4	25.75	102.50	.51
			HOURS									REG				
			UNITS									O.T.				
			HOURS									REG				
			UNITS									O.T.				
			HOURS									REG				
			UNITS									O.T.				
		TOTAL	HOURS	8	8	8	4	8	8			REG	40		700.00	
			UNITS									O.T.	4		102.50	

PAYROLL DEDUCTIONS	F.I.C.A.	INCOME TAXES			HOSPITAL-IZATION			TOTAL DEDUCTIONS	302.49	LESS DEDUCTIONS
		FEDERAL	STATE	CITY/COUNTY						
	$ 56.57	$ 180.62	$ 41.10	$	$ 24.20	$	$	$ 302.49	# 500.01	NET PAY

Figure 16.2

Means Forms

EXTRA WORK ORDER

FROM:
CARL LINDE
C. LINDE ASSOC.
MARSHFIELD, MA.

TO:
R. J. GRANT
OFFICE BUILDERS CORP.
33 SMITH LANE
KINGSTON, MA 02364

EXTRA WORK ORDER NO. 86-1-1
DATE 1 JAN 86
PROJECT R.S.M. OFFICE BUILDING
LOCATION KINGSTON, MA
JOB NO. 86-1
EXTRA WORK ORDER APPROVED

BY

BY

Gentlemen:

This EXTRA WORK ORDER includes all Material, Labor and Equipment necessary to complete the following work;

☐ the work below to be paid for at actual cost of Labor, Materials and Equipment plus ________ percent (____%)

☐ the work below to be completed for the sum of TWELVE THOUSAND SEVEN HUNDRED dollars ($ 12,700.00)

DESCRIPTION

ALTER THE DIMENSIONS AND FINISHES OF THE ENTRY TO THE R.S.M. OFFICE BUILDING AS PER THE PLANS AND SPECIFICATIONS INCLUDED WITH HIS ORDER. WORK TO INCLUDE MOVING EAST PARTITION WALL 5'-0", ALTERING THE STOREFRONT FRAMING DETAILS AND FURNISHING AND INSTALLING TERRAZZO FLOORING AS PRICED IN PROPOSAL NO. 86-10

The work covered by this order shall be performed under the same Terms and Conditions as that included in the original contract unless stated otherwise above.

Signed Carl Linde

By CARL LINDE - C. LINDE ASSOC.

Figure 16.3

Job Site Construction Reports

The Daily Construction Report is one of the most important records that can be kept on a project. Often, this valuable report is held in little esteem by field personnel and managers–until someone needs to prove or disprove something. Crisis conversations often start with, "We wouldn't even be talking about this if the matter had been included in the Daily Report."

A number of different formats can be used for the Job Site Construction Reports. The following information should be recorded in the Daily Report, which should be completed to reflect the special conditions of each day. (Writing "same as yesterday" is unacceptable.)

- Date
- Weather conditions
- Work performed (overall)
- Work performed (own forces)
- Subcontractors working today, and number of workers
- Safety: safety meeting? accidents?
- Equipment rented? written agreement? how long?
- Comments
- Extra work, backcharges, delay. If yes, explain.

The questions listed below should be answered for every report. These questions could be pre-printed on the back of a single-page daily report, with a couple of blank lines on which to make entries. These questions reinforce some of the items that should be included in the Job Progress Report. Do not be concerned if the report cannot fit on one page. As we have said before, paper is one of the cheapest materials with which we build. It is better to have a two-page report with all the information included, than a one-page report that is not readable or lacks crucial information.

- Were you required to do any extra work beyond the requirements of the contract? Was this work directed in writing? What was the cause?
- Did you perform work for a subcontractor or supplier that should be backcharged? Was this confirmed in writing? Why was it necessary?
- Did any condition or circumstance change your method of performing or increase your cost to perform the work?
- Were you prevented from doing work on any part of the project due to circumstances beyond your control, including weather?
- Did anyone perform work for you that will be a charge or backcharge? What was the cause? Was it confirmed in writing?
- Were any operations delayed or suspended by the actions of other contractors or suppliers? Was this situation confirmed in writing?
- Were the work area and conditions as you expected?
- Were there any site visitors? Who? From which company?

Figures 16.4a through 16.4d are several different versions of site reports. Each has an appropriate use. Deciding which to use will depend on many factors, such as the company preferences, type of job, size of the job, and size and make-up of the construction firm or facilities department. Again, these forms should be of the carbonless multi-copy type.

Means Forms

DAILY CONSTRUCTION REPORT

JOB NO.

DATE

PROJECT

SUBMITTED BY

ARCHITECT

WEATHER

TEMPERATURE AM PM

CODE NO.	WORK CLASSIFICATION	FOREMEN	MECHANICS	LABORERS	SUB-CONTR'S	TOTAL HOURS	DESCRIPTION OF WORK
	General Conditions						
	Site Work: Demolition						
	Excavation & Dewatering						
	Caissons & Piling						
	Drainage & Utilities						
	Roads, Walks & Landscaping						
	Concrete: Formwork						
	Reinforcing						
	Placing						
	Precast						
	Masonry: Brickwork & Stonework						
	Block & Tile						
	Metals: Structural						
	Decks						
	Miscellaneous & Ornamental						
	Carpentry: Rough						
	Finish						
	Moisture Protection: Waterproofing						
	Insulation						
	Roofing & Siding						
	Doors & Windows						
	Glass & Glazing						
	Finishes: Lath, Plaster & Stucco						
	Drywall						
	Tile & Terrazzo						
	Acoustical Ceilings						
	Floor Covering						
	Painting & Wallcovering						
	Specialties						
	Equipment						
	Furnishings						
	Special Construction						
	Conveying Systems						
	Mechanical: Plumbing						
	HVAC						
	Electrical						

Page 1 of 2

Figure 16.4a

Means® Forms

EQUIPMENT ON PROJECT	NUMBER	DESCRIPTION OF OPERATION	TOTAL HOURS

EQUIPMENT RENTAL - ITEM	TIME IN	TIME OUT	SUPPLIER	REMARKS

MATERIAL RECEIVED	QUANTITY	DELIVERY SLIP NO.	SUPPLIER	USE

CHANGE ORDERS, BACKCHARGES AND/OR EXTRA WORK

VERBAL DISCUSSIONS AND/OR INSTRUCTIONS

VISITORS TO SITE

JOB REQUIREMENTS

Page 2 of

Figure 16.4a (cont.)

PROGRESS STATUS REPORT

(If yes, elaborate below)	YES	NO	
1. Directed or required to do any extra work	____	____	Job No.____________
2. Backcharges - Against NCC or a Subcontractor	____	____	
3. Delays - Waiting on information or decision	____	____	Week Ending__________
- Subcontractor not performing	____	____	
- Material late	____	____	
- Placed on Hold	____	____	Job Supt.___________
4. Visitors and/or mtgs. held (including safety)	____	____	
5. Accidents - Physical or Equipment	____	____	

MONDAY
Weather__________
Time Lost________
Foreman__________
Carpenters_______
Laborers_________
Others:__________

TUESDAY
Weather__________
Time Lost________
Foreman__________
Carpenters_______
Laborers_________
Others:__________

WEDNESDAY
Weather__________
Time Lost________
Foreman__________
Carpenters_______
Laborers_________
Others:__________

THURSDAY
Weather__________
Time Lost________
Foreman__________
Carpenters_______
Laborers_________
Others:__________

FRIDAY
Weather__________
Time Lost________
Foreman__________
Carpenters_______
Laborers_________
Others:__________

Schedule Status: ______________________________

Figure 16.4b

Site Report

Delta Construction
6065 Mission Gorge Road
San Diego, California 92120
(619) 555-2111

Date: ____________________

Job Name: ____________________ Job No. ____________________

Work Force	No.	Mat'l. Placed – Quantity – Describe Use or Location
Carpenter Foreman		
Carpenters		
Laborers		
Truck Drivers		
Operating Engineers		
Cement Finishers		

Progress Report: (General Scope of Work Accomplished by Subs – Subs Scheduled – Who and When)

Subcontractors	No.	Deliveries From	Type of Materials (Time)

Changes Made:

Directions:

Accidents: (Attach Report Form)	Type Injury	Reason
(Name)		
Weather	Approx. Temp.	Foreman

Figure 16.4c

SUPERINTENDENT: JOB NO.________________ DATE:

Weather: ______________ Lost Time: ___________ Total: ___________

GC ACTIVITIES:		MEN	HOURS
	CARP.		
	LAB.		
	C.M.		
	OTHER		
	TOTAL		

TRADE OR SUB: (M & P) ACTIVITIES:

TRADE OR SUB: (ELEC) ACTIVITIES:

TRADE OR SUB: ________________ ACTIVITIES:

TRADE OR SUB: ________________ ACTIVITIES:

TRADE OR SUB: ________________ ACTIVITIES:

TRADE OR SUB: ________________ ACTIVITIES:

TRADE OR SUB: ________________ ACTIVITIES:

GENERAL NOTES: TOTAL

SIGNED

PAGE______ OF ________

Figure 16.4d

Job Site Summary Report

A periodic Summary Report, either weekly or monthly, is worth the little extra effort required to produce it. This report should not be widely distributed. In fact, it should probably be distributed only to top level management. One reason for this select distribution is that the superintendent or responsible field individual should be able to express himself freely without fear of repercussions.

This report should accomplish three objectives:

1. Make sure the field manager is thinking about the activities for the upcoming time period.
2. Give the readers of the report a quick synopsis of the status of the project from the field level.
3. Allow the field manager the opportunity to express (in some degree of confidence) his opinions about the problem areas that may exist on the project.

The Summary Report is most useful in large organizations with a multitude of projects, although smaller organizations often find it useful as well. Figure 16.5 is a sample form that can be used for a weekly summary.

A monthly narrative of current project status is often put together to keep owners informed. It should be as positive as possible to build up "good cheer" with owners, and should cover in more depth the topics in which he or she has shown an interest.

Short-Term Interval Report

One of the more productive reports is a two-week, or short-term, interval report. This schedule is prepared in the field. It is normally reviewed with all subcontractors on the job site and then presented in the regular job site meeting. The schedule is not intended to look very far downstream. It reflects *current job* status. While the schedule should address the overall job, it should not be concerned with long lead time Submittals. It should be directed to a two- or four-week on-site work schedule.

The short-term interval schedule should not be too fancy or complex. In fact, many field personnel are suspicious of fancy charts. A hand-written bar chart is perfectly adequate. Remember, this type of schedule is created in the field, and used in the field, so do not be too demanding of the form it takes. The important thing about the schedule is that it is well thought-out, timely, reasonably accurate, and utilizes actual conditions and information from the field.

By reviewing a series of short-term interval reports, the manager has a pretty good idea of the report's reliability, as well as how the report coincides with the overall job schedule. The short-term interval schedule is also discussed in Chapter 14.

Figure 16.6 is a sample of a short-term Interval Report, utilizing a Gantt-type chart for this purpose.

FIELD MANAGER'S SUMMARY PROJECT REPORT

JOB NO.______________ PROJECT_________________________ DATE___________

WORK COMPLETED WEEK OF __________________. _______________________

____

____

____

CONSTRUCTION PLANNED FOR WEEK OF __________________. ____________________

____

____

____

____

____

PROJECTED COMPLETION DATE_____________________________.

CONTRACT COMPLETION DATE OR SCHEDULED_______________________________.

CAUSE OF DELAY (IF APPLICABLE): ____

____

____

SCHEDULE PROGRESS (DAYS AHEAD, BEHIND) ____________________________________

PROBLEM AREAS (PRESENT OR FUTURE):_______________________________________

____

____

____

____

____

FIELD MANAGER

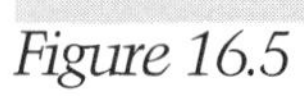

Figure 16.5

Means® Forms

PROJECT SCHEDULE

Delta Construction Company
6678 Mission Road
San Diego, CA 92120
(619) 581-6315

PROJECT Gibson Office Bldg. #321

DATE 10/6/90

BY Bill Hardy

CALENDAR PERIOD Day

NO.	DESCRIPTION	October 10	11	12	13	14	17	18	19	20	21	24	25	26	27	28	COMMENTS
	Drywall 2nd floor																
	3rd floor																
	Pull Elec. Home Run																
	2nd floor																
	Tape 1st floor																
	Tape 2nd floor																
	Elec. Conduit 3rd floor																Need inspection
	4th floor																
	HVAC 3rd floor																
	4th floor																
	Fire Sprinkler																
	3rd floor																
	4th floor																
	5th floor																

Figure 16.6

Backcharges

Backcharges are a major concern in the construction industry. Applied with reckless abandon, they can wipe out an entire species of subcontractors. Like Field Orders, backcharges *should* not occur. When they do, it is generally because someone did not do what they were supposed to do–at least in the eyes of the individual issuing the backcharge. (Backcharges are also used when a subcontractor "borrows" a person or piece of equipment from the general contractor.) Figure 16.7 is an example of a backcharge form. Refer to Chapter 3 for discussion of how to use Transmittals as standard letters sent from the office in regard to backcharges.

Backcharges, if and when necessary, should be issued in numerical order. A log should be kept of all backcharges.

Many times the items being backcharged are "traded out" at some point in the job. This is not necessarily bad, though we do not advocate this practice. It is good business to know that the trade-out of services is performed on the project and not at someone's home.

A pre-printed form with sequential identifying numbers is a preferable format for backcharges. The backcharge should include:

- Job number
- Date
- Backcharge number
- Credit to
- Precise description of activities, equipment, and reasons for charge
- Cost/pricing information
- Space for signatures of involved parties
- Status of backcharge

Filing

The Problem

How many times have we gone to the field office and seen piles of papers on the field superintendent's desk? Or, for that matter, how often have we ourselves been guilty of allowing documents to accumulate in disorganized piles? When asked a question, how many places and piles must we search to find the right document? Too often, significant items, such as Field Reports, Price Requests, Requests for Information, and Transmittals, get lost or misplaced because there is no proper filing system.

Effective management of construction paperwork includes not only creating and sending the proper documents, but also being able to find them. This applies to both the main office and the field office. If a question arises, such as "Why was this not done?", or "Who is responsible?", nothing finalizes a response better than back-up information in the form of letters, communications, and sketches. On the other hand, if you cannot find the appropriate document, then these questions really cannot be answered.

Backcharge

Delta Construction
6065 Mission Gorge Road
San Diego, California 92120
(619) 555-2111

Job Number: ______

Date: ______

Backcharge Number: ______

Backcharge Against: ______

Backcharge Credit: ______

Description	Quantity	Unit	Price	Amount
			Total	

1. Signature: (of company/individual being backcharged)
2. Notice given to: (company/individual being backcharged)
3. ☐ in person ☐ telephone ☐ job meeting
4. Formal letter sent with price? ☐ Yes ☐ No
5. Disposition of backcharge: ☐ traded out ☐ withheld from contract payment ☐ void

Figure 16.7

Correcting the Problem

While this section is directed chiefly to the general contractor and the facility manager, the goals for efficient filing methods are the same for any firm connected with a construction project. Keep in mind that filing procedures, and compliance with these procedures, must have the full support of the president of the construction firm or the facilities manager, as well as all levels of management.

The basic requirements are as follows.

- Set up correspondence handling procedures.
- Make it easier to file.
- Get the filing done.
- Ensure long-term compliance.

Correspondence Handling Procedures

The first step in setting up proper document handling procedures is to establish company policies for each phase and type of correspondence. The filing procedures should include:

- Dates and date stamps
- Entry into logs
- Review and delegation of tasks

Dates and Date Stamps

It is vitally important that every piece of paper received by a firm be date-stamped. Whether it is a set of plans, specifications, bids, proposals, invoices or any other correspondence item, a date should be stamped on it the day it arrives. Do not depend on the date of the correspondence as recorded by the sender. Many contested claims are determined based on the sequence of events and information received or not received. By making date-stamping part of the correspondence handling procedure, you apply a record keeping tool of your own. The following example illustrates the importance of this advice.

> *A Price Request carried a date of September 15, but did not reach the contractor's office until December 1. The document may have been prepared by the architect on September 15, but then the owner spent several weeks reviewing it and making a number of changes. The date of origination, of course, was never changed. The architect then tried to make a case that the general contractor was not doing his job by failing to respond in a "timely manner."*

Note to architects and owners: be reasonable in setting a date by which the contractor is required to respond. Some forms specify that unless a response is received within seven days, it is assumed to be "no charge." That approach will not work. Everyone knows that you want the information as soon as possible. However, if you press for an answer too soon, before the general and subcontractors have had a chance to assess the long-range impact, the quotations you receive will contain "fear factors" (extra costs) built into the number. Every time someone is forced to respond before they have a chance to do their homework, the price will err on the "up" side, to the owner's detriment.

When sending correspondence, be conscientious about accurately recording the outgoing correspondence dates. This is good policy for two reasons: (1) Your reputation for recording the correct date on a piece of correspondence will support you in negotiations and other areas of question on the job. A reputation for back-dating memorandums, or otherwise manipulating correspondence, will only make your case more difficult to support when the true dates really are on your side. People will assume that you misrepresented the date even when they have no proof. (2) Accurate dates will ensure that you have a correct history of events as they occurred in order to evaluate your own performance.

Entry Into Logs

As soon as incoming documents (whether they are sketches, samples, letters, or requests) have been date-stamped, they should be logged–before they have a chance to "disappear" into the system. Chapter 20, "Logs," explains in more detail how and why items should be logged, and shows examples of effective formats. It must be noted that as much condensed information as possible should be entered into the logs. This makes it much easier to trace the cause and course of an event.

Logs should include the following items:

- Item number (If the correspondence refers to a previously numbered item, then use that number. If not, assign it a number.)
- Date received
- From (Be as specific as possible, stating not just "Ace Brick Co." but "Al Brown, Ace Brick Co.")
- Short description (if new) of action taken

Proper documentation of correspondence is the first step to an overall improvement in construction communications. If the time is taken to properly log and document correspondence as it arrives or departs your office, it can be done as a routine, non-demanding activity, and should require no more than a few minutes a day. However, if you delay (waiting until the end of the day or week to do proper logging), this task can become monumental, if it can be done at all, since important documents will have been taken and distributed, possibly without being logged, dated, or getting to all who need to know.

Review and Delegation of Tasks

When the two prior steps are complete, all correspondence should be reviewed, separated, and distributed to the person(s) to whom it is intended. It is recommended that a copy of important correspondence immediately be placed in the proper file for back-up and log identification. Items that should be sent to the field (or their contents made known to field personnel) should be duplicated, and a copy sent to the field.

Items of correspondence should never be allowed to "sit around" until they are uncovered or an ominous "where is it?" telephone call comes through. This is the advantage of using logs, as upon review, it can be seen which items are outstanding and need attention.

Make It Easier To File

When a deal is struck or a contract signed, it is important to immediately set up a complete filing system. These files should include:

- **Construction Documents**
  Copy of Contract
  Subcontracts
  Plans on Rack
  Specifications
  Soils Report
  Permits
  Test Reports
  Submittals
- **Correspondence**
  From Architect/Engineers
  To Architect/Engineers
  From Owner
  To Owner
  Subcontractor Files
- **Financial**
  Budget
  Cost Reports
  Backcharges
  Change Orders
  Price Requests
  Purchase Orders
  Time Sheets, Time Cards
  Field Orders
  Clarifications
- **Miscellaneous**
  Long-Term Schedule
  Short-Term Schedule
  Logs (all)
  Meeting Minutes
  Accident Reports
  Personnel Files
  Equipment Rental Sheets

Note: Originals should be kept in the home office. The field should have copies, as necessary.

It makes life easier if the company policy or "standard operating procedures" include guidelines for setting up files, the proper forms to use, and clerical help, if necessary. Many times, a company's office personnel will visit the field office as the job begins to set up the filing system.

Get the Filing Done

Having set up the proper guidelines and resources, the next step is actually getting the filing done. As a rule, field personnel are not very interested in filing. Depending on the company organization, consider having someone from the office go to the job site once a week, or every other week, and ensure that all files are up to date. A number of field problems can be avoided or resolved if field personnel can find the information they need–documents that they have probably already received, such as Job Minutes, Submittals, Price Requests, Field Orders, and Clarifications.

Ensure Long-Term Compliance

Make it a company policy that the manager check on the field personnel to keep their filing up to date. Once again, "people do well what the boss checks." Every week or every other week, inspect the offices to ensure that filing is not stored in a nook or cranny to be attended to "later." It is a minor task to put a few pieces of paper in the files as they are processed. However, if one waits and stores it up for a "later" time, the paper has a tendency to turn into a paper avalanche and filing becomes a major, all-afternoon project.

Chapter Seventeen

Meetings and Minutes

Meetings before, during, and after the construction project help to create and maintain good communication. That is their purpose. Attendees convey information for the use of others, and receive information from others for their own use. Construction meetings are also held for various specific purposes. This chapter examines the types of construction meetings and the types of minutes that are normally kept for those meetings.

Meetings are necessary. Meetings allow people to get to know others who are working on the job. This personal contact is important to the establishment of a good working relationship. Establishing contact in person makes it easier to conduct business over the phone, which is an integral part of our work in the construction industry.

Meetings should be scheduled on a regular basis. Other than a few special meetings, *all* meetings should be scheduled and held regularly and consistently. Weekly meetings should be held on the same day of the week at the same time, and in the same place. In this way, those connected to the project know when and where to be, without a lot of "meeting coordination." More than likely, project personnel will know what to prepare for as well. Such regular meetings are more productive than the hit or miss approach which "misses more than it hits."

The worst "meeting killers" are:

- Absence of key personnel involved in the subject to be discussed
- Unprepared personnel
- Confusion about where to meet (resulting in late arrivals) or about the purpose of the meeting
- Discussion of other items that are not the intended subject of the meeting
- Meetings that last too long, lack appropriate focus, or do not involve most of the participants

Any of these circumstances can "hamstring" a meeting, and often an entire project. Enforcing regular, reliable meetings is the first step in eliminating these common "meeting (and, consequently, project) killers."

Regularly scheduled meetings–with a published agenda and minutes–will reduce the volume of telephone calls and correspondence required on most construction projects. If all entities connected with the project know that every Tuesday morning at 9:00 AM there is a meeting with the architect (and possibly the owner, project manager, superintendent, and subcontractors), they will know that this is the proper place to raise job-related problems. When a subcontractor calls the project manager to say the design plans are not clear, the response should be:

> *"Show up at the Tuesday meeting and let's review it with the architect afterward. The meeting usually only lasts about 45 minutes, so we should have a good chance to get to the bottom of the problem."*

Note that the discussion is to take place *after the meeting*. Scheduling a detail-oriented side meeting keeps the primary meeting shorter and does not detain all other participants over one very specific problem.

If there is a specific problem, hold a specific meeting. Do not use everyone's time over an isolated problem. If all the specific problems are discussed at general meetings, then the meetings will last three to four hours and very little will be accomplished. Only (scheduled) design-oriented meetings can last three to four hours and actually accomplish something. The general site meeting is best used for talking about scheduling problems, interferences, and progress.

People are interested only in the problems that directly affect themselves. Therefore, meetings often get out of hand, and the loudest (and most persistent) voice gets the attention while everyone else gets bored, goes to sleep, and stops attending the meetings.

Chairing the Meetings

The person who chairs the meeting should be responsible for preparing the minutes and the agenda. This individual should direct the meeting and not allow it to stray too far from the agenda. If the meeting starts to get hung up on one issue, schedule a separate, special meeting at another time to discuss the special items.

The meeting leader must:

- Follow the agenda and get the meeting over as soon as possible.
- Start the meeting on time.
- End the meeting as soon as possible.
- Keep the discussion to the specific topic at hand.
- Cut the meeting short if required.
- Prepare and distribute the minutes.

If the meeting goes on too long, people will not attend the next meeting–the first hint of the decay of job site organization. Meetings can be the most efficient means to disseminate information and resolve problems, so long as they are properly controlled and contained.

The primary task of the meeting leader is to handle the mechanics and the personalities of the meeting so it remains a productive instrument for solving problems. Never, never harass, embarrass, or "chew out" *anyone* at a meeting. If necessary, do it in private later. To chew someone out in a meeting or in public will cause tension on-site, embarrass the victim, and disrupt the productivity of the meeting. It will polarize the work forces in a way that generally hampers productivity.

The first opportunity to set the pace for the project is the pre-construction meeting. The manager should determine when the meeting will be held and who will attend this and future meetings, and that attendance will be mandatory when the project starts. The pre-construction meeting is an ideal opportunity to make these decisions and announcements.

Agendas

All meetings proceed more smoothly if they have an agenda and it is followed. Even single-purpose meetings will proceed in a more logical manner if they are guided in this way. A meeting agenda is like a Critical Path Schedule of the meeting to be held. A handwritten copy of the agenda distributed to everyone at the meeting is sufficient. The primary purpose of an agenda is to let the attendees know what will be discussed and in what order. Additional items may come up that need to be discussed, but if the agenda has a logical pattern, the same topics will not be discussed two and three times in random order, as happens in so many cases.

For job-site meetings, the minutes of the prior meeting can usually serve as an agenda. Discuss the oldest items first and, if possible, put them to rest. If the meeting always starts with new business, it is more difficult to get around to older items, and they continue to hang on.

Keeping Meetings Short

Meetings should be as short as possible. The following is a guideline for meeting length:

Working Design Meetings:	4 hours
Special Meetings:	30 minutes to 1 hour
Job Site Meetings:	30 minutes to 1 hour
Pre-construction Meetings	30 to 45 minutes

Any meeting that lasts longer than these guidelines tends to become ineffective and boring, and will decrease productivity. Long meetings are usually caused by the following:

- The leader of the meeting does not distribute a written agenda.
- The leader of the meeting lets the meeting get out of control.
- The participants in the meeting have not done their homework.

Attendees who come uninformed or unprepared to discuss the items on the agenda can cause a meeting to be excessively long while they get "educated" during the meeting. Sometimes, there is good reason for taking time to inform people. However, if this happens when it could have been avoided, the meeting leader may choose to stop the meeting, and reschedule it for another time in order to give everyone the opportunity to prepare properly. *There is no reason to waste everyone's time.*

Meetings Should Be Specific

Meetings should be held for a specific purpose. If there is a problem over the landscape plans, then have a specific meeting about the landscape plans. Do not invite everyone. Include only those who can contribute to the resolution of the problem for which the meeting was called. Nowhere are "too many chefs" a bigger problem than on a construction project. There are no spectators, only participants–whether they are invited to join the conversation or not. Do not waste their time and do not allow them to confuse or derail the discussion of issues that do not concern them.

Do not allow the job site construction meeting to deteriorate into a design meeting. Call a separate meeting to discuss design issues if the need arises. Often, the architect/engineer has a problem with a particular detail, and everyone sits around while the architect and one subcontractor discuss the problem. In such cases, it is best to arrange for the two or three concerned parties to hold a separate meeting on another time, day, or following the meeting under way.

It is a common courtesy to other participants in the meeting to avoid flights of fancy and to concentrate on areas that directly relate to their work. Many times someone who does not regularly attend meetings will attend one meeting for a single purpose. If it is not a complicated matter, it should be addressed first. In this way, that individual need not stay for the discussion of other items that do not concern him or her. The length of a meeting has a tendency to increase exponentially based on the number of persons attending. Therefore, it pays to:

- Review the list of attendees and weed out those who need not attend.
- Not allow attendees to discuss matters they do not know anything about.

The Minutes of the Meeting

Keeping and distributing the minutes of all meetings is an unwelcome task. No one enjoys the process. Nevertheless, it must be done. The format of the meeting minutes is less important than the fact that minutes are published and distributed. Because the need for distributed information is critical to the success of a construction project, every communication vehicle available must be used. The minutes of the various meetings provide an excellent "newspaper" to inform not only the people who attended the meeting, but *everyone involved.*

The minutes of construction meetings can provide substance and accountability for the items discussed. The minutes should be considered a "construction document" and *are* considered to be a legal document. During the pre-construction meetings, it should be pointed out that the Project Minutes will be considered a legal document. Good minutes can be of great benefit to the management of a project.

It is difficult to "chair" a meeting and keep accurate minutes at the same time. It is much better to have someone else in the firm (a clerk or assistant) keep the minutes. However, be careful to assign that important task to someone who will understand the conversations at the meetings. There is nothing more frustrating than to have a set of minutes that do not reflect the actual contents of the meetings. The utility of the minutes collapses as does the project participants' respect for them. Minutes prepared by a clerk, project engineer, or someone else should be reviewed in draft form before they are issued.

Although difficult, in some cases, it is advantageous for the meeting organizer to keep the minutes for the following reasons:

1. The minutes from prior meetings can be used as the agenda (see sample minutes, Figures 17.2 and 17.3). That makes it easier to keep notes in the proper order.
2. If the record of the meeting needs interpretation, the organizer (or chair) maintains control of that interpretation.
3. The organizer becomes better prepared for the upcoming meeting while preparing its agenda and reviewing the minutes from the past meeting. A few telephone calls or some research may be required; this can usually be accomplished prior to the next meeting.

Distribution of Minutes

There should be a set of minutes taken for each meeting. Distribution of these minutes must be timely, despite the fact that preparation and distribution of the minutes takes time. If there is a job-site meeting every week, then the minutes of each meeting should be prepared before the next meeting is held, and distributed at that next meeting. Mail the minutes to individuals who "need to know" as soon as possible. Mail the minutes to those who do not normally attend the meeting, and then take extra copies to the next meeting. Before any meeting, give everyone a little time to review last week's minutes before starting the meeting.

Veracity in Reporting

All reporting is, to some degree, biased. Everyone sees the item in question from a different perspective. There is a difference between being biased, withholding some information, and misrepresenting the facts. If the person preparing the minutes leaves out some important items that were discussed, the veracity of the minutes is compromised. In this case, the validity of the minutes as a "contract" or "construction document" is destroyed. Presenting the complete facts is essential, even if this means that some important players may be seen in a negative light. Credibility of the minutes is paramount. At times it may be necessary to note that the "chair" is at fault. While that is not so good, admitting one's own error can lend credibility to the job minutes if reported accurately.

As mentioned earlier in this chapter, the meetings and minutes should be regarded as a part of the construction documents. In this way, the need for issuing additional correspondence and communications can be reduced. Other types of communication cannot, or should not, be eliminated, but they should be made more efficient through the effective use of the minutes as a communication vehicle.

Who Should Receive the Minutes

Send a copy to everyone who has a "need to know." Generally, this includes the owner, the architect/engineer, subcontractors and vendors. All those who attend the meetings should receive a copy of the minutes. It is not uncommon to send a copy to the primary subcontractors, the lender, or others who may need to be filled in on the actions and decisions that were reached in the meeting.

What Should Be Included

There are two basic types of meetings: special meetings, and reoccurring, regularly-scheduled ones. Therefore, there are two approaches to recording minutes. A good format is important in the creation of minutes. However, remember the style of the minutes is *secondary* to the fact that there *are* minutes.

Special Meeting Minutes

Special meetings occur once, for one purpose, and are not repeated. Often, a series of meetings will be held on the subject, but still refer to one "single purpose." The following information should be included in the minutes:

- Project
- Location of meeting
- Purpose of meeting
- Date and Time
- Author of the report
- Persons attending and the companies they represent
- Subjects discussed
- Conclusions reached
- Who is responsible for implementing conclusions
- Time target for implementation
- Who received copies of report

The Means form for "Meetings/Trip Report" (Figure 17.1) is well-suited to the purpose of special meetings. On the back, or on attached sheets, the actual items may be discussed in greater detail. The front page serves nicely as a recap of the "five points of Communication": Who, What, Where, Why, and How. If the report is to be typed in a different format, the same information must be addressed to convey adequately what took place at the meeting to all concerned. Someone not attending the meeting should be able to understand the items discussed and the decisions reached.

The following section discusses six types of meetings, including the purpose, time and location, recommended participants, agenda, and minutes of each. The types of meetings covered are:

- Design
- Pre-bid
- Pre-construction meetings with general contractor (in-house)
- Pre-construction meetings with manager (in-house)
- Subcontractors' job kick-off
- Job close-out (in-house)
- Projects, regularly scheduled

Means Forms

MEETING/TRIP REPORT

PROJECT | FILE NO

MEETING/TRIP LOCATION | SHEET NO

REASON FOR MEETING/TRIP | START TIME | DAY

REPORT BY | FINISH TIME | DATE

PERSONS INVOLVED	TITLE	REPRESENTING

SUBJECTS DISCUSSED (SEE REVERSE SIDE)

ITEMS EXCHANGED	FROM	TO

SUMMARY OF DISCUSSION

MEETING/TRIP NOTES (SEE REVERSE SIDE)

QUESTIONS RAISED	BY WHOM

DECISIONS/CONCLUSIONS/ACTIONS TAKEN

RECOMMENDATIONS/TO BE DONE	WHEN	BY WHOM

TO BE CONTINUED/DISCUSSED LATER

NEXT MEETING/TRIP: DATE | TIME | LOCATION

PURPOSE

COPIES OF THIS REPORT TO

Figure 17.1

Design Meetings

Purpose

The purpose of design meetings is to work out details or additions to the design. This is not a "how to" or "when to" meeting. A design meeting should be directed only to problems that cannot be solved in any way other than redesign or requests for additions to the original design. These meetings are notorious for their length, given the uncertainties involved in design. The main functions of the design meeting are:

- To resolve design problems that occur during construction
- To incorporate changes in the scope of the work
- To resolve problems of interference between different subcontractors

Meeting Time and Location

The meeting should be held as soon as possible after there is evidence of the need for a meeting.

For construction-related problems, the job site is the appropriate meeting location. For design-type problems relating to a change in the scope of the work, the architect/engineer's or contractor's office may be a good choice. If it is necessary to inspect any work accomplished already, then the meeting should be held at the job site.

Who Should Attend

- Owner/Representative
- Architect/Engineer
- General Contractor
- Subcontractors, as required
- Others that may have input or money at stake

Items to be Discussed

The architect should prepare the agenda as, most often, he or she has the best overall perspective of the design intent and, therefore, the design difficulty. By preparing an agenda, the architect is forced to think through the problem and address the possible solutions in outlining the sequence of steps that may have to be addressed.

Requirements for Minutes

The requirement for exhaustive minutes may not be necessary, or even possible, as the completed plans and construction documents are, in effect, the minutes. However, general minutes of at least the decisions/revisions should be taken and distributed. It is important that the minutes be distributed, especially if individuals have been assigned some specific responsibility in the course of the meeting. The minutes are normally taken by the individual or firm that "chairs" the meeting.

Pre-bid Meetings

Purpose

Pre-bid meetings inform and assist bidders prior to the actual bidding process. They are intended to familiarize the bidders with the scope of the project and potential obstacles to progress. Generally, this type of meeting is used in connection with renovations to existing buildings or expansion projects.

Meeting Location and Time

Pre-bid meetings should be held two to three weeks in advance of the bid due date in order to be useful to the bidders. This means no earlier than the time when the scope of the bidding requirement is established, and no later than allows time for the bidders to respond to information disseminated at the meeting. Pre-bid meetings should be held at the site of the proposed construction, whenever possible.

Participants

- Owner/Representative
- Architect
- Interested general contractors, if bidding for the general construction contract. If not bidding for the general contract, then the selected general contractor.
- Interested subcontractors

Agenda

- Tour of the project site
- General discussion about such items as:
  Working conditions
  Demolition problems
  Quiet hours
  How payments are to be made to contractors or subcontractors
- Answering questions from bidder and attendees
- Reviewing of the General Conditions
- Possible review of the construction documents
- Method of construction (i.e., phased work)

Requirements for Minutes

Following the meeting, a report should be made listing the attendees, and the items discussed. It is especially important to record any decisions made by the architect or owner, so that all contractors may know what was said. Often, one contractor will ask a very pertinent question, but not everyone will hear the answer. The answer may impact the bidder's prices significantly. The answers should, therefore, be included in the minutes.

Pre-construction Meeting with G.C. and Owner

Purpose

The pre-construction meeting reviews the project and sets the ground rules for responsibilities and communication during the course of construction.

Meeting Time and Location

The pre-construction meeting with the general contractor and owner should be held as soon as possible after notice of Award of the Contract. The site of the meeting is not crucial.

Participants

- Owner
- End user of project, (e.g., department head, or facility manager)
- Architect
- Engineer

- General Contractor
  - Project Manager
  - Estimator (optional)
  - Chief Executive Officer (optional)
  - Accounting Representative (optional)
  - Superintendent for the General Contractor
- On some projects it will be important to involve the major subcontractors (if selected). This decision should be reviewed on a project-by-project basis.

Agenda

- Responsibilities of the Owner
  - To give direction
  - To make decisions promptly
  - To make payments according to a procedure and schedule
  - To specify who has authority to sign for changes in the scope of work
  - To establish a format for billing
  - To attend project meetings (as required)
- Responsibilities of the architect/engineer
  - To provide accurate plans
  - To review and return submittals in a timely way. (A specific time frame should be set at the meeting.)
  - To review scope of work changes in a timely way.
  - To attend project meetings
- Responsibilities of the General Contractor
  - To make prompt submittals to the Architect/Engineer
  - To keep the owner/architect/engineer informed of job progress and problems
  - To create the Critical Path construction schedule. A scheduled date for submittal of the schedule should be set at this meeting.
- "Housekeeping" Items
  - To review the submittal procedure
  - To review the billing procedure
  - To set a time and place for project meetings
  - To specify who will be responsible for job meetings and minutes
  - To review with participants the style and format of the meeting minutes, as well as how any item(s) entered into the minutes may be contested
  - To reach a consensus with all concerned that the project minutes are to be considered a legal document

If the job site minutes are to be considered part of the construction documents, then, for example, if it is recorded in the minutes that the owner has given approval to proceed with Price Request No. 5, that authorization becomes binding.

Requirement for Minutes

The minutes of the pre-construction meeting with the general contractor should be taken and distributed to all attendees and those people and/or firms that may be affected. Distribution is important, especially if individuals have been assigned a specific responsibility as a result of the meeting. This meeting can set a positive tone for the whole project.

Pre-construction Meeting with Manager (In-house)

Purpose

This meeting should be a general review of the project: how it was designed and bid, and any potential problems. It is a good idea for the "boss" of the general contracting firm to sit in on these meetings to help determine if there are items that need his particular expertise. It is also a good time to review the bid and the game plan for completing the project and making money while doing so. An agenda or checklist of items to be discussed is very helpful.

Meeting Time and Location

The in-house meeting should be held as soon as possible after notification of award. It can be held anywhere, but the main office is probably most suitable.

Participants

- Project Manager
- Estimator
- Superintendent
- Accounting Representative
- Chief Executive Officer

Agenda

- Review Construction Contract and General Conditions
- Review Scope of Work of Project
  - Start date
  - Completion date
  - Can completion date be moved up?
- Review the original project design (bid package)
- Can all areas of construction be accessed at once? Phasing?
- Review of critical materials
  - Structural steel and anchor bolts
  - Glue-lam beams and/or trusses
  - Finish hardware
  - Hollow metal
  - Exterior envelope
  - Elevators
  - Switchgear
  - Other finishes
- Review construction schedule per contract
- Prepare in-house construction CPM schedule
- Are lien notices required?
- Subcontractor/vendor list
- Review subcontractor scope of work
- Subcontractors' bonding needs, if any
- Any city or county licenses required?
- Plan distribution: Who gets what plans?
- Review submittal requirements. Any long lead items, specialties, unknowns?
- Do any in-house departments need to relocate? How soon? For how long?
- Permits Required
  - Who is responsible for acquiring permits?
  - Land development
  - Demolition and grading
  - Building permit

  - Engineering permit
  - OSHA permit
- Submittals required?
- Temporary construction facilities
  - Phone
  - Power
  - Trailer
  - Heat
  - Temporary enclosures
- Existing utility disconnects required?
  - Phone
  - Electric
  - Gas
- Has application for permanent utilities been made, and if so, by whom? *If some of these items are the responsibility of the owner, architect or user, write a letter stating what needs to be done and by whom.*
  - Telephone
  - Gas
  - Electric
  - Cable television
  - Water and sewer laterals
  - Fire service
  - Alarm, and elevator alarm monitoring services
- Is special inspection required?
  - Soils and compaction
  - Concrete
  - Masonry
  - Steel, bolt and welding
  - Waterproofing
  - In-house
- Is there a soils report?
  - In main file?
  - Copy for field?
- Are plans complete?

Minutes Required

Brief handwritten notes or minutes of this meeting should be taken and distributed. At a specific in-house meeting like this, one person should be made responsible for the minutes or notes. The checklist shown above can serve as the outline for the notes. It is important that the notes be distributed to all attendees, especially if individuals have been assigned specific responsibilities during the course of the meeting.

Subcontractors' Job "Kick-off" Meeting

Purpose

This "kick-off" meeting is called by the general contractor and should include all subcontractors. It is similar to the pre-construction meeting held with the owner and architect.

Time and Location

This meeting should be held as soon as possible after award of contract and prior to construction. It should be chaired by the general contractor. Hopefully, all subcontracts can be executed at this meeting. However, in fast-track and some design/build-type projects, the subcontracts cannot be executed at this stage due to the nature of their construction practices.

Participants

- General Contractor. This includes representation by the following:
  - Chief Executive Officer
  - Project Manager
  - Superintendent
  - Accounting Representative (optional)
- All subcontractors
- Architect (optional)

Agenda

- General review of the project
- Distribute list of subcontractors
- Review of contract documents
- Review of construction schedule
- Review of billing procedure and payment schedule
- Review of submittal procedure and schedule, if possible
- Set a time for regular job meetings
- Review insurance and bonding requirements
- Review job clean-up procedures
- Review backcharge procedures
- Inform all concerned that the project minutes will be considered a legal document. Inform participants of the style and format of the meeting minutes and review the procedures for contesting or adding something entered into the minutes.

Minutes Required

Written notes or minutes of this meeting should be taken and distributed, not only to the individuals attending the meeting, but also to anyone who did not attend and, in some cases, directly to the Chief Executive Officer of each of the subcontractors. This meeting, like the kick-off meeting with the owner/architect/ engineer, can set the tone and flow of paperwork for the entire project and assist in the completion of the project.

Job Close-out Meeting (In-house)

This meeting is held "in-house" for the purpose of reviewing the project just completed and ensuring that all conditions of the contract have been met and all monies paid and collected.

Time and Location

This meeting should take place before final payments are made to subcontractors and vendors and after all job site construction is complete. All punch list items should have been resolved by this time. The meeting should be held at the contractor's/facility manager's office.

Participants

- Chief Executive Officer of general contracting firm
- Project Manager
- Superintendent
- Estimator
- Accounting representative

Agenda

- Get all Price Requests, Field Orders, and Clarifications signed
- Get all Change Orders signed and processed
- Write all subcontractor change orders that have not already been processed
- Resolve all backcharges before making final payment to subcontractors
- Perform final sign-off of all permits
- File notice of completion
- Prepare final billing
- Retention payout review
- Obtain final lien releases from subcontractors, if not done already
- Give lien release to lender, or owner, as required (at final payout from owner)
- Resolve all quotes and claims from subcontractors
- Submit warranties and guarantees
- Submit as-built drawings
- Submit operations and maintenance manuals
- Transfer elevator registration from general contractor or subcontractor to owner (if any)
- Disconnect temporary utilities (if not done already)
- Transfer permanent power from general contractor to owner
- Turn over keys, get receipts
- Change owner's address to new building
- Finalize subcontractor bonds, if any
- Tie up loose ends

Minutes Required

Brief handwritten notes or minutes of the job close-out should be taken and distributed. At a specific in-house meeting like this, one person should be designated as responsible for the minutes or notes. It is important that the notes be distributed. Attendees will be assigned some specific responsibilities and deadlines through the meeting.

Regularly Scheduled Project Meetings

Purpose

These are general purpose meetings called to handle the problems that occur on a project. They deal with changes in the scope of work, delivery problems, scheduling problems, and other such items. There are two types of regularly scheduled meetings: one with the owner, and one with the subcontractors. The concept of the meeting and the minutes taken is the same. The only difference is the participants. If the project is a major one, it may be advisable to hold separate "owner" and "subcontractor" meetings. The following project meeting description is for a combined owner and subcontractor meeting. This is a common arrangement.

Time and Location

The Project, or Subcontractor's Meeting should be held every week, or at least every other week. It should always be held at the same time and at the same place–usually at the job site.

Participants in Owner Meetings

- Owner or User
- Project Manager or Facility Manager
- Superintendent
- Major Subcontractors
- Architect or Engineer

Participants in Subcontractor Meetings

- Project Manager or Facility Manager
- Project Superintendent
- Subcontractors
- Architect, not usually in attendance, but as required

Agenda

- Review of project progress/summary of work
- Review of project schedule
- Items from prior meeting
- New items for discussion/action, and who is to act
- Problems, potential problems, interferences (from subs)
- *Safety Items*
- Review of all outstanding pricing and change documents

Minutes Required

Structured, clear, typewritten minutes should contain:

- Project name
- Date of meeting
- Location of meeting
- Time of meeting
- Attendees and companies' representatives
- Lost day tabulation
- Condensed statements of items discussed (each item gets its own separate line.)
- Statement allowing attendees to correct mis-stated information

All items not resolved in the previous meeting should remain in the minutes.

All safety items are to remain in the minutes from meeting to meeting as a constant reminder of past problems (no names or companies to be associated with these items).

Like the special meeting minutes, regular project or job site minutes should be taken and distributed to those individuals and companies that require or are affected by the problems and decisions that occur on the job. The same information items included in special meetings also should be included in regular meetings. It is important to identify who is responsible for implementing conclusions.

Summary of Work

A brief summary of the job status, including major activities for the next period, should be included with the minutes. Alternatively, a copy of a two-week, short-term, interval schedule may be attached. (See Chapter 14, "Scheduling.")

Project Completion "Block"

Several statements regarding project completion should appear on the front page of every set of minutes. These are the items that people look for, especially if they do not regularly attend meetings. They are:

- Original contract completion date
- Adjusted contract completion date
- CPM scheduled completion date
- Anticipated completion date as of date of meeting

This is an area where credibility is so important. If a review of the job schedule reveals that the job is behind, extend the completion date. Get the owner and the other participants to realize that the anticipated completion date is just that–the date construction is expected to be completed–not when someone would *like* the project to be completed.

The adjusted contract completion date is the revised completion date when all approved extensions in time are included. The completion date is usually adjusted because of added scope in the work as shown in the pricing documents. This date is not changed until the owner has signed the Change Order or otherwise authorized the increase in contract completion.

Lost Days and Extensions of Time

All lost days, whatever their cause (e.g., weather, strikes), should be included. Refer to Chapter 15, "Special Use Letters," and Chapter 3, "Transmittals," for samples of Time Extension documents. Apply for all extensions in the Request for Change format as shown in Chapter 22. Record the cumulative total of days lost in the minutes for each weekly meeting, as shown in Figure 17.3. The extensions in contract time in this category are for "acts of God." It is important to keep track and record the amount of time lost due to such conditions, so they do not surprise third parties who were not immediately involved in the extension of time.

Extensions caused by a change or changes in the scope of work should be included in the Price Requests or Field Order with a commensurate cost implication.

Include All Items Until Resolved

All items should continue to appear in the minutes until they are resolved. It is sometimes difficult to keep everyone interested in these older issues, as they want to move on to new challenges. However, it is necessary to use the minutes, as well as other tactics, to make sure the old assignments are handled.

Assign Responsibility for Action

This can be accomplished in at least three different ways. Once again, the format is not as important as the information itself, and the fact that the responsibility is assigned. The information should, however, be easy to scan in order to determine who is responsible for each item. The most effective way to identify in the minutes who is responsible for a particular item in the minutes is to have it handwritten or typed in one of the following formats, showing the initials of the company responsible for the item:

- Within the text of the item in the minutes (in bold print)
- In the left hand margin (usually handwritten)
- In the right hand margin (usually typed)

Several sets of initials can be organized in the margin for a particular item. Using a word processing program, where the text can be divided into columns, it is fairly easy for the typist to enter the initials (of responsible parties) in a column to the right of the item.

Use Company Names

If possible, avoid using personal names. Use the company name or initials instead, because the company is the one responsible for the action, decision, or work. If it is necessary to use an individual's name, always refer to "James Jones, for ABC Glass Company."

Include Statements of Review

Include a statement that all outstanding Price Requests, Field Orders, Clarifications, and Requests for Change received thus far by the owner from the subcontractors have or have not been reviewed. Also state if a copy of the logs was left with the owner.

Include a statement that all Outstanding Price Requests, Field Orders, Clarification Letters, and Requests for Change from the general contractor to the subcontractors have or have not been reviewed.

Include a statement regarding the submittal log: whether or not it has been reviewed with the owner, architect, and subcontractors. State whether or not a copy of the log was left with the architect.

Include Date and Time of Next Meeting

Include the date, time, and place of the next meeting.

Minutes as a Construction Document

If it has been agreed that the minutes are to become a part of the construction documents, then a statement similar to the following can (and should) be included in the minutes–usually at the very end of the document.

These minutes will become a part of the project documents. Please inform the writer within five working days if items included are not correct. Any required corrections or additions will appear in the next minutes.

Recommended Paragraph Format

The items in the minutes should be numbered as they occurred in the meeting. Item 8.0, for example, will always be the job status for meeting number 8. Item 9.0 will be the job status for meeting number 9. Item 8.1 will then be the first item discussed in meeting number 8. The following paragraph is a typical entry:

8.4(5.7) There still appear to be some questions regarding the requirement for plastic laminate cabinet fronts. WARA to review and report at next meeting.*

*Note: **WARA informed ABC on 5/7/89 that plastic laminate would not be required.*

** The number in the parentheses indicates this is a recurring item, and was first brought up in the meeting and paragraph indicated in the parentheses. To research the events that occurred since this topic first was raised, one would look in the minutes for meetings number five, six, and seven. The item number (5.7) would be shown following the paragraph number in the order it was discussed in the particular meeting. This approach helps to keep the minutes short and in order. It also makes it easier for the person chairing the meeting to keep the minutes while conducting the meeting.*

*** The "Note" following the paragraph indicates that the "note" was inserted by the individual preparing the minutes after the meeting. This is a way to include pertinent information not available at the time of the meeting.*

Alternative Method of Paragraph Recording

Some firms use a system of paragraph numbering whereby each item, as it is entered into the minutes, is permanently assigned an I.D. number. For example, item 5.7 would have been *permanently* assigned the number 5.7. In this case, the minutes would appear as shown in Figure 17.3.

This method makes it easy to follow the history of the item from each meeting; it also gives a complete overview of the subject matter.

The disadvantage from a practical standpoint is that the minutes can get very long, and as a result, the people who should read them do not! It is also difficult to record the minutes if the sequence of the actual meeting differs from the order of the minutes.

All other items (referred to earlier in this chapter) to be included in the minutes are still necessary.

One format may be more comfortable to a particular organization than another. Once again, it does not matter too much which format is used as long as:

- All information is included
- All information is accurate
- The minutes are distributed in a timely manner

M I N U T E S

March 1, 1990

PROJECT: Job #503, GordOn Mercantile

ATTENDANCE:
John Carrier, Gordon and Associates
Mary Johnson, S.G.T. & Co.
Donna Bailer, Patterson Design
Kevin David, Bildenhope Construction
Lou Mayflower, " "
Michael Louis, " "
Angie Goldman, " "
Dwayne Lehigh, " "

===

4.0 JOB PROGRESS

Now taping on 17th floor, installing t-bar on 17, and will start fire sprinklers on 17th floor March 2, 1990. Working on curtain pockets on 17th floor. Finishing up on second siding of sheetrock on 14, 15, and 16. Rough HVAC complete on 17, 90% on 14 and 16, rough HVAC on 15 will be started on Monday. Rough plumbing is complete on all floors.

Project Completion:

Original Contract Completion Date:	December 22, 1990
Adjusted Contract Completion Date	December 22, 1990
CPM Scheduled Completion Date:	December 22, 1990
Anticipated Completion as of March 1, 1990:	December 22, 1990

APPROVED EXTENSIONS: None to date

* * * * * * * * * *

4.1 Gordon and Associates requested all meetings to be on Thursdays at 2:00 PM, no exceptions.

4.2 (1.10) Bildenhope Construction received price from Peter Piping Co. for moving the pipe - $708.00.

4.3 (2.9) Bildenhope Construction, Gordon and Karle will meet before next Thursday's meeting.

4.4 (3.12) This is a correction - should have read Data and receptacle on West wall in room 1519 to be in wire mold. Receptacles in rooms 1432 and 1433 to be moved to avoid conflict.

Figure 17.2

Minutes
March 1, 1990
Page 2

4.5 (3.13) Bildenhope Construction has a drawing from Faude Sheet Metal for the panel fold door support system, and has submitted it to Patterson Design.

4.6 Gordon has decided to accept an alternate to polished aluminum in lieu of chrome-plated for signage. Alternate has been accepted - $897.00 credit.

4.7 Electrical drawings for insta-hot water locations - Bildenhope Construction stated that there are no electrical outlets shown on drawings to power insta-hots. P.D. TO RESOLVE

4.8 Segmented glass wall on the 14th floor - Patterson Design will supply Bildenhope Construction with detail showing segments. P.D. TO PROVIDE

4.9 G-type wall to be built with drywall head at sidelites, per Patterson Design at job site meeting.

4.10 14th floor wall detail missing at column line G between columns 4 and 7. Bildenhope Construction to align with existing furred columns, per Patterson Design.

4.11 Meeting adjourned at 5:00 PM

NOTE THAT THESE MINUTES BECOME A PART OF THE PROJECT DOCUMENTS. PLEASE INFORM BILDENHOPE CONSTRUCTION, INC. BY THE NEXT MEETING IF ITEMS INCLUDED ARE NOT CORRECT. ANY REQUIRED CORRECTIONS OR ADDITIONS WILL APPEAR IN THE NEXT MINUTES.

/SFC

Figure 17.2 (cont.)

January 5, 1990

JOB #708
Raikow and Associates, Tenant Improvements
Construction Meeting #8

Present	Distribution
David Cleary, Raikow Assoc.	All Present
Mary Carrier, Raikow Assoc.	J. Edward Grimes, Goodenchepe Inc.
Hal Horvat, Gotham, G, & T Co.	Adam Joseph, Gotham Guaranty
Meg Hackett, Goodenchepe, Inc.	Peter Wilford, Kendall Design Assoc.
Karen McCarty, Goodenchepe, Inc.	
Douglas Hull, Kendall Design.	

The following is an overview/summary of the meeting with follow-up comments or action to be taken. These minutes will be deemed accurate and correct unless notified in writing within five (5) working days.

SCHEDULE:

- Plastic laminate cabinets being installed.
- Light fixtures being installed.
- Monocote 50% complete.
- Ceiling grid re-started.
- HVAC trim started.
- Paint, vinyl wallcover started.

- Delays:
  1. 5 working days from actual start date being moved from 10/11/89 to 10/18/89.
  2. 5 working days delay due to lack of permit from 11/14/89 through 11/18/89. Inspection could not be called until 11/21/89.

- Potential Delays:
  1. Steel Stairs - Various elements have contributed to delays. Principal among them was the rejection of the design by Raikow. This design was based on KDA drawings and revised per Bulletin 3, which incorporated Raikow's design criteria, and further revised during a meeting with the subcontractor on December 1, 1989. Revised drawings were submitted simultaneously to KDA (sepia) and Raikow (blueline) on December 19, 1989. KDA claims Raikow not available for coordination during week of 12/19/89 and those individuals at Raikow involved with the project were on vacation week of 12/26/89. Gotham will revise schedule and discuss possible delays after receipt of

Figure 17.3

Job 708
Construction Meeting #8
January 5, 1990
Page Two

- Potential Delays: (Continued)

approved drawings by subcontractor.
Update 1/5/90: KDA sent approved stair drawings to Raikow by messenger on 1/3/90. Not yet received by Gotham.

2. Building skin at manlift must be closed before drywall can be installed at Southeast corner of both floors. Estimated by Skinco to be closed by Christmas. Impact to be determined.
Update 12/29/89: Granite installed to 6th floor.
Update 1/5/90: Granite installed to 15th floor. Two weeks more to complete framing drywall.

3. Curtain pocket hardware must be installed prior to framing and drywall of this area. Ceiling grid must be attached to the completed assembly. ABC is currently working on acceleration of this process. Impact to be determined.
Updated 12/15/89: Hardware due to be delivered 1/3/90, installed 1/4/90 through 1/6/90.
Update 1/5/90: New move-on date 1/9/90 for this subcontractor.

ACTION ITEM OLD BUSINESS

Raikow 2.1 Keying Schedule: Requested from Raikow. M.H. to coordinate with building key schedule.
Update 1/5/90: Schlage FG Keyway throughout building; Raikow/San Diego sent revised schedule to Raikow/Los Angeles.

2.7 COR's: See Attached

KDA 3.2 Anegre Flitch: 11/17/89 GA to verify millwork is intended to match sample.
Update 12/1/89: Sample Flitch supplier has only 2,200 SF available. Quality to provide alternate samples.
Update 12/8/89: KDA to discuss with veneer supplier, meanwhile samples are to be forwarded to Gotham by KDA for review.
Update 12/15/89: Raw samples reviewed by KDA; samples due next week.
Update 12/29/89: Approval by KDA still needed.
Update 1/5/90: Flitch approved and ordered 1/3/90.

Figure 17.3 (cont.)

Job #708
Construction Meeting #8
January 5, 1990
Page Three

ACTION ITEM OLD BUSINESS

4.1 Telephone coordination - KDA to schedule work before title is installed - vendor to work with Steve Scofield. Note: Vendor is required to comply with GA's insurance requirements as though he were a subcontractor (must furnish primary endorsement).
Update 12/29/89: Contract with MEG/Telephone Company signed. No word to Steve from Tel Co.
Update 1/5/90: Revised date to begin installing tile 1/23/90.

Raikow 5.3 Infrared motion detector to be reviewed by Raikow. Subcontractor claimed to have missed this item on the bid, but competitive bid situation called per plans and specifications. Raikow/KDA to review and advise GA.
Update 12/29/89: Motion detectors are to be installed as shown.
Update 1/5/90: DH to verify locations of card readers and sensors.

KDA 5.6 Lights and life safety speaker in elevator lobby are to be re-located by KDA
Update 12/29/89: South and North wall washers to be surface-mounted. KDA to provide specification.

KDA 6.3 Hardware set not shown door #53.
Update 12/29/89: Lockset required, no closer.

KDA 7.1 Herman MUller furniture. GA requests electrical requirements

KDA 7.2 Power distribution unit at Computer Room - GM requests specifications.

GA 7.3 Paint sample P-1 corrected and approved. Benjamin Moore GN-93 Semi-Gloss to be corrected to match flat, which was approved.

ACTION ITEM NEW BUSINESS

GA 8.1 Finishes not shown:
Room 3233 - VWC-1
Room 3234 - VWC-1
Room 3241 - P-1

GA 8.2 Downlights in Room 3235 - Northernmost is in conflict with ductwork running East-West. To be resolved.

GA 8.3 Light fixture in Room 3315 - Westernmost in conflict with duct - to be moved 1 Bay East.

Figure 17.3 (cont.)

Job #708
Construction Meeting #8
January 5, 1990
Page Four

ACTION ITEM NEW BUSINESS (Continued)

GDC 8.4 J-Box in Computer Room per note 44-A 32-2 required as soon as possible from Raikow.

CMC 8.5 Contrasting color at top and bottom of each flight of stairs is required by Building Inspection Department.

CMC 8.6 Type A door hinges may conflict with wood frame. KDA to provide solution.

Next Meeting: Thursday, January 12, 1990 at 9:30 AM at Raikow office.

Submitted by:

I. Bildem
Project Engineer

IB:lh

Attachment

Figure 17.3 (cont.)

Part Four

Pricing Communications

Chapter Eighteen

Concepts of Pricing Documents

There are four basic types of construction correspondence used to disseminate information about a change in the scope of work, and to request or provide a cost quotation for that change. The four categories are:

1. Price Request and Price Request Quote to Owner
2. Field Order and Field Order Quote to Owner
3. Clarification and Clarification Quote to Owner
4. Request for Change and Request for Change Quote to Owner

These four types of *pricing communication* letters convey information from the owner/architect to the general contractor, then to the appropriate subcontractors, and back up the line. The pricing communication to subcontractors and the resulting Quote to Owner (or user in the case of a facilities project) document represent the cost for the scope change as initiated by the owner, architect, or engineer.

Pricing Communications Are Not Change Orders

The four categories of pricing communication documents should not be confused with *Change Orders.* A Change Order is a separate contract document normally issued by the owner, or for the owner, by the architect or engineer. The purpose of a Change Order is to change or modify the original, or existing, *Construction Contract Between Owner and Contractor.*

Within the industry, many of us refer to *any* change in scope as a "Change Order." This is *wrong.* It is not a Change Order until it has been executed by all parties.

A Change Order is a modification of the Construction Contract. It may be concerned with a scope of work change, a bonus for finishing early, or an extension of time to the contract–with or without compensation. The Change Order may go so far as to record a change of owners, or to document any other change to the Construction Contract that is agreeable to both parties. Remember, paper is the cheapest material we build with.

It is strongly recommended that multiple scope changes not be included in a single Change Order. One separate Change Order for each scope change, makes everyone's life easier.

A Change Order is a separate contractual instrument which may incorporate several Price Requests, Field Orders, Clarifications, and a time extension, even though the pricing documents and request for a time extension may originate at different times and for different purposes.

On many projects, "official" Change Orders are not issued. Instead, the "sign and return" part of the form contained in the Quote to Owner letter (see Chapter 22) provides a contractual vehicle for returning a price quote to the owner on a proposed change in the scope of the work. This method is an alternative to issuing a Change Order as an additional document. In this way, the Quote to Owner letter, signed and returned by the owner, serves as a Change Order to the original contract. This one-item/one-document approach allows for tracking of modifications to the original contract.

The alternative system–the owner/architect issuing a formal change (consisting of one or many Price Requests, Field Orders, Clarifications, or Requests for Change), can make it difficult to maintain document system control. In this way, modifications to the original contract can be tracked.

Note: Individual companies may elect not to use every available type of "pricing" concept and document, choosing only the ones that particularly suit their business. Each of the following four chapters (19, 20, 21, and 22) is written to stand alone as an individual reference to one of the pricing documents. In this way, if a company elects to use only the Field Order format for pricing communication, company personnel need only read *this* chapter (18) and Chapter 20, "Field Orders." Remember that each company is unique and will have different requirements and priorities, just as each job is unique.

Timely Response

It is the duty of all parties (owner, architect, facility manager, general contractor, and subcontractors) to make sure that the change in scope description is sent to those who do the work, and to then get that description back to the owner. Often, the architect, engineer, facilities manager, general contractor, and subcontractors do not put a high priority on expediting the various quotes. Everyone assumes that the owner/architect will simply remember that a larger footing was constructed until someone gets around to quoting how much the larger footing actually costs. Unfortunately, that is not the way it happens.

Most owners look only at the monthly billing to check approved changes in the scope of work; they may consider the total adjusted contract price to be their maximum exposure. Managers should avoid situations where owners/users are surprised by a late change of scope increase and where they (contractors) must produce proof that a change was, in fact, directed. No matter how correct the manager's position might be, it will be a rude awakening for the owner/user who discovers suddenly that those costs were not included in the original plans.

The importance of the manager promptly sending the information to the proper subcontractors, vendors, and field personnel, as well as returning the resulting responses to the owner/architect as quickly as possible, cannot be emphasized enough. Every day that the information and response is delayed is another day that the owner hopes he will get the scope change for $100.00 when the real cost will be more like $1,000.00. The longer the delay, the more likely it is that he will spend the remaining $900.00 on other items. By waiting until the end of the job to provide quotes to the owner, the owner's money may already be spent, and the change in scope of work may not be on his list of payables. Now the question becomes: Is the job over when the money is gone, or is the money gone when the job is over?

By sending or giving the owner a copy of the logs on the "open" (unresolved) pricing letters (those items not yet quoted by the manager), the potential of a price increase remains visible to the owner for his budgetary review. Better to let him be concerned about future costs today, than to run out of money and be unable to pay the bills tomorrow.

Logs

A manual log usually shows all pricing items–even those that have been approved. While it can be very thorough, information in this format can also be confusing, and the owner may find it difficult to interpret. A computer-generated list is another option. A computer program can generate a log that shows only the work that is still in the process of being priced, or only the work that has been approved.

Many firms group all of the pricing vehicles (Price Requests, Field Orders, Clarifications, and Requests for Change) together, although this practice is not recommended. It is easier for the owner/architect to grasp these different concepts one type at a time, rather than all jumbled together. The general contractor and subcontractors understand the difference in concept between these four types of scope changes because they deal with them constantly. The owner, on the other hand, will probably build only one, two, or maybe as many as five buildings in a lifetime.

Remember, it is not necessary to use all of the pricing vehicles. A particular company's business activity may require only the use of the Field Order and Field Order Quote to Owner documents and not the Price Request with the Price Request Quote to Owner documents or the clarification documentation. The manager will, however, *always* need to use the Request for Change, as it is the only vehicle available for *initiating* a pricing and approval procedure.

Chapter Nineteen
Price Request and Quote to Owner

A Price Request (PR) is one of the four basic concepts or types of construction correspondence that are used to process the pricing of a change in the scope of work on a project. Other types of pricing letters include Field Orders, Clarifications, and Requests for Change. These are described in Chapters 20, 21, and 22.

The Price Request Quote To Owner (PRQTO) is the response to the original Price Request from the contractor to the owner/architect. Briefly, it should specify the time period for which the quote is valid, and any other stipulations or conditions that might be required. One of the conditions usually accompanying a PRQTO is an extension in contract time. The use of the PRQTO is discussed in detail in the second half of this chapter.

The Price Request

Purpose

A Price Request (PR) is a communication from the owner/architect to the contractor requesting a price quote for a particular task. The work involved may be an adjustment to the existing plans or to the project. It may involve a change in the scope of work. It may also represent "window shopping" on the part of an owner who wishes to investigate a variety of options. A Price Request does not necessarily mean additional cost; it could also result in a credit.

The Price Request function is known by many names, such as:

- Request for Proposal
- Proposal Request
- Bulletin
- Informational Bulletin
- Request for Quotation

Regardless of what they are called, Price Requests are requests (by the owner/user or architect) for a price to perform a specific task. The work has not yet been authorized to start or to complete. It is important to remember that a Price Request is *not* an authorization to proceed with the work described.

Get Price Requests Authorized Before Starting Work

It is possible that a Field Order may be issued that includes the work described in the PR without having the PR itself authorized. Often, a Field Order may be issued to proceed with part of the work described in an unpriced or unapproved PR so that long lead time items or other work that needs immediate attention can proceed. The following are examples of Price Requests that may be issued or requested. Note that these questions are phrased as statements:

- Provide a cost to change the carpet Brand X, Grade 2, to Brand Y, Grade 4.
- Provide a cost to add 2'-0" to the height of the retaining wall at the back of the parking lot.
- Provide a cost to change the cabinet tops from plastic laminate to ceramic tile.

There may be hundreds of these kinds of Price Requests on the same job.

Again, the work will normally not be authorized until the question of cost is answered. The owner must give specific authorization, based on the price submitted, either by signing the quotation or issuing a Change Order, before the work described in the Price Request can proceed. To do otherwise is to proceed at the company's own financial peril. Architects, engineers, facility groups, general contractors, and subcontractors should all follow this procedure.

Some Price Requests are really "fishing trips." In other words, authorization to proceed may be doubtful or the owner may simply be "fishing" for a price to see what it may cost to get something they want, but may not need. Many Price Requests have been severely modified in scope after the first prices are quoted. The PR may retain the same ID number (such as PR #14), although it barely resembles the original PR scope of work. Contractors who proceed with work on an unauthorized PR are treading on dangerous ground until the scope of the proposed work has been clearly identified and approved. Owners have no obligation to pay for unauthorized work. The manager who performs unauthorized work on a regular basis exposes his own company to great risk.

Price Requests should be kept separate from Field Orders and Clarifications because a Price Request is "in limbo" until it is clearly determined whether or not the work it describes will be constructed.

The sequence of events for the paperwork processing of Price Requests is as follows:

1. The owner/user and architect/user determine the scope of additional work that should be included in the Price Request. The document is assigned an identification number (such as PR #15) and is issued to the general contractor with the appropriate documentation, including description, plans, and specifications fully describing the work to be priced. In the case of the facility manager, the department (user) of the project would be the same as the owner. Often, the facility department also serves as the design entity. If the owner makes the request orally, the general contractor/facility manager should document this event, and treat it as a written request, assigning it a PR number.
2. The project manager reviews PR #15 and determines which of the various companies and trades (including his own forces) are affected by the Price Request. The project manager then sends these designated parties a copy of the Price Request, along with

supporting documents, plans, and specifications. The manager should indicate an appropriate and realistic due date for a response from the trades and subcontractors. If the Price Request is to add 15,000 S.F. of new manufacturing area (a significant change in scope), then an answer cannot be expected within ten working days. If the Price Request is to add panic hardware on one door, it is reasonable to require a response in less than ten working days. For more on the amount of time required to obtain complete and accurate pricing, please refer to Chapter 15, "Special Use Letters."

For pricing purposes, managers should treat the work of their own forces just as they would the work of subcontractors. If the manager does the estimating for the PR, he should do so when he sends out the Price Requests for subcontractors. Then, when all of the subcontractors respond, the manager can immediately work up the price for the owner and get the paperwork out. Often, project managers wait until all of the subcontractors'/vendors' responses are in. They then begin to look at the detail of their own work, which can further delay the quote being sent to the owner. This approach will not help job progress.

3. Subcontractors and vendors review the Price Request and return their quotations to the manager.
4. When the manager receives the quotes for the Price Request from all of the trades and subcontractors, the profit and overhead numbers should be added, and a Price Request Quote To Owner sent to the owner/architect/engineer. The owner/architect should be allowed sufficient time to respond; usually ten working days should be adequate.
5. The owner approves the Price Request and returns it to the manager, using all ten of the allocated days.
6. The manager should write a subcontract Change Order to each of the subcontractors involved. Allow four working days for this step.

It is easy to see what might happen at this point in the process. The following events must occur in sequence and, together, require a significant amount of time.

- It usually takes the owner/user two weeks to determine what should be included in the PR. It takes the manager's organization four days to get the information out to the subcontractors.
- The subcontractors take ten days to answer.
- It takes the manager at least four days to do the paperwork in-house.
- The owner takes ten working days to approve the quotation.
- It then takes the manager four days to process and mail the change order to the subcontractor for the work described.

All of a sudden, 42 working days are gone–approximately eight work weeks. At this point, it may be that the original quotes are outdated or no longer applicable.

7. The subcontractor orders the long lead time items, which may take several weeks from the date that the authorization to

proceed is received. He then installs the work when the materials are delivered to the job.

In summary, the steps involved and the approximate time required for each are as follows:

1. Owner to manager	10 days
2. Manager to subcontractor	4 days
3. Subcontractor prices and returns	10 days
4. Manager to owner	4 days
5. Owner reviews and returns to manager	10 days
6. Manager writes subcontractor change orders	4 days
	42 days
7. Subcontractor orders materials	

The steps and time allocation just described may not represent the working procedures of every project. However, they do illustrate the sequence and timing problems that may occur when processing a Price Request. They also show the importance of "timely" responses.

Many firms group all of the pricing vehicles together. This is not recommended. Making information accessible and usable (in a "readable" format) to owners/users is important. Most project owners (or department heads) will build one, two, or perhaps five projects in a lifetime, while a project manager may work on that many in a just a few months. The owner should not be expected to be totally familiar with construction documentation.

Additionally, the owner may have more at risk. He may have every red penny he can gather together invested in the project, and his occasional nervousness is understandable. Combining the pricing vehicles, such as Price Requests, Field Orders, and Clarifications, makes the documentation too confusing for the owner to easily determine the impact and risk on his or her project.

The manager does well to take good care of the owner or, for that matter, anyone who has to pay for changes in the scope of work. Every architect, engineer, project manager, and subcontractor should, at some time in their life, build a project for themselves with their own money. This experience provides a different outlook and a little more appreciation for the trials, uncertainties, checkbook balances, and doubts that face every owner at some time during a project.

It is not necessary to use all of the different types of pricing letters on every job. Some jobs do not lend themselves to the use of Price Requests because of the time involved from inception to completion. Yet, on other jobs, especially government work, all changes in scope may be required to go through the Price Request process. That is one of the reasons that government work costs so much and takes so long.

Each company has different requirements and priorities, just as each job is unique. A particular business operation may be such that only the Field Order and Field Order To Owner pricing vehicles are required. A manager should, however, set up the company's internal system in such a way that it can handle all four different types of pricing correspondence. On one job or another, all four types will be used. Once using the four types becomes normal practice, the advantages will become evident. Each type of pricing correspondence represents a unique function and requires responses tailored to a particular project.

Figure 19.1a is a sample of a computer-generated Price Request to the subcontractors/vendors. Figure 19.1b is a manual Price Request to subcontractors/vendors, using the standard Transmittal form. Both of the formats are effective.

The main advantages of a computer-generated system are:

- Maintaining a consistent format
- Automatically entering the information in the logs when the Price Request letter to subcontractors is written
- Making it easier for new employees to learn (the hard part is getting the old employees to learn)

The key information items that should be included in the Price Request letter to subcontractors and vendors are listed below:

- From
- To
- Job identification
- Date from architect or engineer
- Date sent to subcontractors/vendors
- Date information is due from subcontractor or vendor
- Short description of the work described in the Price Request
- A note if the full Price Request is attached, or if there are plans included
- Special notes to subcontractor or vendor, if necessary

The Price Request Quote To Owner

The Price Request Quote To Owner (PRQTO) is the response to the original Price Request to the owner/architect. The following items represent information that should be included in the PRQTO.

- From
- To
- Date sent
- Job identification
- Price Request number and identification
- Brief description of work involved
- Date decision is due from owner
- Amount of the quotation
- Extension of time, if required
- How long the quote is valid
- Any special notes to the owner
- Breakdown of costs, depending on the contractual arrangement

Figure 19.2a is a sample of a computer-generated Price Request Quote to Owner. Figure 19.2b is a manual version, using the standard Transmittal format.

DELTA CONSTRUCTION COMPANY

6065 Mission Gorge Road, #193
San Diego, CA 92120
(619) 582-5829

***** PRICE REQUEST *****

Date: 2/28/90

Job Number: 321
Price Request Number: 004

TO: QUALITY MASONRY
8795 BLOCK STREET
SAN DIEGO, CA 92212

ATTENTION: NED NEAT

JOB NUMBER: 321 GIBSON OFFICE BUILDING

FROM: J. EDWARD GRIMES

SUBJECT: ADD 2'0" TO RETAINING WALL AT SOUTH END OF PARKING LOT

We are transmitting to you the following "PR" which is DESCRIBED BELOW for your review and RESPONSE. Please refer to Job Number & Price Number.

Please review and return this information to us by 3/14/90.

Price the addition of 2'-0" of retaining wall at the south end of the parking lot. Add 3' x 5' sidewalk section per Architect's plans dated Feb 21.

Note to Sub:

Quality, if this changes the footing, price the footing also.

Copies were sent to the following:

LOT'S OF REBAR

dp: 2/28/90 5:55 **PaperWorks,** a Construction Software Program

Figure 19.1a

Means Forms

LETTER OF TRANSMITTAL

Delta Construction Company
6678 Mission Road
San Diego, CA 92120
(619) 581-6315

FROM: J. Edward Grimes
Project Manager

TO: Quality Masonry
8795 Block Street
San Diego, Ca. 92212

DATE 2/28/90
PROJECT Gibson Office Bldg. #321
LOCATION San Diego, Ca.
ATTENTION Ned Neat
RE: Additional ret'g. wall

Gentlemen:

WE ARE SENDING YOU ☒ HEREWITH ☐ DELIVERED BY HAND ☐ UNDER SEPARATE COVER

VIA ______ THE FOLLOWING ITEMS:

☐ PLANS ☐ PRINTS ☐ SHOP DRAWINGS ☐ SAMPLES ☐ SPECIFICATIONS

☐ ESTIMATES ☐ COPY OF LETTER ☒ Price Request

COPIES	DATE OR NO.	DESCRIPTION

THESE ARE TRANSMITTED AS INDICATED BELOW

☐ FOR YOUR USE ☐ APPROVED AS NOTED ☐ RETURN ______ CORRECTED PRINTS

☐ FOR APPROVAL ☐ APPROVED FOR CONSTRUCTION ☐ SUBMIT ______ COPIES FOR ______

☐ AS REQUESTED ☐ RETURNED FOR CORRECTIONS ☐ RESUBMIT ______ COPIES FOR ______

☐ FOR REVIEW AND COMMENT ☐ RETURNED AFTER LOAN TO US ☐ ~~FOR BIDS~~ DUE 3/14/90

☐ ______

REMARKS: Enclosed is a Price Request for adding 2'-0" of retaining wall at the south end of the parking lot. Add 3'x 5' sidewalk section per architect's plans dated Feb. 21.

CC: Lots of Rebar

SIGNED: J. Edward Grimes

IF ENCLOSURES ARE NOT AS INDICATED, PLEASE NOTIFY US AT ONCE.

Figure 19.1b

DELTA CONSTRUCTION COMPANY

6065 Mission Gorge Road, #193
San Diego, CA 92120
(619) 582-5829

*** PRICE REQUEST QUOTE TO OWNER ***

Date: 3/12/90

Job Number: 321
Price Request Number: 004

TO: FRANCES GIBSON
6109 CAMINITO CLAVO
SAN DIEGO, CA 92120

ATTENTION: FRANCES GIBSON

JOB NUMBER: 321 GIBSON OFFICE BUILDING

FROM: J. EDWARD GRIMES

SUBJECT: ADD 2'0" TO RETAINING WALL AT SOUTH END OF PARKING LOT

We are submitting the following changes in cost of $ 4,377.00
The details and breakdown of this change in cost are enclosed.

A complete description of the scope of work relating to this proposal is enclosed. The summary of the work is as follows:

Price the addition of 2'-0" of retaining wall at the south end of the parking lot. Add 3' x 5' sidewalk section per Architect's plans dated Feb 21.

Remarks to owner/architect:

We will be starting the retaining wall in three weeks, therefore, the price is only good until April 2.

This work will result in an extension to our contract time of 8 days.

Please indicate your approval by signing and returning one copy of this letter to our office or issue a CHANGE ORDER for this work no later than 3/26/90 in order to not delay the project.

Sincerely,

J. EDWARD GRIMES

APPROVED BY: FRANCES GIBSON

SIGNED: ______________________ DATE: __/__/__

dp: 3/12/90 5:55 **PaperWorks, a Construction Software Program**

Figure 19.2a

Means Forms

PROPOSAL

FROM: J. Edward Grimes
Project Manager

TO: Frances Gibson
6109 Caminito Clavo
San Diego, Ca. 92120

PROPOSAL NO. PRQ 321.004
DATE 3/12/90
PROJECT Gibson Office Bldg. #321
LOCATION San Diego

CONSTRUCTION TO BEGIN

COMPLETION DATE

Gentlemen:

The undersigned proposes to furnish all materials and necessary equipment and perform all labor necessary to complete the following work:

Add 2'0" retaining wall at South End of parking lot.
Add 3'x5' sidewalk section per architect's plans dated Feb. 21st.

Note: We will be starting the retaining wall in three weeks; therefore, the price is only good until April 2nd.

This work will result in an extension to our contract time of 8 days.

All of the above work to be completed in a substantial and workmanlike manner

☒ for the sum of four thousand three hundred seventy-seven dollars ($4,377.00)

☐ to be paid for at actual cost of Labor, Materials and Equipment plus ______ percent (______ %)

Payments to be made as follows: ______

______ The entire amount of the contract to be paid within ______ after completion

Any alteration or deviation from the plans and specifications will be executed only upon written orders for same and will be added to or deducted from the sum quoted in this contract. All additional agreements must be in writing.

The Contractor agrees to carry Workmen's Compensation and Public Liability Insurance and to pay all taxes on material and labor furnished under this contract as required by Federal laws and the laws of the State in which this work is performed.

Respectfully submitted,

Contractor Delta Construction

By J. Edward Grimes

ACCEPTANCE

You are hereby authorized to furnish all material, equipment and labor required to complete the work described in the above proposal, for which the undersigned agrees to pay the amount stated in said proposal and according to the terms thereof.

Date ______ 19___ ______

Figure 19.2b

The Impact of Price Requests

The owner (or user), architect, general contractor, facility manager, construction manager, vendors, and subcontractors (and *their* material suppliers and vendors) involved in a project should remember the following regarding Price Requests:

1. They always change the scope of the work.
2. They usually interfere with the originally planned "normal" course of construction.
3. A great deal of effort may be required on the part of the manager and subcontractors, vendors, and material suppliers to accurately estimate the extent of the scope of work being changed in the Price Request.
4. So much time and effort are involved in preparing Price Requests that if the owner/architect issues many of these documents, and then does not authorize the work to be done, contractors may no longer take them seriously. As a result, they may not take the time to prepare thorough estimates. Perhaps they will simply give a high guess, believing that the owner probably will not authorize the work anyway.
5. The amount of time between the manager's receiving the information and authorizing the work can be so long, and the job may have progressed so far, that the price may no longer be accurate.
6. Most legal and arbitration cases involving changes in the scope of work begin because:
   - There has been a lack of proper documentation.
   - There is a misunderstanding regarding the price of the additional work to be completed.
   - The actual effect of the change on the original scope was not fully understood.
   - The work was completed without proper notification and/or authorization.
   - The owner was not given the opportunity to say "no" to a particular item of work.
   - The additional work has affected completion time, and the owner was not informed that it would take "30" extra days to accomplish the work.

Reducing the Negative Impact of Price Requests

Ideally, the owner (or user) and architect/engineer should do a thorough job in preparing the original plans. If there are too many questions, such as "How much is it going to cost to add an additional elevator?", these issues should be included by the architect in the original bid package as an alternate. Handled in this way, not only will the actual price for the work be less than if it were quoted as individual Price Requests, but it also allows for the *additional work to be scheduled in its proper construction sequence.*

Completing Price Request scope changes out of the proper construction sequence is costly in terms of time, money, and effort. Work performed out of sequence also reduces morale, as it lowers the esteem held by the job personnel for the architect and owner. Many scope changes involve destroying and redoing the work of someone else. No one likes to see something they have put together with nails and wood, lath and plaster, brick and mortar, torn down or replaced. If changes in the scope of work are required, the owner/architect should make sure all the ramifications of the change are clear and understandable. Do not cause more confusion by issuing incomplete instructions, drawings, and specifications.

If it is necessary to make significant changes in the scope of work, it is also important to explain why the change is necessary. Many owners, architects, and engineers are reluctant to explain anything to field personnel. This is difficult to understand, as it is the men and women on the job who really make the project work. People can, and will, understand a major revision in the radiology department of a new hospital, for example, if it is explained that the configurations of lead shielding are being modified to accommodate new equipment that was unavailable at the time the hospital was designed. They will, however, be turned off if they are told that it does not concern them. It *is* their business. They are the ones actually building the project, and they want to feel proud of what they build. They want to be proud of the project, the owner, the architect, and the engineer.

Let the journeymen understand and be proud and they will help the project. Even in the case of a mistake, if they feel that most of the rest of the work is worthwhile, they will help rectify the "goof," no matter who is responsible. On the other hand, if they feel that they have not been included "in" the project, they will be more inclined to simply search for and blame the person who caused the goof.

Earlier in this chapter, it was stated that a signed Change Order or Price Request Quote to Owner is required before any of the work described in the PR is performed. Many times, if the minutes are accepted as a part of project documentation (See Chapter 17), those minutes may show that the owner has given his authorization to proceed with the work. The purpose of this approach is not to "trick" the owner into authorizing work, but rather to shorten the time span for authorization–to everyone's benefit. To use this method, however, make sure that everyone involved in the project accepts the fact that the job minutes are a part of the project documents. Establish this agreement at the pre-construction meeting. It also helps to keep the minutes clear and accurate so they retain their credibility and do not create confusion.

Timely Response

The manager must get the Price Request Quote back to the owner/architect/engineer promptly in order to be listed among the payables. Often, the manager and subcontractors fail to provide timely quotes. They think that the owner/architect will remember that a larger footing was installed in the building, and that he or she will take it into account. Unfortunately, that is not the way it happens. Most owners look only at what they have approved and consider the total adjusted contract price as their maximum exposure.

Logs

Figure 19.3 is a Means *Job Progress Form,* adapted to serve as a manual log of pricing all items, even those that are approved. Manual logs can be somewhat difficult to decipher to the untrained eye. Computer-generated lists may be somewhat easier to interpret, as they can be generated to show only the outstanding items.

The advantages of a computer-generated correspondence program are as follows:

- Consistent format
- Key information included in every letter
- Logs updated almost automatically
- More specific logs generated

The system will generate:

- Reference Reports and Logs which are detailed in nature, and valuable for reference purposes
- The PR #3 logs show what has not been done and who did not do the work
- The PR #5 logs are exception reports that show who is delinquent in their responses

Figures 19.4a through 19.4c are computer-generated logs for Price Requests. Figures 19.5a through 19.5c are computer-generated Price Requests Quotes To Owner Logs. Price Requests and Price Request Quote to Owner Logs contain similar information. The PR logs show all of the subcontractor costs, and general contractor costs and fees, while the PRQTO shows only the price quoted to the owner. It should be noted that the subcontractors should never see the PRQTO logs, and the owner should never see the Price Request logs with subcontractor pricing information.

PROJECT Gibson Office Bldg. #321 SHEET 2 OF —

LOCATION San Diego, Ca. JOB NO. 321

	To/Subject	Date Rec'd.	Sent to Subs.	Date Due	Date Rec'd.	Amount	Sent to owner/Arch.	Date Due	Date Rec'd	Amount		Comments
04	Add 2'-0" to Retaining Wall	2/28	2/28									
	Quality Masonry		2/28	3/14	3/10	2979						
	Lot's of Rebar		2/28	3/14	3/11	400						
	G.C. Work		2/28	3/14	3/12	600						
	G.C. Fee		2/28	3/14	3/2	398						
						4377	3/13	3/26				
05	Custom Elevator Cab											
	Up and Down Elevator Co.	2/28	2/28	3/14	3/1	600						
	Western Mill	2/28	2/28	3/14	3/10	3600						
	Sparks Electric	2/28	2/28	3/14								
	G.C. Work / Fee			3/14								
06	Brass Handrails	2/28	2/28									
	Smith and Smith		2/28	(3/14)	3/18	5687						
	G.C. Work/Fee		2/28	(3/14)	3/19	569						
						6256	3/20	4/2				

circled = Late

Figure 19.3

DELTA CONSTRUCTION COMPANY

PaperWorks Version 1.23F

*** PRICE REQUEST LOG ***

.PR1

Job 321 GIBSON OFFICE BUILDING

Date: 03/20/90

Page: 2

pr #: 004 sent: 02/28/90 method: DESCRIBED due: 03/14/90

summary: ADD 2'0" TO RETAINING WALL AT SOUTH END OF PARKING LOT

description: Price the addition of 2'-0" of retaining wall at the south end of the parking lot. Add 3' x 5' sidewalk section per Architect's plans dated Feb 21.

rem to sub: Quality, if this changes the footing, price the footing also.

sent to:	date returned	amount quoted
QUALITY MASONRY	03/10/90	$ 2,979.00
LOT'S OF REBAR	03/11/90	$ 400.00
GENERAL CONTRACTOR WORK	03/12/90	$ 600.00
GENERAL CONTRACTOR FEE	03/12/90	$ 398.00
	sub total:	$ 4,377.00

date due owner	amount quoted	date returned	amount approved
03/26/90	$ 4,377.00		$

rem to owner: We will be starting the retaining wall in three weeks and therefore the price is only good until April 2.

---

pr #: 005 sent: 02/28/90 method: ATTACHED due: 03/14/90

summary: PRICE ELEVATOR CUSTOM CAB

description: Price custom elevator cab for the three passenger elevators, per the attached design from One Line Design SK E-1, E-2, E-3. Cabs to be customized on site.

rem to sub: Value engineering suggestions are welcome.

sent to:	date returned	amount quoted
UP AND DOWN ELEVATOR CO	03/01/90	$ 600.00
WESTERN MILL	03/10/90	$ 3,600.00
SPARKS ELECTRIC COMPANY	NOT QUOTED	$ 0.00
	sub total:	$ 4,200.00

---

pr #: 006 sent: 02/28/90 method: DESCRIBED due: 03/14/90

summary: PRICE FOR BRASS HANDRAILS

description: Price substitution of brass handrails for iron handrails at stair # 3 from the first to second floor at the lobby area. See spec's attached from the architect.

rem to sub: Let me know about delivery time.

sent to:	date returned	amount quoted
SMITH AND SMITH HARDWARE	03/18/90	$ 5,687.00
GENERAL CONTRACTOR FEE	03/19/90	$ 569.00
	sub total:	$ 6,256.00

date due owner	amount quoted	date returned	amount approved
04/02/90	$ 6,256.00		$

rem to owner: Deduct $1,250 if attached alternate detail is acceptable.

Figure 19.4a

```
                         DELTA CONSTRUCTION COMPANY
==================================================PaperWorks Version 1.23F==
                *** PRICE REQUEST LOG DUE FROM SUBS ***                  .PR3
           Job 321      GIBSON OFFICE BUILDING
Date: 03/20/90                                                    Page:    1
=============================================================================
 pr #: 003          sent: 02/21/90      method: ATTACHED         due: 03/07/90
    summary: PRICE ADDITION OF EXIT DOOR AT CORRIDOR #104
description: Price all work in connection with the addition of an exit door at
             the south end of corridor #104 per the attached drawings and
             price request description from Oneline Design.

        sent to:                  not responded:
       SMOOTH                  SMOOTH DRYWALL, INC.
       CONTHDWE                CONTINENTIAL HARDWARE
-----------------------------------------------------------------------------

 pr #: 005          sent: 02/28/90      method: ATTACHED         due: 03/14/90
    summary: PRICE ELEVATOR CUSTOM CAB
description: Price custom elevator cab for the three passanger elevators, per
             the attached design from One Line Design SK E-1, E-2, E-3.  Cabs
             to be customized on site.

        sent to:                  not responded:
       SPARKS                  SPARKS ELECTRIC COMPANY
-----------------------------------------------------------------------------

 pr #: 007          sent: 03/07/90      method: DESCRIBED        due: 03/21/90
    summary: PRICE CHANGE OF EXTERIOR GLASS COLOR
description: Price change of exterior glass color and specs from graylite 40 t
             High Performance PPG # 885, medium blue.

        sent to:                  not responded:
       CLEAR                   CLEAR GLASS COMPANY
-----------------------------------------------------------------------------

 pr #: 008          sent: 03/07/90      method: ATTACHED         due: 03/21/90
    summary: PRICE UPGRADE OF LANDSCAPING PLANS
description: Price upgrade of landscaping per plans prepared by Growbig
             Landscape Architects,  Sheet L-1 through L-10 dated Feb. 28,
             which are attached.

        sent to:                  not responded:
       GREEN                   GREEN LANDSCAPING
-----------------------------------------------------------------------------

 pr #: 009          sent: 03/07/90      method: ATTACHED         due: 03/21/90
    summary: PRICE LOBBY REVISIONS
description: Price millwork lobby revisions per attacted sheets M-1 to M-5 as
             prepared by One Line Design dated Feb 22.

        sent to:                  not responded:
       WESTERN                 WESTERN MILL
       ABC                     ABC PAINTING COMPANY
-----------------------------------------------------------------------------
```

Figure 19.4b

```
                          DELTA CONSTRUCTION COMPANY
==============================================PaperWorks Version 1.23F==
       *** PRICE REQUEST LOG DELINQUENT FROM SUBS as of 03/20/90  *      .PR5
           Job 321      GIBSON OFFICE BUILDING
Date: 03/20/90                                                     Page:   1
=============================================================================
 pr #: 003          sent: 02/21/90      method: ATTACHED         due: 03/07/90
    summary: PRICE ADDITION OF EXIT DOOR AT CORRIDOR #104
description: Price all work in connection with the addition of an exit door at
             the south end of corridor #104 per the attached drawings and
             price request description from Oneline Design.
 rem to sub: Price as an alternate a double rather than a single exit door.

       sent to:
       SMOOTH         SMOOTH DRYWALL, INC.
       CONTHDWE       CONTINENTIAL HARDWARE
-----------------------------------------------------------------------------
 pr #: 005          sent: 02/28/90      method: ATTACHED         due: 03/14/90
    summary: PRICE ELEVATOR CUSTOM CAB
description: Price custom elevator cab for the three passenger elevators, per
             the attached design from One Line Design SK E-1, E-2, E-3.  Cabs
             to be customized on site.
 rem to sub: Value engineering suggestions are welcome.

       sent to:
       SPARKS         SPARKS ELECTRIC COMPANY
-----------------------------------------------------------------------------
```

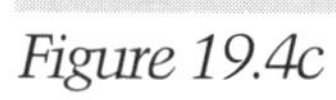

Figure 19.4c

```
                        DELTA CONSTRUCTION COMPANY
=============================================PaperWorks Version 1.23F==
              *** PRICE REQUEST QUOTE TO OWNER LOG ***                .PQ1
         Job 321      GIBSON OFFICE BUILDING
Date: 04/13/90                                                  Page:    1
==========================================================================
 pr #: 001           sent: 02/21/90       method: DESCRIBED       due: 03/07/90
    summary: PRICE UPGRADE IN CARPET
description: Price upgrade in carpet from XYZ Grade 4 to ABC Carpet Mill, Grade
             7, Color steel gray, at lobby # 101 and corridor #104.
             Price upgrade of carpet pad from 3/8" to "5/8", same specificatio
 rem to sub: Provide a color selection book in case owner wants to change colo

 date due owner   amount quoted    date ret by owner  amount approved by owner
    03/22/90    $        52,800.00      04/02/90          $

remarks: This carpet is a long lead item.  Please decide as soon as you can
        if you want to make this change.
                                             STATUS:    PRICE REQUEST QTO REVISED
--------------------------------------------------------------------------

 pr #: 002           sent: 02/21/90       method: ATTACHED        due: 03/07/90
    summary: PRICE ENTRY SIGN AND MONUMENT
description: Price entry sign and monument based on the architect's landscape
             sketches, SK L-1, 2, 3.

 rem to sub: Price both chrome and brass letters

 date due owner   amount quoted    date ret by owner  amount approved by owner
    03/22/90    $         4,463.00      04/12/90          $          4,463.00

remarks: If you want brass letters add $1,800 for this price request.

                                             STATUS:    PRICE REQUEST QTO APPROVED
--------------------------------------------------------------------------

 pr #: 004           sent: 02/28/90       method: DESCRIBED       due: 03/14/90
    summary: ADD 2'0" TO RETAINING WALL AT SOUTH END OF PARKING LOT
description: Price the addition of 2'-0" of retaining wall at the south end of
             the parking lot.  Add 3' x 5' sidewalk section per Architect's
             plans dated Feb 21.
 rem to sub: Quality, if this changes the footing, price the footing also.

 date due owner   amount quoted    date ret by owner  amount approved by owner
    03/26/90    $         4,377.00                        $

remarks: We will be starting the retaining wall in three weeks, there-
        fore, the price is only good until April 2.
                                                 STATUS:    PENDING Q.T.O.
--------------------------------------------------------------------------

 pr #: 006           sent: 02/28/90       method: DESCRIBED       due: 03/14/90
    summary: PRICE FOR BRASS HANDRAILS
description: Price substitution of brass handrails for iron handrails at stair
             # 3 from the first to second floor at the lobby area.  See specs
             attached from the architect.
 rem to sub: Let me know about delivery time.
```

Figure 19.5a

```
                        DELTA CONSTRUCTION COMPANY
═══════════════════════════════════════════════PaperWorks Version 1.23F══
              *** PRICE REQUEST LOG DUE FROM OWNER ***                .PQ3
         Job 321       GIBSON OFFICE BUILDING
Date: 04/13/90                                                  Page:    1
═════════════════════════════════════════════════════════════════════════
 pr #: 004          sent: 02/28/90      method: DESCRIBED      due: 03/14/90
    summary: ADD 2'0" TO RETAINING WALL AT SOUTH END OF PARKING LOT
description: Price the addition of 2'-0" of retaining wall at the south end of
             the parking lot.  Add 3' x 5' sidewalk section per Architect's
             plans dated Feb 21.

   date due owner 03/26/90     amount quoted $        4,377.00
remarks: We will be starting the retaining wall in three weeks, there-
         fore, the price is only good until April 2.
─────────────────────────────────────────────────────────────────────────

 pr #: 008          sent: 03/07/90      method: ATTACHED       due: 03/21/90
    summary: PRICE UPGRADE OF LANDSCAPING PLANS
description: Price upgrade of landscaping per plans prepared by Growbig
             Landscape Architects,  Sheet L-1 through L-10 dated Feb. 28,
             which are attached.

   date due owner 04/09/90     amount quoted $       15,147.00
remarks: This quotation is good for 30 calendar days.

─────────────────────────────────────────────────────────────────────────

 pr #: 010          sent: 03/14/90      method: DESCRIBED      due: 03/28/90
    summary: PRICE REVISION OF CERAMIC TILE IN RESTROOMS
description: Price revision of ceramic tile in both public restrooms from
             DULL # 555 to BRIGHT # 886 and change size from 4 x 4 to 6 x 6
             inches.

   date due owner 04/23/90     amount quoted $        1,980.00
─────────────────────────────────────────────────────────────────────────

 pr #: 014          sent: 03/21/90      method: DESCRIBED      due: 04/04/90
    summary: PRICE REVISION OF BUILDING STANDARD DOORS
description: Price the change of the building standard doors from paint grade
             solid core to stain grade solid core doors.
             Delete paint and add stain and 2 coat varnish on doors.

   date due owner 04/23/90     amount quoted $        4,400.00
─────────────────────────────────────────────────────────────────────────
```

Figure 19.5b

```
                          DELTA CONSTRUCTION COMPANY
==============================================PaperWorks Version 1.23F==
      *** PRICE REQUEST LOG DELINQUENT FROM OWNER as of 04/13/90 ***      .PQ5
           Job 321      GIBSON OFFICE BUILDING
Date: 04/13/90                                                    Page:    1
=========================================================================
 pr #: 004           sent: 02/28/90      method: DESCRIBED       due: 03/14/90
    summary: ADD 2'0" TO RETAINING WALL AT SOUTH END OF PARKING LOT
description: Price the addition of 2'-0" of retaining wall at the south end of
             the parking lot.  Add 3' x 5' sidewalk section per Architect's
             plans dated Feb 21.

  date due owner 03/26/90     amount quoted $        4,377.00
 remarks: We will be starting the retaining wall in three weeks, there-
          fore, the price is only good until April 2.

-------------------------------------------------------------------------

 pr #: 008           sent: 03/07/90      method: ATTACHED        due: 03/21/90
    summary: PRICE UPGRADE OF LANDSCAPING PLANS
description: Price upgrade of landscaping per plans prepared by Growbig
             Landscape Architects,  Sheet L-1 through L-10 dated Feb. 28,
             which are attached.

  date due owner 04/09/90     amount quoted $       15,147.00
 remarks: This quotation is good for 30 calendar days.

-------------------------------------------------------------------------
```

Figure 19.5c

Chapter Twenty

Field Orders and Quotes to Owner

The Price Request (PR) and the Price Request Quote to Owner (PRQTO) were identified previously as a "Tell me how much it will cost, and then I will tell you if I want you to do it" process. The concept of Field Orders is: "This has to be done. Do it and *then* tell me how much it is going to cost."

Purpose

The Field Order is a directive, issued in the field, to perform certain work immediately, without the owner, architect, general contractor, or subcontractor knowing how much it will cost. Generally, Field Orders are used when time is not available to go through the Price Request process.

In some cases, the owner/architect may insist that the general contractor provide a not-to-exceed, or budget, number to put a "cap" on the individual Field Order cost at the time the Field Order is issued. All parties should be wary of trying to "pin" the subcontractor or general contractor to a specific price on a Field Order. By definition, if there is enough time to obtain a specific and accurate price for the work to be accomplished, then a Field Order should not be issued. A Price Request should be issued instead. Field Orders should be used when no one (contractor, architect, or owner) had any idea that the subject would ever arise.

Contractors and subcontractors should be careful about issuing budget numbers and not-to-exceed prices. Owners and architects should not force any manager to give them a dollar figure and then hold them to that number without allowing them enough time to do the required homework. The following example is a case in point.

A project manager, under pressure from the owner, gave a hasty Not-To-Exceed (N-T-E) price on the Field Order. The scope of work required to complete the work was mis-estimated in the Field Order. When the Field Order was finally quoted, it was significantly higher than the N-T-E number. The owner said he would not pay more than the N-T-E amount stated on the Field Order. After many harsh words, the project manager agreed to absorb the cost of the estimating error in the first Field Order.

It happened that another Field Order was needed that same day and, when the owner forced him to produce another not-to-exceed number, this time, the project manager gave an astronomically high figure. He also assured the owner that the cost of the work would come close to that number (no matter how much anyone else thought the work would cost). The owner finally got the point and allowed the previous

Field Order to be set at the increased, realistic price. The project manager than issued a new, more "in-line" number for the second Field Order.

Directive

Many architects and engineers feel that a Field Order should not involve any costs, but should only serve as clarification, or amplification, of the contract documents. As a result, they prefer to use the term, "Directive," for an order that involves a change in the contract scope of work.

The purpose of a Directive is to "direct" an individual or firm to perform work that differs from that described in the contract documents. If everyone on the project agrees to use the "Directive," and it has been discussed in the pre-construction meeting, then this document can be substituted for the Field Order as described in this book. One or the other is appropriate, but not both.

Authority to Sign Field Orders

One of the most important things to accomplish in the pre-construction meeting is to determine who has the authority to sign Field Orders and for how much money. The final responsibility is, of course, the owner's (or user's). Sometimes an owner/user authorizes the architect to sign Field Orders up to a specified limit. It is also important to determine whether telephone authorization is acceptable. *Be sure to act only on approved signatures.*

Managers should make an effort on every job to have the owner or architect sign a Field Order at the first job meeting. Chances are, a review of the plans will reveal something that needs immediate action. Make it routine for the participants to sign from the very beginning so that when the going gets tough, they are accustomed to the procedures for authorizing changes in the scope of the work.

Sample Field Orders

The dialogue leading up to a Field Order or Directive might be as follows:

Owner: *"Please change the color of the living room from blue to pink."*

Contractor: *"I've already bought the blue paint, so there will be a restocking charge."*

Owner: *"How much?"*

Contractor: *"10%, that figures out to be $25.00 plus delivery charges of another $25.00....totalling $50.00."*

Owner: *"O.K."*

Contractor: *"Fine. Please sign this Field Order for $50.00, and I'll include it with my next monthly invoice."*

Most Field Orders are, in fact, more involved than the previous example. Usually, the instructions to proceed with the work are issued. Then each subcontractor and vendor quotes a price to be included in the final price given to the owner for the cost of the work (which may now be in progress or already completed). Being timely with quotes and getting the owner to respond promptly in signing the Field Order as documentation is important in the case of all of the pricing vehicles–but especially in the case of Field Orders.

Another example follows. It is a telephone call from the job superintendent to the architect.

General Contractor: *"Hi, Bob. The building inspector was just here and red-tagged the footing that we're planning to pour this afternoon. He says that it should be 12" deeper. I can get the guys busy and get it done, but we never figured it that way. I sure don't want to miss this pour because I want to be ready to pour the slab on Friday."*

Architect: *"How much do you think it's going to cost?....Never mind, it needs to be done. I'll sign a Field Order the next time I'm there, or at the next job meeting. What was the name of the inspector? I want to check this out with him."*

General Contractor: *"His name is Charlie Smith, but we really can't wait any longer."*

Architect: *"You're right, but I want to make sure this doesn't happen again. I'll have Jerry (the structural engineer) look at all the footings again. Maybe we can add more steel rather than going deeper with the footing. In either case, we'll issue a Price Request for that work."*

General Contractor: *"You had better make that a Field Order for today's pour, and then issue a Price Request for the other footings after you find out what you need to know."*

Architect: *"You're right, I guess. I hate to do the extra paperwork, and I really hate getting 'boxed in' like this."*

General Contractor: *"I know how you feel. I had all those laborers working on Building B, and now I'm going to get behind on that one. Let me know what you find out. Thanks for the Field Order. I sure didn't want to cancel that concrete order. Get me the revised Building B footings as soon as you can...Bye."*

Authorization for instructions issued in the field is required, even if the price of the work cannot be provided at that time. The authorization to proceed must be signed.

If the participants can establish good working relationships (this is known as *trust*, and it works both ways), immediate, signed authorization may not be required. If you have a good working relationship, the architect/owner should not be reluctant to sign the next time he is on the job.

Field Orders and Directives are often issued and approved orally. This may be unavoidable in many cases. However, it is worth remembering the axiom, "no trust without testing."

The next step is to get the completed Field Order Quote to Owner returned as soon as possible. In this case, ten working days would be acceptable. Many contractors wait until the end of the job to submit this quote. At that point, it is a bit of a surprise and everyone starts trying to assign the blame. It is much better to get those prices out into the open early. The problem (for everyone) is that it may take several weeks or even months to get the pricing information back to the owner/architect. By that time, no one remembers that this is actually additional work that was not originally included in the scope of the contract.

The manager can make it a practice to send the owner a copy of the Field Order letter at the same time that it is sent to the subcontractors and vendors. This procedure documents the fact that the work was requested and authorized, even though it was not priced at the time of authorization.

Tracking Field Orders

Using logs for Field Orders helps to keep track of the costs that have not yet been documented. The use of the Exception Logs, showing the items that have not yet been quoted, is especially effective.

Many firms group all of the pricing vehicles together. Price Requests may be filed with Field Orders, Clarifications, and Request for Changes. This is not a good policy, since each of these documents has a particular use. Furthermore, keep in mind the importance of the KISS (Keep It Simple, Silly) principle. It is easier for the owner (or user)/architect to grasp these different concepts one type at a time, rather than jumbled together.

It is easy for the project manager and the subcontractors to understand the difference between these four types of scope changes, because they work with them every day. They are pretty good at it or they would not be in business. On the other hand, most owners of projects do not build new buildings every day, and are not familiar with the documentation. Additionally, the owner probably has quite a lot at stake. Take good care of him/her. Good owners are hard to come by.

It is important to remember that it is not necessary to use all of the different types of pricing letters on every job. Some jobs do not lend themselves to the use of Price Requests, for example, because of the time required from inception to completion. Yet on other jobs, such as government, commercial, and institutional work, all the changes in scope may be required to go through the Price Request process. This is one of the reasons government work is often more costly and time-consuming.

The procedures of each particular company may determine the system used for Field Order and Field Order Quote to Owner identification. An internal system can be set up to handle all four of the different types of pricing correspondence since, sooner or later, all four types will undoubtedly be required on the same job.

Again, each company has its own requirements and priorities, in the same way that each job is different. When the parties to the construction process adjust to using these four formats, their advantages will become evident and their use will spread. Each type of letter represents a unique function and requires a unique response. The following pages are samples of Field Orders to subcontractors/vendors. Figure 20.1a is computer-generated and Figure 20.1b is a manual Field Order to subcontractors/vendors using the standard Transmittal form. Both do the job; the advantage of the computer-generated system is that it maintains a consistent format, and automatically enters the information in the logs.

The following information should be included in the Field Order letter to subcontractors.

- From
- To
- Job identification
- Field Order number and identification
- Date Field Order given
- Date sent to subcontractors, vendors
- Short description of the work described in the Field Order

- A note if the Field Order has a long description or is accompanied by plans or sketches
- Special note to subcontractor or vendor, if necessary

Purpose of Field Order Quote to Owner

The Field Order Quote to Owner (FOQTO) is the response to the original Field Order as signed or authorized by the owner/architect. One of the items usually addressed in the FOQTO is an extension of time.

If an extension of time in the completion date of the project is needed as a result of the Field Order work, by all means include one. Many times, the owner/architect will not agree and will say that an extension of contract time is unnecessary. However, it is still important that the extension of time was *requested* for a change in the scope of the project. Should project-related issues ever go to court, documentation of such a request could be very meaningful. At that point, the general contractor can say that the owner/architect had him "over a barrel" in that if he insisted on an extension of time, the owner might have delayed paying for the monthly progress billing. Keep including those extension of time requests....it is not wasted effort.

Key Information for the Field Order Quote to Owner

The key items to be included in the FOQTO are:

- From
- To
- Date sent
- Job identification
- Field Order number and identification
- Brief description of work involved in Field Order
- Date decision is required from owner
- The amount of the quotation
- The request for extension of time, if required
- Any special notes to the owner
- Breakdown of costs, depending on the contractual arrangement

Included in the FOQTO is a space for the owner's signature, an indication that he/she approves the price of the Field Order. Make it easy for the owner. It may be possible to eliminate the usual Change Order process by using the FOQTO authorization as a Change Order. This eliminates one whole sequence in the authorization process. Additionally, it is less stressful for the owner/user to sign one Field Order Quote to Owner at a time, rather than one large Change Order consisting of many different items.

One institutional administrator boasts that he did a 12 million dollar job with just two Change Orders. One of the Change Orders was for 2.5 million dollars, and one was for 1.4 million dollars, and each was comprised of virtually hundreds of different items. That was not much to brag about. The amount of confusion and headaches caused by those overstuffed Change Orders offset the benefits of the reduction in paperwork, and in all probability, cost more money because the contractors involved had to wait so long for their money that they increased their "quote" for the change work to be accomplished.

DELTA CONSTRUCTION COMPANY

6065 Mission Gorge Road, #193
San Diego, CA 92120
(619) 582-5829

*** FIELD ORDER ***

Date: 2/28/90

Job Number: 321
Field Order Number: 006

TO: ROCKY CONCRETE CONTRACTORS
451 Gravel Road
San Diego, CA 92111

ATTENTION: C. Ment

JOB NUMBER: 321 GIBSON OFFICE BUILDING

FROM: J. EDWARD GRIMES

SUBJECT: DECREASE SLOPE OF RAMP TO GARAGE

We are transmitting to you the following which is DESCRIBED BELOW for your review and RESPONSE. Please refer to Job Number & Field Order Number. Please review and return this information to us by 3/14/90.

Decrease slope of ramp to garage from parking lot. Extend concrete ramp 20'. Add masonry, waterproofing, rebar. This will require realignment of approach to garage.

Please contact me as soon as possible if you need any further information or if you have any questions.

Sincerely,

J. Edward Grimes

Copies were sent to the following:

QUALITY MASONRY
LOT'S OF REBAR
GENERAL CONTRACTOR WORK

dp: 2/28/90 16:47 **PaperWorks,** a Construction Software Program

Figure 20.1a

Means Forms

PROPOSAL

FROM: Delta Construction
6065 Mission Gorge Rd
San Diego, Ca. 92120

TO: Rocky Concrete Contractors
451 Gravel Rd.
San Diego, Ca. 92111

PROPOSAL NO. F.O. - 321.01
DATE 2/28/89
PROJECT Gibson Office Bldg. #321
LOCATION San Diego, Ca.

CONSTRUCTION TO BEGIN

COMPLETION DATE

Gentlemen:

The undersigned proposes to furnish all materials and necessary equipment and perform all labor necessary to complete the following work:

Decrease slope of ramp, to garage from parking lot. Extend concrete ramp 20'. Add masonry, waterproofing, rebar. This will require realignment of approach to garage.

Please contact me as soon as possible if you need further information or have any questions.

All of the above work to be completed in a substantial and workmanlike manner

☐ for the sum of ______ dollars ($______)

☐ to be paid for at actual cost of Labor, Materials and Equipment plus ______ percent (______%)

Payments to be made as follows: ______

______ The entire amount of the contract to be paid within ______ after completion

Any alteration or deviation from the plans and specifications will be executed only upon written orders for same and will be added to or deducted from the sum quoted in this contract. All additional agreements must be in writing.

The Contractor agrees to carry Workmen's Compensation and Public Liability Insurance and to pay all taxes on material and labor furnished under this contract as required by Federal laws and the laws of the State in which this work is performed.

Respectfully submitted,

Contractor Delta Construction

By J. Edward Grimes

ACCEPTANCE

You are hereby authorized to furnish all material, equipment and labor required to complete the work described in the above proposal, for which the undersigned agrees to pay the amount stated in said proposal and according to the terms thereof.

Date ______ 19____ X ______

Figure 20.1b

Sample Field Orders

Figure 20.2a is a sample of a computer-generated Field Order Quote to Owner. Figure 20.2b is a sample of a manual FOQTO, using the standard Transmittal form.

It is important for everyone involved in a project (owner, architect, general contractor, and subcontractors) to remember these points about Field Orders:

1. They are usually unexpected and involve a change in scope of the work.
2. They usually interfere with the planned course of construction, either physically or mentally.
3. They usually cost more per unit than the original estimate.
4. Because the item that caused the Field Order usually comes as a surprise, it does not do a lot of good for any of those involved to complain about the cost. If it is really necessary to do it that way, the work simply has to be done. If too many hassles result, then the general contractor may not handle the next "Field Order situation" in the best interests of the project. He may instead choose to cancel the concrete truck, and the job could be delayed two weeks while the architect issues a Price Request and a firm price is determined for the owner. Most people want to do a good job with as few problems as possible. If their word is constantly questioned, they are likely to become obstinate and do not communicate well. Clearly, getting other people to do their best work is extremely important.

Timely Response

Earlier in this chapter, timely responses were discussed. This concept is so important for Field Orders that it is raised again here. Get the Field Order Quote to Owner back to the owner (or user)/architect promptly in order to get on the list of payables. It is especially important that Field Order responses be timely, as the very reason for the Field Order was an unanticipated requirement. No one likes surprises that cost them money, and they tend to forget about the unplanned expenses.

DELTA CONSTRUCTION COMPANY

6065 Mission Gorge Road, #193
San Diego, CA 92120
(619) 582-5829

*** FIELD ORDER QUOTE TO OWNER ***

Date: 4/25/90

Job Number: 321
Field Order Number: 019

TO: FRANCES GIBSON
6109 CAMINITO CLAVO
SAN DIEGO, CA 92120

ATTENTION: FRANCES GIBSON

JOB NUMBER: 321 GIBSON OFFICE BUILDING

FROM: J. EDWARD GRIMES

SUBJECT: CHANGE LATCH SETS TO LOCKSETS

We are submitting the following changes in cost of $ 880.00
The details and breakdown of this change in cost is enclosed.

A complete description of the scope of work relating to this proposal is enclosed. The summary of the work is as follows:

Change the following doors from latchsets to locksets, Hardware Group 88-44-22, with master key function: 111, 112 118 120 and 221 222 228 and 240.

Remarks to owner/architect:

If the locksets don't get here in time for TCO we can install them later. They are not required for Occupancy Permit.

This work will result in an extension to our contract time of 3 days.

Please indicate your approval by signing and returning one copy of this letter to our office or issue a CHANGE ORDER for this work no later than 5/09/90 in order to not delay the project.

Sincerely

J. EDWARD GRIMES

APPROVED BY: FRANCES GIBSON

SIGNED: ______________________ DATE: __/__/__

dp: 3/21/90 16:50 **PaperWorks, a Construction Software Program**

Figure 20.2a

Means Forms

PROPOSAL

Delta Construction Company
6678 Mission Road
San Diego, CA 92120
(619) 581-6315

FROM: J. Edward Grimes
Project Manager

PROPOSAL NO. FOQ 321.19
DATE 4/25/90
PROJECT Gibson Office Bldg #321
LOCATION San Diego, Ca.

TO: Frances Gibson
6109 Caminito Clavo
San Diego, Ca. 92120

CONSTRUCTION TO BEGIN

COMPLETION DATE

Gentlemen:

The undersigned proposes to furnish all materials and necessary equipment and perform all labor necessary to complete the following work:

Change the following doors from latchsets to locksets, Hardware Group 88-44-22, with master key function: 111, 112, 118, 120, 221, 222, 228, and 240

This work will result in an extension to our contract time of three (3) days.

All of the above work to be completed in a substantial and workmanlike manner

☒ for the sum of Eight-hundred and eighty dollars ($ 880.00)

☐ to be paid for at actual cost of Labor, Materials and Equipment plus ______ percent (______ %)

Payments to be made as follows: ______

______ The entire amount of the contract to be paid within ______ after completion

Any alteration or deviation from the plans and specifications will be executed only upon written orders for same and will be added to or deducted from the sum quoted in this contract. All additional agreements must be in writing.

The Contractor agrees to carry Workmen's Compensation and Public Liability Insurance and to pay all taxes on material and labor furnished under this contract as required by Federal laws and the laws of the State in which this work is performed.

Respectfully submitted,

Contractor Delta Construction Co.

By J Edward Grimes

ACCEPTANCE

You are hereby authorized to furnish all material, equipment and labor required to complete the work described in the above proposal, for which the undersigned agrees to pay the amount stated in said proposal and according to the terms thereof.

Date ______ 19____

Figure 20.2b

Logs

The Field Order logs show all of the subcontractor and general contractor costs and fees. The Field Order Quote to Owner, on the other hand, shows only the price quoted to the owner. The subcontractors should never see the FOQTO logs, and the owner should never see the Price Request logs with subcontractor pricing information.

Figure 20.3a is a manually-generated Field Order log and Figure 20.3b is a computer-generated Field Order log. Figures 20.3c and 20.3d are examples of a Field Order Due log and a Field Order Delinquent log, respectively.

Figure 20.4a is a computer-generated Quote to Owner log and Figures 20.4b and 20.4c are a Field Order Quote to Owner Due log and a Field Order Quote to Owner Delinquent log.

PROJECT Gibson Office Bldg. SHEET 1 OF —

LOCATION San Diego, Ca. JOB NO. 321

	To/Subject	Date Rec'd	Sent to Subs	Date Due	Date Rec'd	Amount	Sent to Owner/Arch.	Date Due	Date Rec'd	Amount		Comments
06	Decrease of Slope @ Ramp	2/22	2/28	3/14								
	Quality Masonry		2/28	3/14	3/19	1100						
	Rocky Concrete		2/28	3/14	3/19	1890						
	Lot's of Rebar		2/28	3/14	3/19	765						
	G.C. Work/Fees		2/28	3/14	3/21	375						
						4130	3/21	4/4 (circled)	4/11	4130		Approved
07	Ground Breaking Party Prep.	3/7	–	–								
	G.C. Work		–	4/11	3/28	1557						
	G.C. Fee		–	4/11	3/28	156						
						1713	3/28	4/11 (circled)				
08	Add Fire Hydrant											
	Quick Plumbing	3/7	3/7	3/21 (circled)								
	G.C. Work/Fee	3/7	3/7	3/21 (circled)								

circled = Late

Figure 20.3a

```
                         DELTA CONSTRUCTION COMPANY
===================================================PaperWorks Version 1.23F==
                          *** FIELD ORDER LOG ***                        .FO1
         Job 321          GIBSON OFFICE BUILDING
Date: 04/14/90                                                    Page:   2
=============================================================================
 FO #: 006           sent: 02/28/90      method: DESCRIBED       due: 03/14/90
    summary: DECREASE SLOPE OF RAMP TO GARAGE
description: Decrease slope of ramp to garage from parking lot.  Extend
             concrete ramp 20'.  Add masonry, waterproofing, rebar.  This will
             require realignment of approach to garage.
 rem to sub:

       sent to:                          date returned     amount quoted
    QUALITY MASONRY                        03/19/90     $         1,100.00
    ROCKY CONCRETE CONTRACTORS             03/19/90     $         1,900.00
    LOT'S OF REBAR                         03/19/90     $           765.00
    GENERAL CONTRACTOR WORK                03/21/90     $           375.00
                                           subtotal:    $         4,130.00

   date due owner     amount quoted     date returned       amount approved
     04/04/90       $        4,130.00      04/11/90        $          4,130.00

remarks:


                    STATUS:   FIELD ORDER APPROVED
-----------------------------------------------------------------------------
 FO #: 007           sent: 03/07/90      method: DESCRIBED       due: 03/21/90
    summary: PREPARE AREA FOR GROUND-BREAKING PARTY
description: Partially grade and brush area for parking.  Clean up.  Provide
             extra Port-a-potty and parking assistance for ground breaking
             party
 rem to sub:

       sent to:                          date returned     amount quoted
    GENERAL CONTRACTOR WORK                03/28/90     $         1,557.00
    GENERAL CONTRACTOR FEE                 03/28/90     $           156.00
                                           subtotal:    $         1,713.00

   date due owner     amount quoted     date returned       amount approved
     04/11/90       $        1,713.00                      $

remarks:


                    STATUS:   FIELD ORDER APPROVED
-----------------------------------------------------------------------------
 FO #: 008           sent: 03/07/90      method: DESCRIBED       due: 03/21/90
    summary: ADD FIRE HYDRANT AND 120' 6" LINE
description: Add one fire hydrant and 120' of 6" fire line as required by City
             Fire Marshall on evaluation of building occupancy and use.  Work
             not shown on bid plans or specifications.
 rem to sub:

       sent to:                          date returned     amount quoted
    QUICK PLUMBING COMPANY                 NOT QUOTED   $             0.00
    GENERAL CONTRACTOR WORK                NOT QUOTED   $             0.00
                                           subtotal:    $             0.00

-----------------------------------------------------------------------------
```

Figure 20.3b

```
                  DELTA CONSTRUCTION COMPANY
==============================================PaperWorks Version 1.23F==
               *** FIELD ORDER LOG DUE FROM SUBS ***               .FO3
          Job 321     GIBSON OFFICE BUILDING
Date: 04/14/90                                                Page:   2
========================================================================
 FO #: 012          sent: 03/14/90     method: DESCRIBED      due: 03/28/90
    summary: REVISE EXIT HARDWARE
description: Revise exit hardware at door 1001 from hardware group 11-98 to
             hardware group 11-98-8888.

 rem to sub:

        sent to:
       CONTHDWE       CONTINENTAL HARDWARE
       GFEE           GENERAL CONTRACTOR FEE
------------------------------------------------------------------------

 FO #: 013          sent: 03/21/90     method: DESCRIBED      due: 04/04/90
    summary: ADD RESTROOM SIGNS
description: Add international restroom signs to all restrooms.  Color to be
             silver on black plastic in approved sizes and shapes.

 rem to sub:

        sent to:
       SSHDWE         SMITH AND SMITH HARDWARE
------------------------------------------------------------------------

 FO #: 017          sent: 03/28/90     method: DESCRIBED      due: 04/11/90
    summary: CLEAN UP AREA FOR RIBBON CUTTING
description: Clean up general work area and grounds for ribbon cutting
             party.  Project not to complete for 14 more working days.  Work
             out of sequence to complete desired areas
 rem to sub:

        sent to:
       GCW            GENERAL CONTRACTOR WORK
       ABC            ABC PAINTING COMPANY
       SMOOTH         SMOOTH DRYWALL, INC.
       SHAGGY         SHAGGY CARPET COMPANY
------------------------------------------------------------------------

 FO #: 020          sent: 04/04/90     method: DESCRIBED      due: 04/18/90
    summary: CHANGE SIZE OF SPECIMEN TREES
description: Change size of boxed trees from 36" box to 48" box (18) along
             El Cajon Blvd, and from 15 gallon to 24" box (12) along
             54th Street.
 rem to sub: Your field rep has been advised of this, the trees are available.

        sent to:
       GREEN          GREEN LANDSCAPING
------------------------------------------------------------------------
```

Figure 20.3c

```
                         DELTA CONSTRUCTION COMPANY
==========================================================PaperWorks Version 1.23F==
      *** FIELD ORDER LOG DELINQUENT FROM SUBS as of 04/14/90  *          .FO5
          Job 321      GIBSON OFFICE BUILDING
Date: 04/14/90                                                       Page:    2
=================================================================================
 FO #: 009            sent: 03/14/90      method: DESCRIBED       due: 03/28/90
    summary: ADD MIRRORS AT RESTROOMS
description: Add 3x8' mirrors at women's restrooms and 3x6' mirrors at men's


             restrooms.  Mirrors not show on bid set of plans and specifications


             at time of bid.
 rem to sub: Submit edge moldings available

       sent to:
      CLEAR          CLEAR GLASS COMPANY
---------------------------------------------------------------------------------

 FO #: 012            sent: 03/14/90      method: DESCRIBED       due: 03/28/90
    summary: REVISE EXIT HARDWARE
description: Revise exit hardware at door 1001 from hardware group 11-98 to
             hardware group 11-98-8888.

 rem to sub:

       sent to:
      CONTHDWE       CONTINENTAL HARDWARE
      GFEE           GENERAL CONTRACTOR FEE
---------------------------------------------------------------------------------

 FO #: 013            sent: 03/21/90      method: DESCRIBED       due: 04/04/90
    summary: ADD RESTROOM SIGNS
description: Add international restroom signs to all restrooms.  Color to be
             silver on black plastic in approved sizes and shapes.

 rem to sub:

       sent to:
      SSHDWE         SMITH AND SMITH HARDWARE
---------------------------------------------------------------------------------

 FO #: 017            sent: 03/28/90      method: DESCRIBED       due: 04/11/90
    summary: CLEAN UP AREA FOR RIBBON CUTTING
description: Clean up general work area and grounds for ribbon cutting
             party.  Project not to be complete for 14 more working days.  Work


             out of sequence to complete desired areas
 rem to sub:

       sent to:
      GCW            GENERAL CONTRACTOR WORK
      ABC            ABC PAINTING COMPANY
      SMOOTH         SMOOTH DRYWALL, INC.
      SHAGGY         SHAGGY CARPET COMPANY
---------------------------------------------------------------------------------
```

Figure 20.3d

```
                    DELTA CONSTRUCTION COMPANY
==============================================PaperWorks Version 1.23F==
              *** FIELD ORDER QUOTE TO OWNER LOG ***                .FQ1
          Job 321     GIBSON OFFICE BUILDING
Date: 04/14/90                                                 Page:   2
=========================================================================
 fo #: 007          sent: 03/07/90    method: DESCRIBED       due: 03/21/90
    summary: PREPARE AREA FOR GROUND BREAKING PARTY
description: Partially grade and brush area for parking.  Clean up.  Provide
             extra Port-a-potty and parking assistance for ground-breaking
             party
 rem to sub:

 date due owner   amount quoted   date ret by owner  amount approved by owner
    04/11/90    $       1,713.00                        $

remarks:

                                    STATUS:    PENDING Q.T.O.
-------------------------------------------------------------------------

 fo #: 010          sent: 03/14/90    method: DESCRIBED       due: 03/28/90
    summary: ADD DRAFT STOPS IN WAREHOUSE AREA
description: Add 3 draft stops in warehouse area as required by City of San
             Diego Building Inspector.  Draft stops not shown on plans.

 rem to sub:

 date due owner   amount quoted   date ret by owner  amount approved by owner
    04/18/90    $       8,656.00      04/25/90          $        8,600.00

remarks:

                                    STATUS:    FIELD ORDER QTO APPROVED
-------------------------------------------------------------------------

 fo #: 014          sent: 03/21/90    method: ENCLOSED        due: 04/04/90
    summary: ADD EXIT SIGNS
description: Add exit signs per the exiting drawing prepared by One Line Design
             and approved by the City of San Diego Fire Department 3/10.

 rem to sub:

 date due owner   amount quoted   date ret by owner  amount approved by owner
    04/25/90    $       1,980.00      05/02/90          $        1,980.00

remarks:

                                    STATUS:    FIELD ORDER QTO APPROVED
-------------------------------------------------------------------------
```

Figure 20.4a

```
                       DELTA CONSTRUCTION COMPANY
=======================================================PaperWorks Version 1.23F==
                 *** FIELD ORDER LOG DUE FROM OWNERS ***                    .FQ3
         Job 321       GIBSON OFFICE BUILDING
Date: 04/14/90                                                         Page:   1
================================================================================
 FO #: 001             sent: 02/21/90       method: DESCRIBED         due: 03/07/90
    summary: MOVE UNDERGROUND WATERLINE
description: Move approximately 180' of waterline that was not shown on plans
             and interfered with the new construction.

   date due owner 03/28/90     amount quoted $        3,960.00

--------------------------------------------------------------------------------

 FO #: 004             sent: 02/28/90       method: DESCRIBED         due: 03/14/90
    summary: ELIMINATE VISQUEEN UNDER SLAB ON GRADE
description: Eliminate visqueen under slab on grade.  Add 2" gravel for a total
             of 6" of gravel.  Lower finish grade by 2" to adjust for added
             gravel.  Finish floor will remain as shown on plans
   date due owner 04/04/90     amount quoted $         -587.00

--------------------------------------------------------------------------------

 FO #: 007             sent: 03/07/90       method: DESCRIBED         due: 03/21/90
    summary: PREPARE AREA FOR GROUND BREAKING PARTY
description: Partially grade and brush area for parking.  Clean up.  Provide
             extra Port-a-potty and parking assistance for ground breaking
             party
   date due owner 04/11/90     amount quoted $        1,713.00

--------------------------------------------------------------------------------

 FO #: 015             sent: 03/21/90       method: DESCRIBED         due: 04/04/90
    summary: CITY UTILITY CONNECTION FEES
description: Reimburse Quick Plumbing for required utility connection fees
             paid on behalf of the owner

   date due owner 04/25/90     amount quoted $       22,000.00
remarks: Note that we have not included any GC fee on this.  Please pay
         this as soon as possible.

--------------------------------------------------------------------------------

 FO #: 016             sent: 03/28/90       method: ENCLOSED          due: 04/11/90
    summary: REVISED STRIPING PLAN
description: Revise striping plan.  Re layout garage.  Plan given to your men
             in the field and authori them to proceed.

   date due owner 05/02/90     amount quoted $          440.00
remarks: ABC was able to add three more handicap stalls as well as increase
         the total parking spaces from 140 to 148 including handicap.

--------------------------------------------------------------------------------
```

Figure 20.4b

```
                          DELTA CONSTRUCTION COMPANY
==========================================================PaperWorks Version 1.23F==
        *** FIELD ORDER LOG DELINQUENT FROM OWNERS as of 04/14/90  *        .FQ5
            Job 321      GIBSON OFFICE BUILDING
Date: 04/14/90                                                        Page:    1
==================================================================================
 FO #: 001           sent: 02/21/90      method: DESCRIBED       due: 03/07/90
     summary: MOVE UNDERGROUND WATERLINE
 description: Move approximately 180' of waterline that was not shown on plans
              and interfered with the new construction.

   date due owner 03/28/90     amount quoted $        3,960.00
----------------------------------------------------------------------------------
 FO #: 004           sent: 02/28/90      method: DESCRIBED       due: 03/14/90
     summary: ELIMINATE VISQUEEN UNDER SLAB ON GRADE
 description: Eliminate visqueen under slab on grade.  Add 2" gravel for a total
              of 6" of gravel.  Lower finish grade by 2" to adjust for added
              gravel.  Finish floor will remain as shown on plans

   date due owner 04/04/90     amount quoted $         -587.00
----------------------------------------------------------------------------------
 FO #: 007           sent: 03/07/90      method: DESCRIBED       due: 03/21/90
     summary: PREPARE AREA FOR GROUND BREAKING PARTY
 description: Partially grade and brush area for parking.  Clean up.  Provide
              extra Port-a-potty and parking assistance for ground breaking
              party

   date due owner 04/11/90     amount quoted $        1,713.00
----------------------------------------------------------------------------------
```

Figure 20.4c

Chapter Twenty-One
Clarifications and Quotes to Owner

Chapter 19, "Price Requests," deals with the issue, "How much will it cost?" and Chapter 20 is about Field Orders, which convey the message: "Go ahead and do it and we'll handle the cost later." Clarifications are intended to clarify a construction item or detail in the contract documents, General Conditions, Specifications, or plan detail. Clarifications are not intended to be a Change in the Scope of Work.

Purpose

The purpose of the Clarification is to clarify, not to modify or change the cost of the project. Therefore, it should be treated as a *separate function*. The Clarification letter is used in the same way as a Price Request or a Field Order. It gets the information contained in the Clarification into the hands of the people who are doing the work so that they can implement the work that has been clarified.

Because the intent is to *clarify*, there is theoretically no reason for the architect/engineer to expect a response from the general contractor, nor for the project manager to anticipate a response from the subcontractor. In fact, Clarifications are never issued without some cost involved. It may not be directly charged to the project, but it is a cost, nevertheless.

For example, suppose a wood frame detail is not clear to the field personnel who are required to build it (according to the plans). The following events will then take place:

- The foreman on the job cannot figure it out, but spends 30 minutes trying.
- He then talks to the job superintendent about the detail (add another 30 minutes x 2) and, together, they still cannot make heads or tails of the detail.
- The job superintendent then calls the architect and relates his confusion (add 10 more minutes x 2). The architect reviews the original detail with the draftsman (add 15 minutes x 2).
- The draftsman draws the Clarification (30 minutes) and issues it.
- The general contractor receives the Clarification and then sends it to the involved subcontractors and to the field (20 minutes).

- Field personnel and subcontractors review the Clarification and respond to the general contractor that there is no change in cost (2 x 30 minutes).
- The general contractor responds to the architect or engineer that there is no additional contractual cost (30 minutes).

Notice that although there was no additional chargeable cost involved in the previous hypothetical example, it actually requires over five hours of someone's time. At an average rate of $30 an hour (which does not include overhead or mailing costs, typing, opening the mail, and logging both the Clarification and the response), the total ends up at $150.00. Add in an overhead number for a total of $195.00, and this is a conservative estimate. Most importantly, this number does not reflect the importance of what these individuals could have been doing if they were not involved in reviewing and researching a detail and the required Clarification.

In practice, the Clarification format is used for the following reasons:

First, it is intended to clarify the original intent of the construction documents. In other words, it is used to interpret a detail in the construction documents.

Second, it is intended to clarify the details which may not be included in the original construction documents (such as the colors of paint in the finish schedule). Neither the architect nor the manager may be aware that the Clarification could have some downstream costs to one or more of the parties involved in the project. For example:

A Clarification may be issued to change the finish of the hardware from brushed brass to "old brass." The cost for the two items are the same. When an order to change color is sent from the architect, contractor, vendor, or distributor to the hardware manufacturer, it is discovered that while "old brass" is a standard finish with their line of hardware, they are all out of it and will not be making another production run for six months. However, for a price ($1,000), they could make a special run.

The owner is now faced with additional costs and needs to know what they are. He also needs to know whether a six-month wait for this preferred style of hardware will interfere with the construction schedule. Also, how great will the time impact be, and what is the cost of the impact on the construction schedule?

The third situation in which the Clarification format is used is in an attempt to "bury" a change in the scope of the work in hopes that the general contractor or subcontractor will overlook it. Architects and engineers should realize that this use of the Clarification vehicle is like shooting oneself in the foot. The general contractor or subcontractor will see the item *and* the attempt to hide it. From that point on, the relationship will be soured. In the construction business, where so much of an individual's/company's reputation hangs on their "word" and personal integrity, it pays dividends to play it straight. Everyone makes mistakes. However, sympathy and cooperation will evaporate if anyone on the job attempts to pull the wool over someone else's eyes.

The fourth use for Clarifications is to convey a scope change when the owner has said "no more changes," yet the architect/engineer realizes that a particular change should be implemented. Naturally, this absolute statement ("no more changes") on the part of the owner is shortsighted. In any case, the architect issues the Clarification and then the general contractor submits a price for the change in scope of work, as called for in the Clarification. In this case, the architect may incur a round of angry negotiations due to a misguided attempt to keep the job looking "clean," with few changes, for the owner's benefit. Regardless of the architect's actual motive, the general contractor is apt to believe that the architect is trying to pass the blame. Again, misusing the Clarification disables an otherwise useful tool.

Typical Clarification Contents

Clarifications are instructions intended to clarify issues between the architect/engineer and the general contractor. The contractor then distributes the Clarification to the affected subcontractors. For example:

- The dimension of the second floor bathroom wall should be 10'-6", not 10'-1" as shown on the plans.
- The paint color of the living room should be ABC pink, grade 4, not ABC green, grade 4.
- The height of the balcony rail shall be 42", not 36" as shown on the plans.

Clarifications and Field Orders both give direction to proceed with the work described. Do not take a chance by assuming that the Clarification does not involve a significant sum of money on someone's part. Get a response from each of the subcontractors/vendors stating that there "is no additional cost to the Contract" if the directions in the Clarification are followed.

Many subcontractors, caught up in their efforts to keep up with the project manager's schedule, do not review the Clarifications very closely. Then, when it comes time to do the portion of the work described in the Clarification (which may be months later), they discover (to everyone's anguish) that there is, in fact, a real cost to perform the work described in the Clarification, and that it may be a substantial sum.

Often, the architect/engineer's standard Clarification form will include a statement such as, "Do not proceed with this work if there is any cost involved. If no response is made to us within seven days, no costs will be allowed." The fact is, it can take seven days to get a letter across town, let alone to respond properly to the architect. In such cases, it is best to reach an agreement with the architect or engineer (in writing), as to the allowed turnaround time. This issue should be resolved in the pre-construction meeting. If the architect or engineer is unwilling to adjust the time constraints on the response, the contractor or subcontractor (upon receiving a Clarification) should immediately send a Transmittal (see Chapter 3) to the architect, with the following statement:

> *"We have reviewed your Clarification # ____________ (Field Order) and we are proceeding with the work as described.*
>
> *We believe that the work described in the Clarification will require additional cost to be expended.*
>
> *We are currently reviewing the Clarification # ____________ for any cost impact and will advise you of the additional costs when we are able to determine the extent of compensation and the extension of contract time required. We are proceeding with the work per your directions."*

This letter informs the architect or engineer that the manager is unable to accomplish what is being requested in the time dictated. The architect will not receive the answer within the stated time frame because there is not enough time to review the Clarification document properly. The recommended statement should say the following:

1. A Clarification was received.
2. It will probably cost money.
3. The amount of the additional cost is not yet known.
4. The owner/architect will be informed when the costs are determined.
5. The work described in the Clarification has been performed or is proceeding.

It is now up to the architect to say "No, if it is going to cost money, do not proceed." If the architect/owner does not respond and later tries to make a case for not paying for the work, the response should be: "I gave you the opportunity to stop me from continuing and you did nothing. I told you that I thought it was going to add an extra cost to the contract. I am, therefore, entitled to be paid for the work that was completed on behalf of your project and per your instructions."

By giving notice that the Clarification (or Field Order) work is going to cost extra, and if neither the owner nor the architect says not to proceed, they are, in effect, prevented from later claiming that the manager was not authorized to proceed with the work.

Subcontractors take note: If a subcontractor responds in a similar manner to the general contractor (i.e., that there is an extra cost and time involved in the Clarification), and the general contractor does not relay that information to the owner, then the general contractor will be responsible for payment–even if he does not receive payment from the owner.

All Clarifications should be responded to, even if they do not involve money. The mere issue of Clarification direction may have caused a change in the scope of the project or project documents. Clarifications that do not cost anything can be quoted simply as "no change in contract cost." Again, this does not mean that there is no cost involved with the Clarification. It only means that there is no change in the *contract* cost.

By following this procedure, you can make sure that the sub-trades involved also have the most current information, and that the owner does not have to face any surprises as the project comes to an end. Every job goes along more smoothly if all of those involved have full knowledge of the changes in the scope of the work as the job progresses.

Some architects may misuse the Clarification as a vehicle to issue minor changes without drawing attention to the fact that a "hole" exists in the construction documents (that will cost money). For protection, the manager may choose to send the Clarification letter to the sub-trades, and send a copy to the owner. The owner is then aware that while it may not be intended to cost him money, there is, or has been, a change in scope. Whether or not a copy of the Clarification is sent to the owner will depend on the kind of relationships that exist on a particular project.

Timely Response

Always respond as soon as possible in order to get listed among current payables. Many times the general contractor and subcontractors fail to provide timely quotes because they count on the owner/architect to automatically provide new funds for work. That is not how it works. Most owners just look at the approved changes and consider the total adjusted contract price to be their maximum exposure. There are, however, a few ways to warn the owner of storms on the horizon.

By sending the owner a copy of the logs on "open" pricing letters (showing those that have been quoted by the general contractor and those that have not), it is possible to keep the owner aware of the potential upward price movement so that he/she is aware of the latest developments. It is better to let the owner be concerned than to (potentially) run out of funds.

Remember: respond to all Clarifications immediately. Otherwise, it will be assumed that there is no change in contract cost, and that the Clarification has no impact on the schedule.

The sequence of events for processing a Clarification will normally be as follows:

1. The owner (or user)/architect determines the scope of work that should be included in the Clarifications. The document is then assigned an identification number (such as CL #23). The Clarification is issued to the general contractor with the appropriate documentation, including description, plans, and specifications to fully describe the work to be priced. Usually, the general contractor is instructed to proceed with the Clarification.
2. The general contractor views CL #23 and determines which trades, including his own, are affected by it. He sends each of them a copy of the Clarification with the supporting documents, plans, and specifications. The general contractor should include an appropriate (and realistic) due date for the response from subcontractors and vendors. Generally, Clarifications should be returned to the general contractor within 14 calendar days.
3. When the general contractor receives a statement of "no change in contract price," or a quotation for the Clarification from all trades and subcontractors, then he can add his own profit and overhead numbers (if necessary) and send a Clarification Quote to Owner, again giving the owner/architect sufficient time to respond–usually 10 working days, or 14 calendar days. If the response is "no change in contract price," this response should be noted. (Note the difference between "no change in cost," and "no change in contract price.")
4. Since the architect has instructed the general contractor to proceed with the Clarification work, the general contractor can wait a little while for the owner to approve a change in contract price, if required. However, do not wait until the owner forgets the whole issue.

   Be aware that since the owner did not expect to have a contract increase on the Clarification, he or she will probably be very slow in signing for an increase. Follow up and follow through. Otherwise, the owner may wait until the end of the job, and then debate whether or not the Clarification should have cost any money.

Items to Include

The key information to include in Clarification letters to subcontractors is as follows.

- From
- To
- Job identification
- Clarification number and identification
- Date from architect or engineer
- Date sent to subcontractors, vendors
- Date information is due from subcontractor or vendor
- Short description of the work as stated in the Clarification
- Indication of whether or not the original Clarification is attached or there is any other documentation included
- Special note to subcontractor or vendor

Clarifications should be kept and identified separately from Field Orders. Clarifications are issued with a different concept in mind and should not be confused with the other pricing documents. It is easier for the owner/architect to deal with these different issues one type at a time. Remember, it is much easier for the general contractor and subcontractor to understand the difference between these four types of pricing documents for scope changes because they deal with them every day.

Sample Clarification Letters

Figure 21.1a is a sample computer-generated Clarification. Figure 21.1b is a manual Clarification, using the standard Transmittal form.

DELTA CONSTRUCTION COMPANY
6065 Mission Gorge Road, #193
San Diego, CA 92120
(619) 582-5829

*** CLARIFICATION ***

Date: 2/21/90

Job Number: 321
Field Order Number: 001

TO: SMOOTH DRYWALL, INC.
7676 TAPE AVENUE
SAN DIEGO, CA 92110

ATTENTION: SAM SAND

JOB NUMBER: 321 GIBSON OFFICE BUILDING

FROM: J. EDWARD GRIMES

SUBJECT: CLARIFY WIDTH OF CORRIDOR #101

We are transmitting to you the following which is DESCRIBED BELOW for your

review and RESPONSE. Please refer to Job Number & Clarification Number

Please review and return this information to us by 3/07/90.

The width of corridor 101 should be 48" not 4"-8" as shown on the plans, Sheet A-3.

Note to Sub:

Please advise your field personnel so layout is correct.

Please contact me as soon as possible if you need any further information or if you have any questions.

Sincerely,

J. Edward Grimes

Copies were sent to the following:

SHAGGY CARPET COMPANY

dp: 2/21/90 15:29 **PaperWorks,** a Construction Software Program

Figure 21.1a

Means Forms

LETTER OF TRANSMITTAL

Delta Construction Company

6678 Mission Road
San Diego, CA 92120
(619) 581-6315

FROM: J. Edward Grimes

DATE 2/21/90
PROJECT Gibson Office Bldg. #321
LOCATION San Diego, Ca.
ATTENTION Sam Sand
RE: Width of corridor 101

TO: Smooth Drywall Inc.
7676 Tape Avenue
San Diego, Ca. 92110

Gentlemen:

WE ARE SENDING YOU ☐ HEREWITH ☐ DELIVERED BY HAND ☐ UNDER SEPARATE COVER

VIA ______ THE FOLLOWING ITEMS:

☐ PLANS ☐ PRINTS ☐ SHOP DRAWINGS ☐ SAMPLES ☐ SPECIFICATIONS

☐ ESTIMATES ☐ COPY OF LETTER ☐ ______

COPIES	DATE OR NO.	DESCRIPTION

THESE ARE TRANSMITTED AS INDICATED BELOW

☐ FOR YOUR USE ☐ APPROVED AS NOTED ☐ RETURN ______ CORRECTED PRINTS

☐ FOR APPROVAL ☐ APPROVED FOR CONSTRUCTION ☐ SUBMIT ______ COPIES FOR ______

☐ AS REQUESTED ☐ RETURNED FOR CORRECTIONS ☐ RESUBMIT ______ COPIES FOR ______

☒ FOR REVIEW AND COMMENT ☐ RETURNED AFTER LOAN TO US ☐ FOR BIDS DUE ______

☐ ______

REMARKS: The width of corridor 101 should be 48" not 4'-8" as shown on plan sheet A-3.

Please advise your field personnel so the layout is correct.

IF ENCLOSURES ARE NOT AS INDICATED, PLEASE NOTIFY US AT ONCE.

SIGNED: J. Edward Grimes

Figure 21.1b

Clarification Quote to Owner

The Clarification Quote to Owner is the response to the original Clarification sent to the owner/architect. It should include a time period during which the quote is valid, and any other stipulations or conditions that might be required. One of the conditions usually accompanying a Clarification Quote to Owner is an extension in contract time.

The following items should be included in the Clarification Quote to Owner.

- From
- To
- Date sent
- Job identification
- Clarification number and identification
- Brief description of work involved
- Date approval is due from owner
- Amount of the Clarification quotation
- Extension of time, if required
- Any special note to the owner
- Breakdown of costs for the Clarification, depending on the contractual arrangement

Figure 21.2a is a sample computer-generated Clarification Quote to Owner. Figure 21.2b is a manual (either typed or handwritten) Clarification Quote to Owner, using the standard Transmittal form.

Logs

Figure 21.3a is a manually-generated combined Clarification and Clarification Quote to Owner log. Figure 21.3b is a computer-generated Clarification log. Figure 21.3c is a computer-generated Clarification Due from Subs log. Notice that the logs for Clarifications and Clarification Quotes To Owner contain similar information, but they are not exactly the same. The Clarification logs show all of the subcontractor and general contractor costs and fees, while the Clarification Quotes to Owner show the price quoted to the owner. The subcontractors should never see the Clarification Quote to Owner logs, and the owner should never see the Clarification logs with subcontractor pricing information. Note that the Clarification log and Clarification Quote to Owner log can be combined for easier tracking in the manual mode. The computer version's versatility eliminates the need for a combined log.

Figure 21.3d is a Clarification Delinquent from Subs log, Figure 21.3e is a Clarification Quote to Owner log, Figure 21.3f is a Clarification Quote Due from Owner log, and Figure 21.3g is a Clarification Quote to Owner Delinquent log. All are generated by the computer tracking system for better expediting of the Clarifications.

DELTA CONSTRUCTION COMPANY

6065 Mission Gorge Road, #193
San Diego, CA 92120
(619) 582-5829

*** CLARIFICATION QUOTE TO OWNER ***

Date: 3/14/90

Job Number: 321
Field Order Number: 001

TO: FRANCES GIBSON
6109 CAMINITO CLAVO
SAN DIEGO, CA 92120

ATTENTION: FRANCES GIBSON

JOB NUMBER: 321 GIBSON OFFICE BUILDING

FROM: J. EDWARD GRIMES

SUBJECT: CLARIFY WIDTH OF CORRIDOR #101

We are submitting the following changes in cost of $ -400.00
The details and breakdown of this change in cost is enclosed.

A complete description of the scope of work relating to this proposal is enclosed. The summary of the works is as follows:

The width of corridor 101 should be 48" not 4"-8" as shown on the plans, Sheet A-3.

Remarks to owner/architect:

The credit has been reviewed and is appropriate.

This work will result in an extension to our contract time of 0 days.

Please indicate your approval by signing and returning one copy of this letter to our office or issue a CHANGE ORDER for this work no later than 3/28/90 in order to not delay the project.

Sincerely

J. EDWARD GRIMES

APPROVED BY: FRANCES GIBSON

SIGNED: ______________________________ DATE:

dp: 3/14/90 15:34 **PaperWorks, a Construction Software Program**

Figure 21.2a

Means Forms

PROPOSAL

Delta Construction Company
6678 Mission Road
San Diego, CA 92120
(619) 581-6315

FROM: J. Edward Grimes
Project Manager

TO: Frances Gibson
6109 Caminito Clavo
San Diego, Ca. 92120

PROPOSAL NO. CLQ 321.01
DATE 3/14/90
PROJECT Gibson Office Bldg. #321
LOCATION San Diego, Ca.

CONSTRUCTION TO BEGIN

COMPLETION DATE

Gentlemen:

The undersigned proposes to furnish all materials and necessary equipment and perform all labor necessary to complete the following work:

Width of corridor 101 should be 48" not 4'-8" as shown on plan sheet A-3.

All of the above work to be completed in a substantial and workmanlike manner

☒ for the sum of Credit Total Four hundred dollars ($ -400.00)

☐ to be paid for at actual cost of Labor, Materials and Equipment plus ______ percent (______ %)

Payments to be made as follows: ______

______ The entire amount of the contract to be paid within ______ after completion

Any alteration or deviation from the plans and specifications will be executed only upon written orders for same and will be added to or deducted from the sum quoted in this contract. All additional agreements must be in writing.

The Contractor agrees to carry Workmen's Compensation and Public Liability Insurance and to pay all taxes on material and labor furnished under this contract as required by Federal laws and the laws of the State in which this work is performed.

Respectfully submitted,

Contractor Delta Construction Co.

By J Edward Grimes

ACCEPTANCE

You are hereby authorized to furnish all material, equipment and labor required to complete the work described in the above proposal, for which the undersigned agrees to pay the amount stated in said proposal and according to the terms thereof.

Date ______ 19___ ______

Figure 21.2b

Means® Forms

Clarification/Quote to owner

PROJECT

SHEET OF

JOB NO.

LOCATION

PREPARED BY

YEARS

CL#	ITEM	Date Sent to Arch.	Due	Rec'd	Action Required	Sent	Due	Rec'd	Amount	Sent to Owners/Accl	Due	Rec'd	Amount Approved/Action
01	Width of Corridor 101	2/16	3/1	2/23	Send to subs								
	Smooth drywall					2/23	2/25	2/25	(100)				
	Shaggy Carpet					2/23	2/25	2/25	200				
	AC Work								0				
	AC OH & P								50				
									150	2/25	3/1		
02	Hardware Finish	2/16	3/1	2/23	✓w/subs								
	Continental Hdwe					2/23	3/1	3/2	0	3/2			ok
03	Curtain Wall Glass Color	2/16	3/1	2/25	Send to subs	2/25	3/7	3/7					
	Clear Glass Co.						3/7	3/7	5600				
	G.C. Work						3/7	3/7					
	G.C. OH & P						3/7	3/7	560				
									6160	3/7	3/17	3/17	$6,160 ✓

Figure 21.3a

```
                    DELTA CONSTRUCTION COMPANY
==========================================PaperWorks Version 1.23F==
                       *** CLARIFICATION ***                    .CL1
          Job 321     GIBSON OFFICE BUILDING
Date: 03/28/90                                            Page:    1
====================================================================
 CL #: 001          sent: 02/21/90     method: DESCRIBED      due: 03/07/90
    summary: CLARIFY WIDTH OF CORRIDOR #101
description: The width of corridor 101 should be 48" not 4"-8" as shown on the
             plans, Sheet A-3.

 rem to sub: Please advise your field personnel so layout is correct.

       sent to:                         date returned     amount quoted
   SMOOTH DRYWALL, INC.                   03/04/90     $              0.00
   SHAGGY CARPET COMPANY                  03/14/90     $           -400.00
                                        sub total:     $           -400.00

  date due owner     amount quoted     date returned      amount approved
    03/28/90       $         -400.00                      $

remarks: The credit has been reviewed and is appropriate.

--------------------------------------------------------------------
 CL #: 002          sent: 02/21/90     method: DESCRIBED      due: 03/07/90
    summary: FINISH HARDWARE FINISH
description: The finish of all finish hardware groups shall be US 10.

 rem to sub: Check to make sure that the US 10 is available.

       sent to:                         date returned     amount quoted
   CONTINENTAL HARDWARE                 NOT QUOTED    $              0.00
                                        sub total:    $              0.00
--------------------------------------------------------------------
 CL #: 003          sent: 02/21/90     method: DESCRIBED      due: 03/07/90
    summary: CURTAIN WALL GLASS COLOR
description: The curtain wall glass color shall be 345 Blue by PPG

 rem to sub: Please advise me if there are any delivery problems

       sent to:                         date returned     amount quoted
   CLEAR GLASS COMPANY                    03/14/90     $          5,867.00
   GENERAL CONTRACTOR FEE                 03/14/90     $            588.00
                                        sub total:     $          6,455.00

  date due owner     amount quoted     date returned      amount approved
    03/28/90       $        6,455.00                      $

remarks: The blue glass cost a lot more.  We are not proceeding until
         you confirm approval of this additional cost.
--------------------------------------------------------------------
 CL #: 004          sent: 02/21/90     method: DESCRIBED      due: 03/07/90
    summary: BASE BUILDING PAINT COLOR
description: The paint color P-1 shall be Dutch Boy 2-112, Navajo White.
             No substitutions without completing the submittal substitution
             process.
 rem to sub: No EXCEPTIONS!
```

Figure 21.3b

```
                         DELTA CONSTRUCTION COMPANY
=================================================PaperWorks Version 1.23F=
                    *** CLARIFICATION LOG DUE FROM SUBS ***             .CL3
             Job 321     GIBSON OFFICE BUILDING
Date: 03/28/90                                                    Page:    1
===========================================================================
 CL #: 002          sent: 02/21/90     method: DESCRIBED      due: 03/07/90
    summary: FINISH HARDWARE FINISH
description: The finish of all finish hardware groups shall be US 10.


 rem to sub: Check to make sure that the US 10 is available.

       sent to:
      CONTHDWE       CONTINENTAL HARDWARE
---------------------------------------------------------------------------

 CL #: 004          sent: 02/21/90     method: DESCRIBED      due: 03/07/90
    summary: BASE BUILDING PAINT COLOR
description: The paint color P-1 shall be Dutch Boy 2-112, Navajo White.
             No substitutions without completing the submittal substitution
             process.
 rem to sub: No EXCEPTIONS!

       sent to:
      ABC            ABC PAINTING COMPANY
---------------------------------------------------------------------------

 CL #: 006          sent: 02/21/90     method: DESCRIBED      due: 03/07/90
    summary: LOCATION OF LIGHT SWITCHES
description: As the building standard the location of all light switches will
             be 42" above the floor  and 12" from the edge of door, if located
             next to a door.
 rem to sub:

       sent to:
      SPARKS         SPARKS ELECTRIC COMPANY
---------------------------------------------------------------------------

 CL #: 007          sent: 02/28/90     method: DESCRIBED      due: 03/14/90
    summary: LOCATION OF THERMOSTATS
description: The placement and location of thermostats shall be as noted on
             HVAC Dwg. AC-1-8 and shall be located 68" from finish floor.
             The height was not noted on plans
 rem to sub:

       sent to:
      COOL           COOL AIR CONDITIONING
---------------------------------------------------------------------------

 CL #: 010          sent: 03/07/90     method: ENCLOSED       due: 03/21/90
    summary: LOCATION OF LANDSCAPE SPRINKLER CONTROLS
description: Per sketch L-001 the location of the landscape sprinkler controls
             shall be at the NW corner of the building.  See Sketch.

 rem to sub:

       sent to:
      GREEN          GREEN LANDSCAPING
```

Figure 21.3c

```
                          DELTA CONSTRUCTION COMPANY
═══════════════════════════════════════════════════PaperWorks Version 1.23F══
      *** CLARIFICATION LOG DELINQUENT FROM SUBS as of 03/28/90  *       .CL5
          Job 321      GIBSON OFFICE BUILDING
Date: 03/28/90                                                     Page:    2
═════════════════════════════════════════════════════════════════════════════

 CL #: 006             sent: 02/21/90       method: DESCRIBED       due: 03/07/90
    summary: LOCATION OF LIGHT SWITCHES
description: As the building standard the location of all light switches will
             be 42" above the floor  and 12" from the edge of door, if located
             next to a door.
 rem to sub:

       sent to:
      SPARKS          SPARKS ELECTRIC COMPANY
─────────────────────────────────────────────────────────────────────────────

 CL #: 007             sent: 02/28/90       method: DESCRIBED       due: 03/14/90
    summary: LOCATION OF THERMOSTATS
description: The placement and location of thermostats shall be as noted on
             HVAC Dwg. AC-1-8 and shall be located 68" from finish floor.
             The height was not noted on plans
 rem to sub:

       sent to:
      COOL            COOL AIR CONDITIONING
─────────────────────────────────────────────────────────────────────────────

 CL #: 010             sent: 03/07/90       method: ENCLOSED        due: 03/21/90
    summary: LOCATION OF LANDSCAPE SPRINKLER CONTROLS
description: Per sketch L-001 the location of the landscape sprinkler controls
             shall be at the NW corner of the building.  See Sketch.

 rem to sub:

       sent to:
      GREEN           GREEN LANDSCAPING
```

Figure 21.3d

```
                    DELTA CONSTRUCTION COMPANY
=====================================================PaperWorks Version 1.23F==
          *** CLARIFICATION QUOTE TO OWNER LOG ***                    .CQ1
             Job 321      GIBSON OFFICE BUILDING
Date: 04/14/90                                                     Page:    1
===============================================================================
 cl #: 001            sent: 02/21/90      method: DESCRIBED       due: 03/07/90
    summary: CLARIFY WIDTH OF CORRIDOR #101
description: The width of corridor 101 should be 48" not 4"-8" as shown on the
             plans, Sheet A-3.

 rem to sub: Please advise your field personnel so layout is correct.

 date due owner   amount quoted    date ret by owner  amount approved by owner
    03/28/90     $          -400.00                     $

remarks: The credit has been reviewed and is appropriate.

                                          STATUS:    PENDING Q.T.O.
-------------------------------------------------------------------------------
 cl #: 003            sent: 02/21/90      method: DESCRIBED       due: 03/07/90
    summary: CURTAIN WALL GLASS COLOR
description: The curtain wall glass color shall be 345 Blue by PPG


 rem to sub: Please advise me if there are any delivery problems

 date due owner   amount quoted    date ret by owner  amount approved by owner
    03/28/90     $         6,455.00      04/02/90       $          6,000.00

remarks: The blue glass cost a lot more.  We are not proceeding until
         you confirm approval of this additional cost.
                                          STATUS:    CLARIFICATION QTO APPROVED
-------------------------------------------------------------------------------
 cl #: 005            sent: 02/21/90      method: DESCRIBED       due: 03/07/90
    summary: CLARIFY DOOR SWING ON DOORS 21, 31, 41, 51, 61
description: Door swings on doors 21, 31, 41, 51, 61 should be RIGHT HAND
             not LEFT HAND shown on the door schedule.

 rem to sub: Be sure to follow up

 date due owner   amount quoted    date ret by owner  amount approved by owner
    04/11/90     $                                      $
                      NO CHANGE
remarks:

                                          STATUS:    PENDING Q.T.O.
-------------------------------------------------------------------------------
 cl #: 008            sent: 02/28/90      method: ENCLOSED        due: 03/14/90
    summary: CLARIFY DETAIL OF STEPPED FOOTING
description: Note details 3 and 4 on sheet S-2 Revision 1 for clarification
             of stepped footings

 rem to sub:

 date due owner   amount quoted    date ret by owner  amount approved by owner
    04/11/90     $           550.00                     $

remarks: The work was not shown on original drawings.

                                          STATUS:    PENDING Q.T.O.
-------------------------------------------------------------------------------
```

Figure 21.3e

```
                              DELTA CONSTRUCTION COMPANY
=====================================================PaperWorks Version 1.23F==
                  *** CLARIFICATION LOG DUE FROM OWNERS ***                .CQ3
             Job 321      GIBSON OFFICE BUILDING
Date: 04/14/90                                                        Page:   1
================================================================================
 CL #: 001           sent: 02/21/90      method: DESCRIBED        due: 03/07/90
    summary: CLARIFY WIDTH OF CORRIDOR #101
description: The width of corridor 101 should be 48" not 4"-8" as shown on the
             plans, Sheet A-3.

    date due owner 03/28/90     amount quoted  $          -400.00
remarks: The credit has been reviewed and is appropriate.

--------------------------------------------------------------------------------

 CL #: 005           sent: 02/21/90      method: DESCRIBED        due: 03/07/90
    summary: CLARIFY DOOR SWING ON DOORS 21, 31, 41, 51, 61
description: Door swings on doors 21, 31, 41, 51, 61 should be RIGHT HAND
             not LEFT HAND shown on the door schedule.

    date due owner 04/11/90     amount quoted  $             0.00  NO CHANGE

--------------------------------------------------------------------------------

 CL #: 008           sent: 02/28/90      method: ENCLOSED         due: 03/14/90
    summary: CLARIFY DETAIL OF STEPPED FOOTING
description: Note details 3 and 4 on sheet S-2 Revision 1 for clarification
             of stepped footings

    date due owner 04/11/90     amount quoted  $           550.00
remarks: The work was not shown on original drawings.

--------------------------------------------------------------------------------

 CL #: 010           sent: 03/07/90      method: ENCLOSED         due: 03/21/90
    summary: LOCATION OF LANDSCAPE SPRINKLER CONTROLS
description: Per sketch L-001 the location of the landscape sprinkler controls
             shall be at the NW corner of the building.  See Sketch.

    date due owner 04/18/90     amount quoted  $             0.00  NO CHANGE

--------------------------------------------------------------------------------

 CL #: 014           sent: 03/21/90      method: ENCLOSED         due: 04/04/90
    summary: PLASTER DRIP DETAIL
description: Refer to SK-1121 for details of plaster drip not shown on original
             plans.

    date due owner 04/18/90     amount quoted  $           165.00

--------------------------------------------------------------------------------
```

Figure 21.3f

```
                    DELTA CONSTRUCTION COMPANY
==================================================PaperWorks Version 1.23F==
    *** CLARIFICATION LOG DELINQUENT FROM OWNERS as of 04/14/90 ***     .CQ5
          Job 321      GIBSON OFFICE BUILDING
Date: 04/14/90                                                     Page:    1
=============================================================================
 CL #: 001          sent: 03/14/90     method: DESCRIBED         due: 03/28/90
    summary: CLARIFY WIDTH OF CORRIDOR #101
description: The width of corridor 101 should be 48" not 4"-8" as shown on the
             plans, Sheet A-3.

   date due owner 03/28/90    amount quoted  $         -400.00

 remarks: The credit has been reviewed and is appropriate.

_____________________________________________________________________________

 CL #: 005          sent: 03/28/90     method: DESCRIBED         due: 04/11/90
    summary: CLARIFY DOOR SWING ON DOORS 21, 31, 41, 51, 61
description: Door swings on doors 21, 31, 41, 51, 61 should be RIGHT HAND
             not LEFT HAND shown on the door schedule.

   date due owner 04/11/90    amount quoted  $            0.00  NO CHANGE

_____________________________________________________________________________

 CL #: 008          sent: 03/28/90     method: ENCLOSED          due: 04/11/90
    summary: CLARIFY DETAIL OF STEPPED FOOTING
description: Note details 3 and 4 on sheet S-2 Revision 1 for clarification
             of stepped footings

   date due owner 04/11/90    amount quoted  $          550.00

 remarks: The work was not shown on original drawings.

_____________________________________________________________________________
```

Figure 21.3g

Summary

It is important for everyone–owner, architect, general contractor, and subcontractors–involved in a project to remember the following regarding Clarifications.

1. They may change the scope of the work.
2. They may interfere with, or take time away from, the normal course of construction.
3. It may take a great deal of effort on the part of the general contractor and the subcontractors to put together an accurate estimate of the difficulty of the proposed Clarification.
4. If the owner/architect issues a lot of Clarifications (generally the result of poor plans), the general contractor and subcontractors may "up" their estimates to cover their additional costs.

Chapter Twenty-Two

Requests for Changes and Quotes to Owners

In Chapters 19, 20 and 21, we have reviewed Price Requests, Field Orders, and Clarifications, all of which are responses to directions from the owner/user, architect, or engineer. These directions are for work that was not specified in the original contract scope of work. The directions might also be a response to an item that required clarification.

A Request for Change (RC) is the contractor's only *official* vehicle for initiating documentation when there is, or has been, a change in the scope of work. The general contractor may also initiate an RC for his own work, or for the work required of a subcontractor.

Typical examples of Requests for Change include:

- Changes required by the municipal Building Department that did not appear on the construction bid documents
- Items (incorporated in a new issue of plans) that were not on the plans when the job was bid
- A delay in the project due to circumstances beyond the contractor's control, e.g., rain, floods, acts of God

If the documentation procedures recommended in this manual are implemented, the Request for Change becomes an important instrument for documenting and obtaining payment for items beyond the scope of the original contractual agreement.

The Need for RC's

It is important to remember that every construction company, subcontractor, or facilities department needs to use the Request for Change procedure because it is the *only* true documentation vehicle that allows the manager to properly initiate, transmit, and document a legitimate claim for work above and beyond the scope of the contract. It may be the only way in which the manager can go on record that something is happening on the project that was not included in the scope of the original contract documents.

The owner /user or architect/engineer may not be willing to issue a Field Order for certain work that is beyond the scope of the project, but which *must* be accomplished in order to complete the project. In such instances, a claim should be documented by submitting a *Request for Change.* This should be done promptly in order to establish the claim, so that the owner/user or architect cannot claim, at a later date, that the contractor was searching for ways to make an extra profit on the job after the project was finished.

At the pre-construction meeting, or at least early in the job, the Request for Change should be established as an acceptable and proper pricing vehicle. It is much easier for the owner and architect to accept this concept at the beginning of the project, rather than as a surprise midway through the project. On some jobs, the owner/architect/engineer will want the general contractor to initiate and submit *all* prices as a Request for Change. This reduces the paperwork for the architect. This approach requires more diligence on the part of the manager, but it can be quite efficient for smaller jobs. Many custom homes, for example, are built using only Field Orders (directives) and Requests for Change.

Because the Request for Change is normally initiated by the manager, it does not mean the subcontractors cannot use this important documentation tool as well. It is just as important for the subcontractors to use Requests for Change to document the fact that they have a valid claim for an item of work that was not a part of their contractual agreement with the contractor.

The following steps are typically taken when changes are required by the Building Department.

1. The general contractor and subcontractors bid and receive a contract to build Project ABC according to a set of plans and documents.
2. The architect submits plans to the Building Department, which reviews them and directs a number of changes. These changes may never be incorporated into the formal plans, but are simply made by the architect at the Building Department.
3. The Building Department's "Approved" set of drawings (one set only) must remain at the job site.
4. The architect convinces the owner that it would be too costly to update and reissue all of the plans. The architect may also feel (and tell the owner) that the changes are insignificant and do not require a Price Request or a Field Order. Building Department changes are, in fact, usually insignificant to the *design* of the building, which is what interests the architect. However, these "insignificant" changes may represent a major cost impact to a particular subcontractor. The following example illustrates this point.

   Building ABC is designed to have one fire sprinkler head for every 400 S.F. The Building Department then requires one fire head for every 300 S.F. due to the occupancy classification of the building. This is not a major concern of the architect because it does not affect the design. However, it may have a major cost impact overall because it significantly affects:
   - *the fire sprinkler subcontractor*
   - *the HVAC subcontractor*
   - *placement of the light fixtures*
   - *layout of the "T" bar and ceiling tile*
   - *possibly, the piping size for the water supply from the street and in the building*

This is one reason why the Request for Change must be handled just like one of the other pricing documents. The ramifications of seemingly minor scope changes can boggle the mind and drain the pocketbook. These items need to be addressed promptly because they will come as a total surprise to the owner. The owner deserves and needs as much time as possible to deal with the situation. He may want to appeal the Building Department's decision to change building occupancy classifications, or he may even ask the architect to redesign a portion, or all, of the project.

Again, *always* give the owner a chance to say "no." Do not just automatically follow what the Building Department says is right or what the architect dictates. Each person involved in the project has an individual agenda and motivation. Recognize this and monitor activity accordingly.

As a manager, keep an eye out for cost changes on the project and try to respond in the best interests of the entire project, even if it is not a pure design issue. Remember the Golden Rule ("he who has the gold gets to rule"); adhering to it will be a great help in staying out of trouble.

5. Every general contractor, facilities manager, or subcontractor knows that significant scope changes may be required by the Building Department and incorporated into the project documents. In these cases, the Request for Change should be sent to the involved subcontractors as soon as possible. In the case of our example, a copy of the RC should be initiated and sent to subcontractors just as if it was a Field Order. Copies should also be sent to the owner and the architect at this time. In the event that the owner wants to appeal, or chooses not to proceed with a portion of the work, he or she should be given as much time and notice as possible to make that decision.
6. After receiving quotes from the subcontractors and/or vendors, the RC is processed in the same way as the other pricing vehicles. (Refer to Chapters 18 through 21.)

What to Include in RC's to Subcontractors

The following items should be included in the RC letter to subcontractors.

- From
- To
- Date sent to subcontractors
- Job identification
- RC # (assigned by project manager)
- Reason for RC (such as Building Department changes not shown on bid documents)
- Brief description of the work involved
- Date information is due from subcontractors, vendors
- The location of documents that caused the issue of the RC
- Special note to subcontractor or vendor, if necessary

Sample RC Letters to Subcontractors

Figure 22.1a is a sample computer-generated RC letter to a subcontractor/vendor. Figure 22.1b is a manual RC to subcontractors/vendors using the standard Transmittal form. The advantage of the computer-generated system is its ability to maintain a consistent format and to automatically enter all the information into the logs.

Purpose of Request for Change Quote to Owner

The Request for Change Quote To Owner (RCQTO) is the formal quotation of the RC, as originated by the general contractor and sent to the owner/architect. Like the other pricing mechanisms, this document should include statements as to how long the quote will remain valid and any other stipulations or conditions that might be required. One of the conditions that may be a part of an RCQTO is an extension of contract time.

Timely Response

Because the Request for Change will be a surprise to the owner, the manager or subcontractor should send the RCQTO to the owner/architect as soon as possible, thereby providing the owner with as much time as possible to review the situation. It is unwise to trap the owner into doing something because he/she did not have the necessary information at the right time. The right time is defined as "as soon as possible."

What to Include in the RCQTO

The following items should be included in the RCQTO:

- From
- To
- Date sent
- Job identification
- RC number and identification
- Brief description of the work involved
- Reason for the RC, and date it became evident that the RC needed to be submitted
- Cost of the RC
- Alternate methods of solving the subject of the RC, if any
- How long the quote is valid
- Extension of time, if applicable
- Any special notes to owner
- Cost breakdown

DELTA CONSTRUCTION COMPANY

6065 Mission Gorge Road, #193
San Diego, CA 92120
(619) 582-5829

*** REQUEST FOR CHANGE ***

Date: 2/28/90

Job Number: 321
Request for Change Number: 003

TO: CLEAR GLASS COMPANY
8888 MAIN STREET
SAN DIEGO, CA 92342

ATTENTION: WILL BREAK

JOB NUMBER: 321 GIBSON OFFICE BUILDING

FROM: J. EDWARD GRIMES

SUBJECT: VARIOUS CHANGES BY BUILDING DEPARTMENT PLAN CHECK

We are transmitting to you the following which is DESCRIBED below for your review and response. Please refer to the Job Number as well as the Request for Change. Return this information to us by 3/14/90.

Changes in the scope of work bid by Building Department for the following firms. Clear Glass. Continental Hardware. Sparks Electric. Green Landscaping. Rocky Concrete

Please contact me as soon as possible if you need any further information or if you have any questions.

Sincerely,

J. Edward Grimes

Copies were sent to the following:

SPARKS ELECTRIC COMPANY
ROCKY CONCRETE CONTRACTORS
CONTINENTAL HARDWARE
GREEN LANDSCAPING

dp: 2/28/90 13:29 **PaperWorks,** a Construction Software Program

Figure 22.1a

Means Forms

LETTER OF TRANSMITTAL

Delta Construction Company
6678 Mission Road
San Diego, CA 92120
(619) 581-6315

FROM: J. Edward Grimes
Project Manager

DATE 2/28/90
PROJECT Gibson Office Bldg. #321
LOCATION San Diego, Ca.
ATTENTION William Break
RE: Building department changes

TO: Clear Glass Company
888 Main St.
San Diego, Ca. 92312

Gentlemen:

WE ARE SENDING YOU ☒ HEREWITH ☐ DELIVERED BY HAND ☐ UNDER SEPARATE COVER

VIA ______ THE FOLLOWING ITEMS:

☐ PLANS ☐ PRINTS ☐ SHOP DRAWINGS ☐ SAMPLES ☐ SPECIFICATIONS

☐ ESTIMATES ☐ COPY OF LETTER ☐ ______

COPIES	DATE OR NO.	DESCRIPTION
1		Plans with bldg. department changes
CC		Continental Glass
CC		Sparks Electric
CC		Green Landscaping
CC		Rocky Group

THESE ARE TRANSMITTED AS INDICATED BELOW

☒ FOR YOUR USE ☐ APPROVED AS NOTED ☐ RETURN ______ CORRECTED PRINTS

☐ FOR APPROVAL ☐ APPROVED FOR CONSTRUCTION ☐ SUBMIT ______ COPIES FOR ______

☐ AS REQUESTED ☐ RETURNED FOR CORRECTIONS ☐ RESUBMIT ______ COPIES FOR ______

☒ FOR REVIEW AND COMMENT ☐ RETURNED AFTER LOAN TO US ☒ ~~FOR BIDS~~ DUE 3/14/90

☐ ______

REMARKS: Enclosed are Building Department changes as noted on plans.

IF ENCLOSURES ARE NOT AS INDICATED, PLEASE NOTIFY US AT ONCE.

SIGNED: J Edward Grimes

Figure 22.1b

Sample RCQTO Letters

Figure 22.2a is a sample computer-generated Request for Change Quote to Owner. Figure 22.2b is a manual RCQTO, using the standard Transmittal form.

Many firms group all of the pricing vehicles together, mixing Price Requests, Field Orders, Clarifications, and Requests for Change. This is wrong, based on the often-mentioned, but worth repeating, KISS (Keep It Simple, Seymour) principle. It is easier for the owner/architect to grasp each of these concepts individually, rather than jumbled together. Again, it is worth noting that architects, engineers, managers, general contractors, and subcontractors find it easy to understand the differences between these four types of scope changes because it is part of their work. Owners, on the other hand, may oversee only one or two construction projects in a lifetime.

Figures 22.3 through 22.8 are three types of computer-generated logs for both Request for Change and Request for Change Quote to Owner. Notice that the logs for each of these documents contain similar, but not identical, information. The RC logs show all of the subcontractors' and general contractor's costs and fees, while the RCQTO shows only the price quoted to the owner. The subcontractor should never see the RCQTO logs, and the owner should never see the Price Request logs with subcontractor pricing information.

Figure 22.9 shows a manual type of log that combines the Request for Change log and Request for Change Quote to Owner log. This type is for the eyes of the general contractor only.

Summary

It is important for everyone (owner, architect, general contractor, and subcontractors) involved in a project to remember the following about Requests for Change:

1. They always represent a change in the scope of the work.
2. They are almost always a surprise (i.e., not planned) to the project manager, general contractor, and subcontractor.
3. They are almost always a surprise to the owner.
4. No one likes surprises that cost money.
5. They will almost always interfere with, or affect, the normal course of construction.
6. Some effort is required on the part of the general contractor and the subcontractors to accurately estimate the difficulty of the proposed RC. This is largely due to the fact that all of the information must be pieced together. This is one way in which the manager really earns his salary, by defining the limits of the RC, and seeing that all appropriate trades, subcontractors, and overhead costs are included.
7. Because a Request for Change comes as a surprise, the general contractor and subcontractors should expect critical scrutiny from the owner/architect. They should, therefore, make it a point to be diligent, precise, timely, and fair in the pricing. Suggestions for alternate methods are usually very welcome, and do not receive the same scrutiny directed toward the original submittal (particularly if the price is lower).

DELTA CONSTRUCTION COMPANY

6065 Mission Gorge Road, #193
San Diego, CA 92120
(619) 582-5829

REQUEST FOR CHANGE QUOTE TO OWNER

Date: 3/14/90

Job Number: 321
Request for Change Number: 003

TO: FRANCES GIBSON
6109 CAMINITO CLAVO
SAN DIEGO, CA 92120

ATTENTION: FRANCES GIBSON
JOB NUMBER: 321 GIBSON OFFICE BUILDING

FROM: J. EDWARD GRIMES

SUBJECT: VARIOUS CHANGES BY BUILDING DEPARTMENT PLAN CHECK

We are submitting the following changes in cost of $ 9,342.00
The details and breakdown of this change in cost is enclosed.
A complete description of the scope of work relating to this proposal is enclosed. The summary of the works is as follows:

Changes in the scope of work bid by Building Department for the following firms. Clear Glass. Continental Hardware. Sparks Electric. Green Landscaping. Rocky Concrete

Remarks to owner/architect:

No Price Request or Field Order was issued and all work was required by the city.

This work will result in an extension to our contract time of 10 days.

Please indicate your approval by signing and returning one copy of this letter to our office or issue a CHANGE ORDER for this work no later than 3/28/90 in order to not delay the project.

Sincerely,

J. EDWARD GRIMES

APPROVED BY: FRANCES GIBSON

SIGNED: ______________________ DATE: ______________

dp: 3/14/90 13:27 **PaperWorks**, a Construction Software Program

Figure 22.2a

Means Forms

PROPOSAL

Delta Construction Company

6678 Mission Road
San Diego, CA 92120
(619) 581-6315

FROM: J. Edward Grimes
Project Manager

PROPOSAL NO. RCQ - 321.03
DATE 3/14/90
PROJECT Gibson Office Bldg. #321
LOCATION San Diego, Ca.

TO: Frances Gibson
6109 Caminito Clavo
San Diego, Ca. 92120

CONSTRUCTION TO BEGIN

COMPLETION DATE

Gentlemen:

The undersigned proposes to furnish all materials and necessary equipment and perform all labor necessary to complete the following work:

Various changes due to Building Department Plan check.

This work will result in an extension to our contract time of 10 (ten) days.

All of the above work to be completed in a substantial and workmanlike manner

☒ for the sum of Nine Thousand Three Hundred and Forty-Two dollars ($ 9,342.00)

☐ to be paid for at actual cost of Labor, Materials and Equipment plus ______ percent (______ %)

Payments to be made as follows: ______

______ The entire amount of the contract to be paid within ______ after completion

Any alteration or deviation from the plans and specifications will be executed only upon written orders for same and will be added to or deducted from the sum quoted in this contract. All additional agreements must be in writing.

The Contractor agrees to carry Workmen's Compensation and Public Liability Insurance and to pay all taxes on material and labor furnished under this contract as required by Federal laws and the laws of the State in which this work is performed.

Respectfully submitted,

Contractor Delta Construction Co.

By J Edward Grimes

ACCEPTANCE

You are hereby authorized to furnish all material, equipment and labor required to complete the work described in the above proposal, for which the undersigned agrees to pay the amount stated in said proposal and according to the terms thereof.

Date ______ 19____ ______

Figure 22.2b

```
                       DELTA CONSTRUCTION COMPANY
=================================================PaperWorks Version 1.23F==
                     *** REQUEST FOR CHANGE LOG ***                  .RC1
          Job 321     GIBSON OFFICE BUILDING
Date: 04/04/90                                                  Page:   1
===========================================================================
 RC #: 001          sent: 02/21/90      method: DESCRIBED      due: 03/07/90
    summary: INCREASED SCOPE OF WORK DUE TO BUILDING DEPARTMENT CORRECTIONS.
description: The SD Building Department increase rebar requirements at the
             footings that were not on the bid documents.  These additions only
             show on the corrected set as approved by the Building Dept.
 rem to sub:

      sent to:                          date returned     amount quoted
   LOT'S OF REBAR                         03/05/90      $            999.00
   GENERAL CONTRACTOR FEE                 03/14/90      $            100.00
                                         sub total:     $          1,099.00

   date due owner     amount quoted    date returned       amount approved
     03/28/90       $        1,099.00    03/28/90        $            999.00

remarks: This was required and we needed to get the rebar in the ground so
         that we could pour footings

                          STATUS:   REQUEST CHANGE APPROVED
---------------------------------------------------------------------------
 RC #: 002          sent: 02/21/90      method: DESCRIBED      due: 03/07/90
    summary: INCREASE IN SCOPE OF WORK FOR STRUCTURAL WORK
description: The SD Building Department has made significant scope changes in
             the approved set of drawings.  These changes only show on the
             approved set of drawing as the job site
 rem to sub: Include these changes in your shop drawings.

      sent to:                          date returned     amount quoted
   STRONG STRUCTURAL STEEL               NOT QUOTED     $              0.00
                                         sub total:     $              0.00

---------------------------------------------------------------------------
 RC #: 003          sent: 02/28/90      method: DESCRIBED      due: 03/14/90
    summary: VARIOUS CHANGES BY BUILDING DEPARTMENT PLAN CHECK
description: Changes in the scope of work bid by Building Department for the
             following firms.  Clear Glass.  Continental Hardware.  Sparks
             Electric.  Green Landscaping.  Rocky Concrete
 rem to sub:

      sent to:                          date returned     amount quoted
   CLEAR GLASS COMPANY                    03/12/90      $          2,000.00
   SPARKS ELECTRIC COMPANY                03/08/90      $            467.00
   ROCKY CONCRETE CONTRACTORS             03/14/90      $          4,800.00
   CONTINENTAL HARDWARE                   03/01/90      $            876.00
   GREEN LANDSCAPING                      03/14/90      $            350.00
   GENERAL CONTRACTOR FEE                 03/14/90      $            849.00
                                         sub total:     $          9,342.00

   date due owner     amount quoted    date returned       amount approved
     03/28/90       $        9,342.00                    $

remarks: No Price Request or Field Order was issued and all work was
         required by the city.
```

Figure 22.3

```
                         DELTA CONSTRUCTION COMPANY
=============================================PaperWorks Version 1.23F==
              *** REQUEST FOR CHANGE LOG DUE FROM SUBS ***          .RC3
           Job 321      GIBSON OFFICE BUILDING
Date: 04/04/90                                                 Page:    1
==========================================================================
 RC #: 002            sent: 02/21/90      method: DESCRIBED       due: 03/07/90
    summary: INCREASE IN SCOPE OF WORK FOR STRUCTURAL WORK
description: The SD Building Department has made significant scope changes in
             the approved set of drawings.  These changes only show on the
             approved set of drawing as the job site
 rem to sub: Include these changes in your shop drawings.

       sent to:
      STRONG          STRONG STRUCTURAL STEEL
--------------------------------------------------------------------------

 RC #: 008            sent: 03/21/90      method: DESCRIBED       due: 04/04/90
    summary: ELECTRICAL CHANGES REQUIRED BY PERMIT DRAWINGS
description: During plan check various corrections were required by the
             building department and have not been incorporated into any
             Price Requests or Field Orders.  This work is required
 rem to sub:

       sent to:
      SPARKS          SPARKS ELECTRIC COMPANY
--------------------------------------------------------------------------

 RC #: 009            sent: 03/21/90      method: DESCRIBED       due: 04/04/90
    summary: HVAC PLAN CHECK REVISIONS
description: During plan check the building department required some changes
             in the scope of the work.  This work was not originally included
             in the bid.  No Price Request or Field Order has been issued.
 rem to sub:

       sent to:
      COOL            COOL AIR CONDITIONING
--------------------------------------------------------------------------

 RC #: 010            sent: 03/28/90      method: DESCRIBED       due: 04/11/90
    summary: EXTRA WORK FOR HUNT COMPANY
description: It was necessary to perform extra work in order that the Hunt
             Company could occupy their space.

 rem to sub: Hunt Company signed for the work but we need to charge you.

       sent to:
      GCW             GENERAL CONTRACTOR WORK
--------------------------------------------------------------------------

 RC #: 011            sent: 03/28/90      method: DESCRIBED       due: 04/11/90
    summary: EXTRA WORK FOR CONSTRUCTION CONSULTING GROUP
description: We were authorized to perform extra work for CCG so they could
             run their computer hardware.  We do not have a contract with
             CCG, however we do have the authorization.
 rem to sub:

       sent to:
      SPARKS          SPARKS ELECTRIC COMPANY
--------------------------------------------------------------------------
```

Figure 22.4

```
                         DELTA CONSTRUCTION COMPANY
==========================================================PaperWorks Version 1.23F==
      *** REQUEST FOR CHANGE LOG DELINQUENT FROM SUBS as of 04/04/89  *     .RC5
                 Job 321      GIBSON OFFICE BUILDING
Date: 04/04/89                                                        Page:    1
====================================================================================
 RC #: 002           sent: 02/21/89      method: DESCRIBED        due: 03/07/89
    summary: INCREASE IN SCOPE OF WORK FOR STRUCTURAL WORK
description: The SD Building Department has made significant scope changes in
             the approved set of drawings.  These changes only show on the
             approved set of drawing as the job site
 rem to sub: Include these changes in your shop drawings.

       sent to:
      STRONG         STRONG STRUCTURAL STEEL
------------------------------------------------------------------------------------
 RC #: 008           sent: 03/21/89      method: DESCRIBED        due: 04/04/89
    summary: ELECTRICAL CHANGES REQUIRED BY PERMIT DRAWINGS
description: During plan check various corrections were required by the
             building department and have not been incorporated into any
             Price Requests or Field Orders.  This work is required
 rem to sub:

       sent to:
      SPARKS         SPARKS ELECTRIC COMPANY
------------------------------------------------------------------------------------
 RC #: 009           sent: 03/21/89      method: DESCRIBED        due: 04/04/89
    summary: HVAC PLAN CHECK REVISIONS
description: During plan check the building department required some changes
             in the scope of the work.  This work was not originally included
             in the bid.  No Price Request or Field Order has been issued.
 rem to sub:

       sent to:
      COOL           COOL AIR CONDITIONING
------------------------------------------------------------------------------------
```

Figure 22.5

```
                        DELTA CONSTRUCTION COMPANY
=======================================================PaperWorks Version 1.23F==
                *** REQUEST FOR CHANGE QUOTE TO OWNER LOG ***                .RQ1
          Job 321      GIBSON OFFICE BUILDING
Date: 04/14/90                                                          Page:   1
==================================================================================
 RC #: 001            sent: 02/21/90     method: DESCRIBED       due: 03/07/90
    summary: INCREASED SCOPE OF WORK DUE TO BUILDING DEPARTMENT CORRECTIONS.
description: The SD Building Department increase rebar requirements at the
             footings that were not on the bid documents.  These additions only
             show on the corrected set as approved by the Building Dept.
 rem to sub:

   date due owner     amount quoted    date returned      amount approved
     03/28/90       $         1,099.00     03/28/90      $             999.00

remarks: This was required and we needed to get the rebar in the ground so
         that we could pour footings
                                         STATUS:     REQUEST CHANGE QTO APPROVE
----------------------------------------------------------------------------------
 RC #: 003            sent: 02/28/90     method: DESCRIBED       due: 03/14/90
    summary: VARIOUS CHANGES BY BUILDING DEPARTMENT PLAN CHECK
description: Changes in the scope of work bid by Building Department for the
             following firms.  Clear Glass.  Continental Hardware.  Sparks
             Electric.  Green Landscaping.  Rocky Concrete
 rem to sub:

   date due owner     amount quoted    date returned      amount approved
     03/28/90       $         9,342.00                   $

remarks: No Price Request or Field Order was issued and all work was
         required by the city.
                                         STATUS:     REQUEST CHANGE QTO APPROVE
----------------------------------------------------------------------------------
 RC #: 005            sent: 03/07/90     method: DESCRIBED       due: 03/21/90
    summary: SPECIFICATION CHANGE FOR CARPET
description: We received a new specification for the carpet.  The carpet was
             bid on the previous specification.  A Price Request was not issue
             for this work.  The new specification is an upgrade in carpet.
 rem to sub:

   date due owner     amount quoted    date returned      amount approved
     04/04/90       $           979.00                   $

remarks: This will confirm our conference call of 3/20 with you and
         One Line Design.
                                         STATUS:     REQUEST CHANGE QTO APPROVE
----------------------------------------------------------------------------------
 RC #: 007            sent: 03/14/90     method: DESCRIBED       due: 03/28/90
    summary: CORRECTIONS OF WORK
description: Due to incomplete plans we have had to rework plumbing lines in
             order to insure that the sewer goes downhill.  This rework was
             kept to a minimum but required.
 rem to sub:

   date due owner     amount quoted    date returned      amount approved
     04/11/90       $           841.00     04/04/90      $             841.00

remarks: This was urgent, and we could not get in touch with One Line for
         a Field Order
                                         STATUS:     REQUEST CHANGE QTO APPROVE
----------------------------------------------------------------------------------
```

Figure 22.6

```
                        DELTA CONSTRUCTION COMPANY
==================================================PaperWorks Version 1.23F==
          *** REQUEST FOR CHANGE LOG DUE FROM OWNER ***                .RQ3
          Job 321      GIBSON OFFICE BUILDING
Date: 04/14/90                                                    Page:    1
=============================================================================
 RC #: 003           sent: 02/28/90      method: DESCRIBED        due: 03/14/90
    summary: VARIOUS CHANGES BY BUILDING DEPARTMENT PLAN CHECK
description: Changes in the scope of work bid by Building Department for the
             following firms.  Clear Glass.  Continental Hardware.  Sparks
             Electric.  Green Landscaping.  Rocky Concrete

   date due owner 03/28/90     amount quoted $          9,342.00
remarks: No Price Request or Field Order was issued and all work was
         required by the city.

-----------------------------------------------------------------------------

 RC #: 005           sent: 03/07/90      method: DESCRIBED        due: 03/21/90
    summary: SPECIFICATION CHANGE FOR CARPET
description: We received a new specification for the carpet.  The carpet was
             bid on the previous specification.  A Price Request was not issue
             for this work.  The new specification is an upgrade in carpet.

   date due owner 04/04/90     amount quoted $            979.00
remarks: This will confirm our conference call of 3/20 with you and
         One Line Design.

-----------------------------------------------------------------------------

 RC #: 009           sent: 03/21/90      method: DESCRIBED        due: 04/04/90
    summary: HVAC PLAN CHECK REVISIONS
description: During plan check the building department required some changes
             in the scope of the work.  This work was not originally included
             in the bid.  No Price Request or Field Order has been issued.

   date due owner 04/27/90     amount quoted $          1,990.00
remarks: The changes involve some long lead equipment items.  Per our
         telephone conversation we have placed those orders.

-----------------------------------------------------------------------------

 RC #: 012           sent: 04/04/90      method: DESCRIBED        due: 04/18/90
    summary: REWORK AT ONE HOUR CORRIDOR
description: The one hour corridor detail per the original plans was rejected
             in the field by the building inspector.  It was necessary to
             demo some of the work and rework other portions of the corridors.

   date due owner 04/27/90     amount quoted $            865.00
remarks: It was necessary to get this work done now or it would have
         cost more money later.

-----------------------------------------------------------------------------
```

Figure 22.7

```
                           DELTA CONSTRUCTION COMPANY
=================================================PaperWorks Version 1.23F==
     *** REQUEST FOR CHANGE LOG DELINQUENT FROM OWNERS as of 04/14/90  *    .RQ5
             Job 321      GIBSON OFFICE BUILDING
Date: 04/14/90                                                        Page:    1
==============================================================================
 RC #: 003              sent: 02/28/90       method: DESCRIBED        due: 03/14/90
    summary: VARIOUS CHANGES BY BUILDING DEPARTMENT PLAN CHECK
description: Changes in the scope of work bid by Building Department for the
             following firms.  Clear Glass.  Continental Hardware.  Sparks
             Electric.  Green Landscaping.  Rocky Concrete

   date due owner 03/28/90     amount quoted $          9,342.00
remarks: No Price Request or Field Order was issued and all work was
         required by the city.

------------------------------------------------------------------------------

 RC #: 005              sent: 03/07/90       method: DESCRIBED        due: 03/21/90
    summary: SPECIFICATION CHANGE FOR CARPET
description: We received a new specification for the carpet.  The carpet was
             bid on the previous specification.  A Price Request was not issue
             for this work.  The new specification is an upgrade in carpet.

   date due owner 04/04/90     amount quoted $            979.00
remarks: This will confirm our conference call of 3/20 with you and
         One Line Design.

------------------------------------------------------------------------------
```

Figure 22.8

PROJECT Gibson Office Bldg. SHEET OF

LOCATION San Diego, Ca. JOB NO. 321

#	Description	Info Rec'd Req. From Owner	Sent to Subcontractor	Due	Rec'd	Amounts	Sent to owner	Due	Rec'd	Amount Approved		Total Approved Changes
01	Building Department changes	2/21	2/21	3/7								
	Lot's of Rebar		2/21	3/7	3/5	999						
	G.C. Fee				3/14	100						
						1099	3/14	3/28	3/28	999		999
02	Increased Structural Wk	2/21	2/21	(3/7)								
	Strong Steel		2/21	(3/7)								
	G.C. Work / Fee		2/21									
03	Bldg. Dept. Req'd changes	2/28		3/4								
	Clear Glass		2/28	3/14	3/12	2000						
	Sparks Electric		2/28	3/14	3/8	467						
	Rocky Concrete		2/28	3/14	3/14	4800						
	Continental Hdwe.		2/28	3/14	3/1	876						
	Green Landscaping		2/28	3/14	3/4	350						
	G.C. Work / Fee				3/14	849						
						9342	3/14	(3/28)				

Circled = late

Page 1 of 2

Figure 22.9

Appendix

AGC Contract Documents

THE ASSOCIATED GENERAL CONTRACTORS

STANDARD FORM OF DESIGN-BUILD AGREEMENT AND GENERAL CONDITIONS BETWEEN OWNER AND CONTRACTOR

This Document has important legal and insurance consequences; consultation with an attorney and insurance consultants and carriers is encouraged with respect to its completion or modification.

AGREEMENT

Made this day of in the year of Nineteen Hundred and

BETWEEN the Owner, and

the Contractor.

For services in connection with the following described Project: (Include complete Project location and scope)

SAMPLE

REPRESENTED WITH THE PERMISSION OF THE ASSOCIATED GENERAL CONTRACTORS OF AMERICA (AGC). COPIES OF CURRENT FORMS MAY BE OBTAINED FROM AGC'S PUBLICATIONS DEPARTMENT, 1957 E ST. N.W., WASHINGTON, D.C. 20006.

The Owner and the Contractor agree as set forth below:

Certain provisions of this document have been derived, with modifications, from the following documents published by The American Institute of Architects AIA Document A111 Owner-Contractor Agreement, © 1976, AIA Document A201 General Conditions, © 1976 by The American Institute of Architects Usage made of AIA language, with the permission of AIA, does not imply AIA endorsement or approval of this document Further reproduction of copyrighted AIA materials without separate written permission from AIA is prohibited

© 1982 Associated General Contractors of America

INDEX

SAMPLE

©1982 Associated General Contractors of America

ARTICLE 1

The Construction Team and Extent of Agreement

THE CONTRACTOR accepts the relationship of trust and confidence established between him and the Owner by this Agreement. He agrees to furnish the architectural, engineering and construction services set forth herein and agrees to furnish efficient business administration and superintendence, and to use his best efforts to complete the Project in the best and soundest way and in the most expeditious and economical manner consistent with the interests of the Owner.

1.1 *The Construction Team:* The Contractor, the Owner, and the Architect/Engineer called the "Construction Team" shall work from the beginning of design through construction completion. The services of
, as the Architect/Engineer, will be furnished by the Contractor pursuant to an agreement between the Contractor and the Architect/Engineer.

1.2 *Extent of Agreement:* This Agreement represents the entire agreement between the Owner and the Contractor and supersedes all prior negotiations, representations or agreements. When the Drawings and Specifications are complete, they shall be identified by amendment to this Agreement. This Agreement shall not be superseded by any provisions of the documents for construction and may be amended only by written instrument signed by both Owner and Contractor.

1.3 *Definitions:* The Project is the total construction to be designed and constructed of which the Work is a part. The Work comprises the completed construction required by the Drawings and Specifications. The term day shall mean calendar day unless otherwise specifically designated.

ARTICLE 2

Contractor's Responsibilities

2.1 Contractor's Services

2.1.1 The Contractor shall be responsible for furnishing the Design and for the construction of the Project. The Owner and Contractor shall develop a design and construction phase schedule and the Owner shall be responsible for prompt decisions and approvals so as to maintain the approved schedule.

2.1.2 The Owner and Contractor shall develop a design phase schedule. *PHASE 1:* Based upon the Owner's Project requirements, schematic Design Studies will be prepared by the Architect/Engineer. These Schematics are for the purpose of assisting the Owner in determining the feasibility of the Project. *PHASE 2:* Upon approval of Schematic Designs and authorization from the Owner to proceed, the Architect/Engineer shall prepare Design Development Documents to fix the size and character of the Project as to structural, mechanical and electrical systems, materials and other appropriate essential items in the Project. These Development Documents are the basis to establish a Guaranteed Maximum Price for the design and construction of the Project. *PHASE 3:* From approved Design Development Documents the Architect/Engineer will prepare working Drawings and Specifications setting forth in detail the requirements for the construction of the Project, and based upon codes, laws or regulations which have been enacted at the time of their preparation. These Working Drawings and Specifications will be used to confirm the Guaranteed Maximum Price. Construction of the Project shall be in accordance with these Drawings and Specifications as approved by the Owner. The Drawings and Specifications shall remain the property of the Contractor and are not to be used by the Owner on other projects without the written consent of the Contractor. If the working Drawings and Specifications have not been completed and a Guaranteed Maximum Price has been established prior to the completion of the working Drawings and Specifications, the Contractor, the Architect/Engineer and the Owner will work closely together to monitor the design in accordance with prior approvals so as to ensure that the Project can be constructed within the Guaranteed Maximum Price. As these working Drawings and Specifications are being completed, the Contractor will keep the Owner advised of the effects of any Owner requested changes on the Contract Time Schedule and/or the Guaranteed Maximum Price.

SAMPLE

2.1.3 The Contractor will assist the Owner in securing permits necessary for the construction of the Project.

2.2 Responsibilities With Respect to Construction

2.2.1 The Contractor will provide all construction supervision, inspection, labor, materials, tools, construction equipment and subcontracted items necessary for the execution and completion of the Project.

2.2.2 The Contractor will pay all sales, use, gross receipts and similar taxes related to the Work provided by the Contractor which have been legally enacted at the time of execution of this Agreement and for which the Contractor is liable.

2.2.3 The Contractor will prepare and submit for the Owner's approval an estimated progress schedule for the Project. This schedule shall indicate the dates for the starting and completion of the various stages of the design and construction. It shall be revised as required by the conditions of the Work and those conditions and events which are beyond the Contractor's control.

2.2.4 The Contractor shall at all times keep the premises free from the accumulation of waste materials or rubbish caused by his operations. At the completion of the Work, he shall remove all of his waste material and rubbish from and around the Project as well as all his tools, construction equipment, machinery and surplus materials.

2.2.5 The Contractor will give all notices and comply with all laws and ordinances legally enacted at the date of execution of the Agreement, which govern the proper execution of the Work.

2.2.6 The Contractor shall take necessary precautions for the safety of his employees on the Work, and shall comply with all applicable provisions of federal, state and municipal safety laws to prevent accidents or injury to persons on, about or adjacent to the Project site. He shall erect and properly maintain, at all times, as required by the conditions and progress of Work, necessary safeguards for the protection of workmen and the public. It is understood and agreed, however, that the Contractor shall have no responsibility for the elimination or abatement of safety hazards created or otherwise resulting from Work at the job site carried on by other persons or firms directly employed by the Owner as separate contractors or by the Owner's tenants, and the Owner agrees to cause any such separate contractors and tenants to abide by and fully adhere to all applicable provisions of federal, state and municipal safety laws and regulations and to comply with all reasonable requests and directions of the Contractor for the elimination or abatement of any such safety hazards at the job site.

2.2.7 The Contractor shall keep such full and detailed accounts as may be necessary for proper financial management under this Agreement. The system shall be satisfactory to the Owner, who shall be afforded access to all the Contractor's records, books, correspondence, instructions, drawings, receipts, vouchers, memoranda and similar data relating to this Agreement. The Contractor shall preserve all such records for a period of three years after the final payment or longer where required by law.

SAMPLE

2.3 Royalties and Patents

2.3.1 The Contractor shall pay all royalties and license fees for materials, methods and systems incorporated in the work. He shall defend all suits or claims for infringement of any patent rights and shall save the Owner harmless from loss on account thereof except when a particular design, process or product is specified by the Owner. In such case the Contractor shall be responsible for such loss only if he has reason to believe that the design, process or product so specified is an infringement of a patent, and fails to give such information promptly to the Owner.

2.4 Warranties and Completion

2.4.1 The Contractor warrants to the Owner that all materials and equipment furnished under this Agreement will be new, unless otherwise specified, and that all Work will be of good quality, free from improper workmanship and defective materials and in conformance with the Drawings and Specifications. The Contractor agrees to correct all Work performed by him under this Agreement which proves to be defective in material and workmanship within a period of one year from the Date of Substantial Completion as defined in Paragraph 5.2, or for such longer periods of time as may be set forth with respect to specific warranties contained in the Specifications.

2.4.2 The Contractor will secure required certificates of inspection, testing or approval and deliver them to the Owner.

2.4.3 The Contractor will collect all written warranties and equipment manuals and deliver them to the Owner.

2.4.4 The Contractor with the assistance of the Owner's maintenance personnel, will direct the checkout of utilities and operations of systems and equipment for readiness, and will assist in their initial start-up and testing.

2.5 Additional Services

2.5.1 The Contractor will provide the following additional services upon the request of the Owner. A written agreement between the Owner and Contractor shall define the extent of such additional services and the amount and manner in which the Contractor will be compensated for such additional services.

2.5.2 Services related to investigation, appraisals or evaluations of existing conditions, facilities or equipment, or verification of the accuracy of existing drawings or other Owner-furnished information.

2.5.3 Services related to Owner-furnished equipment, furniture and furnishings which are not a part of this Agreement.

2.5.4 Services for tenant or rental spaces not a part of this Agreement.

2.5.5 Obtaining and training maintenance personnel or negotiating maintenance service contracts.

ARTICLE 3

Owner's Responsibilities

3.1 The Owner shall provide full information regarding his requirements for the Project.

3.2 The Owner shall designate a representative who shall be fully acquainted with the Project, and has authority to approve changes in the scope of the Project, render decisions promptly, and furnish information expeditiously and in time to meet the dates set forth in Subparagraph 2.2.3.

3.3 The Owner shall furnish for the site of the Project all necessary surveys describing the physical characteristics, soils reports and subsurface investigations, legal limitations, utility locations, and a legal description.

3.4 The Owner shall secure and pay for necessary approvals, easements, assessments and charges required for the construction, use, or occupancy of permanent structures or for permanent changes in existing facilities.

3.5 The Owner shall furnish such legal services as may be necessary for providing the items set forth in Paragraph 3.4, and such auditing services as he may require.

3.6 If the Owner becomes aware of any fault or defect in the Project or non-conformance with the Drawings or Specifications, he shall give prompt written notice thereof to the Contractor.

3.7 The Owner shall provide the insurance for the Project as provided in Paragraph 12.4.

3.8 The Owner shall bear the costs of any bonds that may be required.

3.9 The services and information required by the above paragraphs shall be furnished with reasonable promptness at the Owner's expense and the Contractor shall be entitled to rely upon the accuracy and the completeness thereof.

3.10 The Owner shall furnish reasonable evidence satisfactory to the Contractor, prior to commencing Work and at such future times as may be required, that sufficient funds are available and committed for the entire Cost of the Project. Unless such reasonable evidence is furnished, the Contractor is not required to commence or continue any Work, or may, if such evidence is not presented within a reasonable time, stop Work upon 15 days notice to the Owner. The failure of the Contractor to insist upon the providing of this evidence at any one time shall not be a waiver of the Owner's obligation to make payments pursuant to this Agreement nor shall it be a waiver of the Contractor's right to request or insist that such evidence be provided at a later date.

3.11 The Owner shall have no contractual obligation to the Contractor's Subcontractors and shall communicate with such Subcontractors only through the Contractor.

SAMPLE

ARTICLE 4

Subcontracts

4.1 All portions of the Work that the Contractor does not perform with his own forces shall be performed under subcontracts.

4.2 A Subcontractor is a person or entity who has a direct contract with the Contractor to perform any Work in connection with the Project. The term Subcontractor does not include any separate contractor employed by the Owner or the separate contractors' subcontractors.

4.3 No contractual relationship shall exist between the Owner and any Subcontractor. The Contractor shall be responsible for the management of the Subcontractors in the performance of their Work.

ARTICLE 5

Contract Time Schedule

5.1 The Work to be performed under this Agreement shall be commenced on or about and shall be substantially completed on or about

5.2 The Date of Substantial Completion of the Project or a designated portion thereof is the date when construction is sufficiently complete in accordance with the Drawings and Specifications so the Owner can occupy or utilize the Project or designated portion thereof for the use for which it is intended. Warranties called for by this Agreement or by the Drawings and Specifications shall commence on the Date of Substantial Completion of the Project or designated portion thereof. This date shall be established by a Certificate of Substantial Completion signed by the Owner and Contractor and shall state their respective responsibilities for security, maintenance, heat, utilities, damage to the Work and insurance. This Certificate shall also list the items to be completed or corrected and fix the time for their completion and correction.

5.3 If the Contractor is delayed at any time in the progress of the Project by any act or neglect of the Owner or by any separate contractor employed by the Owner, or by changes order in the Project, or by labor disputes, fire, unusual delay in transportation, adverse weather conditions not reasonably anticipatable, unavoidable casualties, or any causes beyond the Contractor's control, or a delay authorized by the Owner pending arbitration, then the Date for Substantial Completion shall be extended by Change Order for the period of time caused by such delay.

SAMPLE

Guaranteed Maximum Price

6.1 The Contractor guarantees that the maximum price to the Owner for the Cost of the Project as set forth in Article 8, and the Contractor's Fee as set forth in Article 7, will not exceed
Dollars ($), which sum shall be called the Guaranteed Maximum Price.

6.2 The Guaranteed Maximum Price is based upon laws, codes, and regulations in existence at the date of its establishment and upon criteria, Drawings, and Specifications as set forth below:

6.3 The Guaranteed Maximum Price will be modified for delays caused by the Owner and for Changes in the Project, all pursuant to Article 9.

6.4 Allowances included in the Guaranteed Maximum Price are as set forth below:

6.5 Whenever the cost is more than or less than the Allowance, the Guaranteed Maximum Price shall be adjusted by Change Order.

ARTICLE 7

Contractor's Fee

7.1 In consideration of the performance of the Agreement, the Owner agrees to pay to the Contractor in current funds as compensation for his services a Fee as follows:

7.2 Adjustment in Fee shall be made as follows:

7.2.1 For Changes in the Project as provided in Article 9, the Contractor's Fee shall be adjusted as follows:

7.2.2 For delays in the Project not the responsibility of the Contractor, there will be an equitable adjustment in the fee to compensate the Contractor for his increased expenses.

7.2.3 In the event the Cost of the Project plus the Contractor's Fee shall be less than the Guaranteed Maximum Price as adjusted by Change Orders, the resulting savings will be shared by the Owner and the Contractor as follows:

7.2.4 The Contractor shall be paid an additional fee in the same proportion as set forth in 7.2.1 if the Contractor is placed in charge of managing the replacement of insured or uninsured loss.

7.3 The Contractor shall be paid monthly that part of his Fee proportionate to the percentage of Work completed, the balance, if any, to be paid at the time of final payment.

7.4 Including in the Contractor's Fee are the following:

7.4.1 Salaries or other compensation of the Contractor's employees at the principal office and branch offices, except employees listed in Subparagraph 8.2.3.

7.4.2 General operating expenses of the Contractor's principal and branch offices other than the field office.

7.4.3 Any part of the Contractor's capital expenses, including interest on the Contractor's capital employed for the Project.

7.4.4 Overhead or general expenses of any kind, except as may be expressly included in Article 8.

7.4.5 Costs in excess of the Guaranteed Maximum Price.

ARTICLE 8

Cost of the Project

8.1 The term Cost of the Project shall mean costs necessarily incurred in the design and construction of the Project and shall include the items set forth below in this Article. The Owner agrees to pay the Contractor for the Cost of the Project as defined in this Article. Such payment shall be in addition to the Contractor's Fee stipulated in Article 7.

8.2. Cost Items

8.2.1 All architectural, engineering and consulting fees and expenses incurred in designing and constructing the Project.

SAMPLE

8.2.2 Wages paid for labor in the direct employ of the Contractor in the performance of the Work under applicable collective bargaining agreements, or under a salary or wage schedule agreed upon by the Owner and the Contractor, and including such welfare or other benefits, if any, as may be payable with respect thereto.

8.2.3 Salaries of Contractor's employees when stationed at the field office, in whatever capacity employed, employees engaged on the road expediting the production or transportation of material and equipment and employees from the main or branch office performing the functions listed below:

8.2.4 Cost of all employee benefits and taxes for such items as unemployment compensation and social security, insofar as such cost is based on wages, salaries, or other remuneration paid to employees of the Contractor and included in the Cost of the Project under Subparagraphs 8.2.1, 8.2.2 and 8.2.3.

8.2.5 Reasonable transportation, traveling and hotel and moving expenses of the Contractor or of his officers or employees incurred in discharge of duties connected with the Project.

8.2.6 Cost of all materials, supplies and equipment incorporated in the Project, including costs of transportation and storage thereof.

8.2.7 Payments made by the Contractor to Subcontractors for Work performed pursuant to contract under this Agreement.

8.2.8 Cost, including transportation and maintenance, of all materials, supplies, equipment, temporary facilities and hand tools not owned by the workmen, which are employed or consumed in the performance of the Work, and cost less salvage value on such items used, but not consumed, which remain the property of the Contractor.

8.2.9 Rental charges of all necessary machinery and equipment, exclusive of hand tools, used at the site of the Work, whether rented from the Contractor or others, including installations, repairs and replacements, dismantling, removal, costs of lubrication, transportation and delivery costs thereof, at rental charges consistent with those prevailing in the area.

8.2.10 Cost of the premiums for all insurance which the Contractor is required to procure by this Agreement or is deemed necessary by the Contractor.

8.2.11 Sales, use, gross receipts or similar taxes related to the Project, imposed by any governmental authority, and for which the Contractor is liable.

8.2.12 Permit fees, licenses, tests, royalties, damages for infringement of patents and costs of defending suits therefor for which the Contractor is responsible under Subparagraph 2.3.1 and deposits lost for causes other than the Contractor's negligence.

8.2.13 Losses, expenses or damages to the extent not compensated by insurance or otherwise (including settlement made with the written approval of the Owner), and the cost of corrective work.

8.2.14 Minor expenses such as telegrams, long-distance telephone calls, telephone service at the site, expressage, and similar petty cash items in connection with the Project.

8.2.15 Cost of removal of all debris.

8.2.16 Costs incurred due to an emergency affecting the safety of persons and property.

8.2.17 Cost of data processing services required in the performance of the services outlined in Article 2.

8.2.18 Legal costs reasonably and properly resulting from prosecution of the Project for the Owner.

8.2.19 All costs directly incurred in the performance of the Project and not included in the Contractor's Fee as set forth in Paragraph 7.4

SAMPLE

ARTICLE 9

Changes in the Project

9.1 The Owner, without invalidating this Agreement, may order Changes in the Project within the general scope of this Agreement consisting of additions, deletions or other revisions, the Guaranteed Maximum Price, if established, the Contractor's Fee, and the Contract Time Schedule being adjusted accordingly. All such Changes in the Project shall be authorized by Change Order.

9.1.1 A Change Order is a written order to the Contractor signed by the Owner or his authorized agent and issued after the execution of this Agreement, authorizing a Change in the Project and/or an adjustment in the Guaranteed Maximum Price, the Contractor's Fee or the Contract Time Schedule. Each adjustment in the Guaranteed Maximum Price resulting from a Change Order shall clearly separate the amount attributable to the Cost of the Project and the Contractor's Fee.

9.1.2 The increase or decrease in the Guaranteed Maximum Price resulting from a Change in the Project shall be determined in one or more of the following ways:

9.1.2.1 by mutual acceptance of a lump sum properly itemized and supported by sufficient substantiating data to permit evaluation; or

9.1.2.2 by unit prices stated in this Agreement or subsequently agreed upon; or

9.1.2.3 by cost to be determined as defined in Article 8 and a mutual acceptable fixed or percentage fee; or

9.1.2.4 by the method provided in Subparagraph 9.1.3.

9.1.3 If none of the methods set forth in Clauses 9.1.2.1 through 9.1.2.3 is agreed upon, the Contractor, provided he receives a written order signed by the Owner, shall promptly proceed with the Work involved. The cost of such Work shall then be determined on the basis of the reasonable expenditures and savings of those performing the Work attributed to the change, including, in the case of an increase in the Guaranteed Maximum Price, a reasonable increase in the Contractor's Fee. In such case, and also under Clauses 9.1.2.3 and 9.1.2.4 above, the Contractor shall keep and present, in such form as the Owner may prescribe, an itemized accounting together with appropriate supporting data of the increase in the Cost of the Project as outlined in Article 8. The amount of decrease in the Guaranteed Maximum Price to be allowed by the Contractor to the Owner for any deletion or change which results in a net decrease in cost will be the amount of the actual net decrease. When both additions and credits are involved in any one change, the increase in Fee shall be figured on the basis of net increase, if any.

9.1.4 If unit prices are stated in this Agreement or subsequently agreed upon, and if the quantities originally contemplated are so changed in a proposed Change Order or as a result of several Change Orders that application of the agreed unit prices to the quantities of Work proposed will cause substantial inequity to the Owner or the Contractor, the applicable unit prices and the Guaranteed Maximum Price shall be equitably adjusted.

9.1.5 Should concealed conditions encountered in the performance of the Work below the surface of the ground or should concealed or unknown conditions in an existing structure be at variance with the conditions indicated by the Drawings, Specifications, or Owner-furnished information or should unknown physical conditions below the surface of the ground or should concealed or unknown conditions in an existing structure of an unusual nature, differing materially from those ordinarily encountered and generally recognized as inherent in work of the character provided for in this Agreement, be encountered, the Guaranteed Maximum Price and the Contract Time Schedule shall be equitably adjusted by Change Order upon claim by either party made within a reasonable time after the first observance of the conditions.

9.2 Claims for Additional Cost or Time

9.2.1 If the Contractor wishes to make a claim for an increase in the Guaranteed Maximum Price, or increase in his Fee or an extension in the Contract Time Schedule, he shall give the Owner written notice thereof within a reasonable time after the occurrence of the event giving rise to such claim. This notice shall be given by the Contractor before proceeding to execute the Work, except in an emergency endangering life or property in which case the Contractor shall act, at his discretion, to prevent

SAMPLE

threatened damage, injury or loss. Claims arising from delay shall be made within a reasonable time after the delay. Increases based upon design and estimating costs with respect to possible changes requested by the Owner, shall be made within a reasonable time after the decision is made not to proceed with the change. No such claim shall be valid unless so made. If the Owner and the Contractor cannot agree on the amount of the adjustment in the Guaranteed Maximum Price, the Contractor's Fee or Contract Time Schedule, it shall be determined pursuant to the provisions of Article 16. Any change in the Guaranteed Maximum Price, the Contractor's Fee or Contract Time Schedule resulting from such claim shall be authorized by Change Order.

9.3 Minor Changes in the Project

9.3.1 The Owner will have authority to order minor Changes in the Work not involving an adjustment in the Guaranteed Maximum Price or an extension of the Contract Time Schedule and not inconsistent with the intent of the Drawings and Specifications. Such Changes may be effected by written order and shall be binding on the Owner and the Contractor.

9.4 Emergencies

9.4.1 In any emergency affecting the safety of persons or property, the Contractor shall act, at his discretion, to prevent threatened damage, injury or loss. Any increase in the Guaranteed Maximum Price or extension of time claimed by the Contractor on account of emergency work shall be determined as provided in this Article.

ARTICLE 10

Discounts

All discounts for prompt payment shall accrue to the Owner to the extent the Cost of the Project is paid directly by the Owner or from a fund made available by the Owner to the Contractor for such payments. To the extent the Cost of the Project is paid with funds of the Contractor, all cash discounts shall accrue to the Contractor. All trade discounts, rebates and refunds, and all returns from sale of surplus materials and equipment, shall be credited to the Cost of the Project.

ARTICLE 11

Payments to the Contractor

11.1 Payments shall be made by the Owner to the Contractor according to the following procedure:

11.1.1 On or before the   day of each month after Work has commenced, the Contractor shall submit to the Owner an Application for Payment in such detail as may be required by the Owner based on the Work completed and materials stored on the site and/or at locations approved by the Owner along with a proportionate amount of the Contractor's Fee for the period ending on the   day of the month.

11.1.2 Within ten (10) days after his receipt of each monthly Application for Payment, the Owner shall pay directly to the Contractor the appropriate amounts for which Application for Payment is made therein. This payment request shall deduct the aggregate of amounts previously paid by the Owner.

11.1.3 If the Owner should fail to pay the Contractor at the time the payment of any amount becomes due, then the Contractor may, at any time thereafter, upon serving written notice that he will stop Work within five (5) days after receipt of the notice by the Owner, and after such five (5) day period, stop the Project until payment of the amount owing has been received. Written notice shall be deemed to have been duly served if sent by certified mail to the last business address known to him who gives the notice.

11.1.4 Payments due but unpaid shall bear interest at the rate the Owner is paying on his construction loan or at the legal rate, whichever is higher.

11.2 The Contractor warrants and guarantees that title to all Work, materials and equipment covered by an Application for Payment whether incorporated in the Project or not, will pass to the Owner upon receipt of such payment by the Contractor free and clear of all liens, claims, security interests or encumbrances hereinafter referred to as Liens.

11.3 No Progress Payment nor any partial or entire use or occupancy of the Project by the Owner shall constitute an acceptance of any Work not in accordance with the Drawings and Specifications.

SAMPLE

11.4 Final payment constituting the unpaid balance of the Cost of the Project and the Contractor's Fee shall be due and payable when the Project is delivered to the Owner, ready for beneficial occupancy, or when the Owner occupies the Project, whichever event first occurs, provided that the Project be then substantially completed and this Agreement substantially performed. If there should remain minor items to be completed, the Contractor and the Owner shall list such items and the Contractor shall deliver, in writing, his guarantee to complete said items within a reasonable time thereafter. The Owner may retain a sum equal to 150% of the estimated cost of completing any unfinished items, provided that said unfinished items are listed separately and the estimated cost of completing any unfinished items is likewise listed separately. Thereafter, the Owner shall pay to the Contractor, monthly, the amount retained for incomplete items as each of said items is completed.

11.5 Before issuance of Final Payment, the Owner may request satisfactory evidence that all payrolls, materials bills and other indebtedness connected with the Project have been paid or otherwise satisfied.

11.6 The making of Final Payment shall constitute a waiver of all claims by the Owner except those rising from:

11.6.1. Unsettled Liens.

11.6.2 Improper workmanship or defective materials appearing within one year after the Date of Substantial Completion.

11.6.3 Failure of the Work to comply with the Drawings and Specifications.

11.6.4 Terms of any special guarantees required by the Drawings and Specifications.

11.7 The acceptance of Final Payment shall constitute a waiver of all claims by the Contractor except those previously made in writing and unsettled.

SAMPLE

ARTICLE 12

Insurance, Indemnity and Waiver of Subrogation

12.1 Indemnity

12.1.1 The Contractor agrees to indemnify and hold the Owner harmless from all claims for bodily injury and property damage (other than the Work itself and other property insured under Paragraph 12.4) that may arise from the Contractor's operations under this Agreement.

12.1.2 The Owner shall cause any other contractor who may have a contract with the Owner to perform work in the areas where Work will be performed under this Agreement, to agree to indemnify the Owner and the Contractor and hold them harmless from all claims for bodily injury and property damage (other than property insured under Paragraph 12.4) that may arise from that contractor's operations. Such provisions shall be in a form satisfactory to the Contractor.

12.2 Contractor's Liability Insurance

12.2.1 The Contractor shall purchase and maintain such insurance as will protect him from the claims set forth below which may arise out of or result from the Contractor's operations under this Agreement whether such operations be by himself or by any Subcontractor or by anyone directly or indirectly employed by any of them, or by anyone for whose acts any of them may be liable:

12.2.1.1 Claims under workers' compensation, disability benefit and other similar employee benefit acts which are applicable to the Work to be performed.

12.2.1.2 Claims for damages because of bodily injury, occupational sickness or disease, or death of his employees under any applicable employer's liability law.

12.2.1.3 Claims for damages because of bodily injury, or death of any person other than his employees.

12.2.1.4 Claims for damages insured by usual personal injury liability coverage which are sustained (1) by any person as a result of an offense directly or indirectly related to the employment of such person by the Contractor or (2) by any other person.

12.2.1.5 Claims for damages, other than to the Work itself, because of injury to or destruction of tangible property, including loss of use therefrom.

12.2.1.6 Claims for damages because of bodily injury or death of any person or property damage arising out of the ownership, maintenance or use of any motor vehicle.

12.2.2 The Comprehensive General Liability Insurance shall include premises-operations (including explosion, collapse and underground coverage) elevators, independent contractors, completed operations, and blanket contractual liability on all written contracts, all including broad form property damage coverage.

12.2.3 The Contractor's Comprehensive General and Automobile Liability Insurance, as required by Subparagraphs 12.2.1 and 12.2.2 shall be written for not less than limits of liability as follows:

a. Comprehensive General Liability
   1. Bodily Injury -$________________ Each Occurrence (Completed Operations)
      $________________ Aggregate
   2. Property Damage -$________________ Each Occurrence
      $________________ Aggregate

b. Comprehensive Automobile Liability
   1. Bodily Injury -$________________ Each Person
      $________________ Each Occurrence
   2. Property Damage -$________________ Each Occurrence

12.2.4 Comprehensive General Liability Insurance may be arranged under a single policy for the full limits required or by a combination of underlying policies with the balance provided by an Excess or Umbrella Liability policy.

12.2.5 The foregoing policies shall contain a provision that coverages afforded under the policies will not be cancelled or not renewed until at least sixty (60) days' prior written notice has been given to the Owner. Certificates of Insurance showing such coverages to be in force shall be filed with the Owner prior to commencement of the Work.

12.3 Owner's Liability Insurance

12.3.1 The Owner shall be responsible for purchasing and maintaining his own liability insurance and, at his option, may purchase and maintain such insurance as will protect him against claims which may arise from operations under this Agreement.

12.4 Insurance to Protect Project

12.4.1 The Owner shall purchase and maintain property insurance in a form acceptable to the Contractor upon the entire Project for the full cost of replacement as the time of any loss. This insurance shall include as named insureds the Owner, the Contractor, Subcontractors and Subsubcontractors and shall insure against loss from the perils of Fire, Extended Coverage, and shall include "All Risk" insurance for physical loss or damage including, without duplication of coverage, at least theft, vandalism, malicious mischief, transit, collapse, flood, earthquake, testing, and damage resulting from defective design, workmanship or material. The Owner will increase limits of coverage, if necessary, to reflect estimated replacement cost. The Owner will be responsible for any co-insurance penalties or deductibles. If the Project covers an addition to or is adjacent to an existing building, the Contractor, Subcontractors and Subsubcontractors shall be named as additional insureds under the Owner's Property Insurance covering such building and its contents.

SAMPLE

12.4.1.1 If the Owner finds it necessary to occupy or use a portion or portions of the Project prior to Substantial Completion thereof, such occupancy shall not commence prior to a time mutually agreed to by the Owner and the Contractor and to which the insurance company or companies providing the property insurance have consented by endorsement to the policy or policies. This insurance shall not be cancelled or lapsed on account of such partial occupancy. Consent of the Contractor and of the insurance company or companies to such occupancy or use shall not be unreasonably withheld.

12.4.2 The Owner shall purchase and maintain such boiler and machinery insurance as may be required or necessary. This insurance shall include the interests of the Owner, the Contractor, Subcontractors and Subsubcontractors in the Work.

12.4.3 The Owner shall purchase and maintain such insurance as will protect the Owner and the Contractor against loss of use of Owner's property due to those perils insured pursuant to Subparagraph 12.4.1. Such policy will provide coverage for expediting expenses of materials, continuing overhead of the Owner and the Contractor, necessary labor expense including overtime, loss of income by the Owner and other determined exposures. Exposures of the Owner and the Contractor shall be determined by mutual agreement and separate limits of coverage fixed for each item.

12.4.4 The Owner shall file a copy of all policies with the Contractor before an exposure to loss may occur. Copies of any subsequent endorsements will be furnished to the Contractor. The Contractor will be given sixty (60) days notice of cancellation, non-renewal, or any endorsements restricting or reducing coverage. If the Owner does not intend to purchase such insurance, he shall inform the Contractor in writing prior to the commencement of the Work. The Contractor may then effect insurance which will protect the interest of himself, the Subcontractors and their Subsubcontractors in the Project, the cost of which shall be a Cost of the Project pursuant to Article 8, and the Guaranteed Maximum Price shall be increased by Change Order. If the Contractor is damaged by failure of the Owner to purchase or maintain such insurance or to so notify the Contractor, the Owner shall bear all reasonable costs properly attributable thereto.

12.5 Property Insurance Loss Adjustment

12.5.1 Any insured loss shall be adjusted with the Owner and the Contractor and made payable to the Owner and Contractor as trustees for the insureds, as their interests may appear, subject to any applicable mortgagee clause.

12.5.2 Upon the occurrence of an insured loss, monies received will be deposited in a separate account and the trustees shall make distribution in accordance with the agreement of the parties in interest, or in the absence of such agreement, in accordance with an arbitration award pursuant to Article 16. If the trustees are unable to agree between themselves on the settlement of the loss, such dispute shall also be submitted to arbitration pursuant to Article 16.

12.6 Waiver of Subrogation

12.6.1 The Owner and Contractor waive all rights against each other, the Architect/Engineer, Subcontractors and Subsubcontractors for damages caused by perils covered by insurance provided under Paragraph 12.4, except such rights as they may have to the proceeds of such insurance held by the Owner and Contractor as trustees. The Contractor shall require similar waivers from all Subcontractors and Subsubcontractors.

12.6.2 The Owner and Contractor waive all rights against each other and the Architect/Engineer, Subcontractors and Subsubcontractors for loss or damage to any equipment used in connection with the Project which loss is covered by any property insurance. The Contractor shall require similar waivers from all Subcontractors and Subsubcontractors.

12.6.3 The Owner waives subrogation against the Contractor, Architect/Engineer, Subcontractors, and Subsubcontractors on all property and consequential loss policies carried by the Owner on adjacent properties and under property and consequential loss policies purchased for the Project after its completion.

12.6.4 If the policies of insurance referred to in this Paragraph require an endorsement to provide for continued coverage where there is a wavier of subrogation, the owners of such policies will cause them to be so endorsed.

SAMPLE

ARTICLE 13

Termination of the Agreement And Owner's Right to Perform Contractor's Obligations

13.1 Termination by the Contractor

13.1.1 If the Project is stopped for a period of thirty (30) days under an order of any court or other public authority having jurisdiction, or as a result of an act of government, such as a declaration of a national emergency making materials unavailable, through no act or fault of the Contractor or if the Project should be stopped for a period of thirty (30) days by the Contractor for the Owner's failure to make payment thereon, then the Contractor may, upon seven days' written notice to the Owner, terminate this Agreement and recover from the Owner payment for all Work executed, the Contractor's Fee earned to date, and for any proven loss sustained upon any materials, equipment, tools, construction equipment and machinery, including reasonable profit and damages.

13.2 Owner's Right to Perform Contractor's Obligations and Termination by the Owner for Cause

13.2.1 If the Contractor fails to perform any of his obligations under this Agreement, including any obligation he assumes to perform Work with his own forces, the Owner may, after seven days' written notice, during which period the Contractor fails to perform such obligation, make good such deficiencies. The Guaranteed Maximum Price, if any, shall be reduced by the cost to the Owner of making good such deficiencies.

13.2.2 If the Contractor is adjudged a bankrupt, or if he makes a general assignment for the benefit of his creditors, or if a receiver is appointed on account of his insolvency, or if he persistently or repeatedly refuses or fails, except in cases for which extension of time is provided, to supply enough properly skilled workmen or proper materials, or if he fails to make proper payment to Subcontractors or for materials or labor, or persistently disregards laws, ordinances, rules, regulations or orders of any public authority having jurisdiction, or otherwise is guilty of a substantial violation of a provision of this Agreement, then the Owner may, without prejudice to any right or remedy and after giving the Contractor and his surety, if any, seven (7) days' written notice, during which period the Contractor fails to cure the violation, terminate the employment of the Contractor and take possession of the site and of all materials, equipment, tools, construction equipment and machinery thereon owned by the Contractor and may finish the Work by whatever reasonable method he may deem expedient. In such case, the Contractor shall not be entitled to receive any further payment until the Work is finished nor shall he be relieved from his obligations assumed under Article 6.

13.3 Termination by Owner Without Cause

13.3.1 If the Owner terminates the Agreement other than pursuant to Article 13.2.2, he shall reimburse the Contractor for any unpaid Cost of the Project due him under Article 8, plus the unpaid balance of the Contractor's Fee. If the Contractor's Fee is based upon a percentage of the Cost of the Project, the Fee shall be calculated upon the adjusted Guaranteed Maximum Cost, if any, otherwise to a reasonable estimated Cost of the Project when completed. The Owner shall also pay to the Contractor fair compensation, either by purchase or rental at the election of the Owner, for any equipment retained. In case of such termination of this Agreement the Owner shall further assume and become liable for obligations, commitments and unsettled claims that the Contractor has previously undertaken or incurred in good faith in connection with said Work. The Contractor shall, as a condition of receiving the payments referred to in this Article 13, execute and deliver such papers and take all such steps, including the legal assignment of his contractual rights, as the Owner may require for the purpose of fully vesting in the Owner the rights and benefits of the Contractor under such obligations or commitments.

ARTICLE 14

Assignment and Governing Law

14.1 Neither the Owner nor the Contractor shall assign his interest in this Agreement without the written consent of the other except as to the assignment of proceeds.

14.2 This Agreement shall be governed by the law in effect at the location of this Project.

SAMPLE

ARTICLE 15

Miscellaneous Provisions

ARTICLE 16

Arbitration

16.1 All claims, disputes and other matters in question arising out of, or relating to, this Agreement or the breach thereof, except with respect to the Architect/Engineer's decision on matters relating to artistic effect, and except for claims which have been waived by the making or acceptance of Final Payment shall be decided by arbitration in accordance with the Construction Industry Arbitration Rules of the American Arbitration Association then obtaining unless the parties mutually agree otherwise. This agreement to arbitrate shall be specifically enforceable under the prevailing arbitration law.

16.2 Notice of the demand for arbitration shall be filed in writing with the other party to this Agreement and with the American Arbitration Association. The demand for arbitration shall be made within a reasonable time after the claim, dispute or other matter in question has arisen, and in no event shall it be made when institution of legal or equitable proceedings based on such claim, dispute or other matter in question would be barred by the applicable statute of limitations.

16.3 The award rendered by the arbitrators shall be final and judgment may be entered upon it in accordance with applicable law in any court having jurisdiction thereof.

16.4 Unless otherwise agreed in writing, the Contractor shall carry on the Work and maintain the Contract Time Schedule during any arbitration proceedings and the Owner shall continue to make payments in accordance with this Agreement.

16.5 All claims which are related to or dependent upon each other shall be heard by the same arbitrator or arbitrators, even though the parties are not the same, unless a specific contract prohibits such consolidation.

16.6 These provisions relating to mandatory arbitration shall not be applicable to a claim asserted in an action in a state or federal court by a person who is under no obligation to arbitrate such claim with either of the parties to this Agreement insofar as the parties to this Agreement may desire to assert any rights of indemnity or contribution with respect to the subject matter of such action.

This Agreement entered into as of the day and year first written above.

ATTEST: OWNER:

ATTEST: CONTRACTOR:

SAMPLE

THE ASSOCIATED GENERAL CONTRACTORS

STANDARD FORM OF DESIGN-BUILD AGREEMENT AND GENERAL CONDITIONS BETWEEN OWNER AND CONTRACTOR

(WHERE THE BASIS OF COMPENSATION IS A LUMP SUM)

This Document has important legal and insurance consequences; consultation with an attorney and insurance consultants and carriers is encouraged with respect to its completion or modification.

SAMPLE

AGREEMENT

Made this day of in the year of Nineteen Hundred and **BETWEEN** the Owner, and the Contractor.

For services in connection with the following described Project: (Include complete Project location and scope)

REPRESENTED WITH THE PERMISSION OF THE ASSOCIATED GENERAL CONTRACTORS OF AMERICA (AGC). COPIES OF CURRENT FORMS MAY BE OBTAINED FROM AGC'S PUBLICATIONS DEPARTMENT, 1957 E ST. N.W., WASHINGTON, D.C. 20006.

The Owner and the Contractor agree as set forth below:

Certain provisions of this document have been derived, with modifications, from the following documents published by The American Institute of Architects: AIA Document A111, Owner-Contractor Agreement, © 1976; AIA Document A201, General Conditions, ©1976 by The American Institute of Architects. Usage made of AIA language, with the permission of AIA, does not imply AIA endorsement or approval of this document. Further reproduction of copyrighted AIA materials without separate written permission from AIA is prohibited.

AGC DOCUMENT NO. 415 • STANDARD FORM OF DESIGN-BUILD AGREEMENT AND GENERAL CONDITIONS BETWEEN OWNER AND CONTRACTOR (LUMP SUM) • FEBRUARY 1986
©1986 Associated General Contractors of America

INDEX

SAMPLE

AGC DOCUMENT NO. 415 • STANDARD FORM OF DESIGN-BUILD AGREEMENT AND GENERAL CONDITIONS BETWEEN OWNER AND CONTRACTOR (LUMP SUM) • FEBRUARY 1986
©1986 Associated General Contractors of America

ARTICLE 1

The Construction Team and Extent of Agreement

1.1 *THE CONSTRUCTION TEAM:* The Contractor, the Owner, and the Architect/Engineer called the "Construction Team" shall work from the beginning of design through construction completion. The services of ______________________ as the Architect/Engineer will be furnished by the Contractor pursuant to an agreement between the Contractor and the Architect/Engineer.

1.2 *EXTENT OF AGREEMENT:* This Agreement represents the entire agreement between the Owner and the Contractor and supersedes all prior negotiations, representations or agreements. When the Drawings and Specifications are complete, they shall be identified by amendment to this Agreement. This Agreement shall not be superseded by any provisions of the documents for construction and may be amended only by written instrument signed by both Owner and Contractor.

1.3 *DEFINITIONS:* The Project is the total construction to be designed and constructed of which the Work is a part. The Work comprises the completed construction required by the Drawings and Specifications. The term day shall mean calendar day unless otherwise specifically designated.

ARTICLE 2

Contractor's Responsibilities

2.1 Contractor's Services

2.1.1 The Contractor shall be responsible for furnishing the Design and for the Construction of the Project. The Contractor shall develop a design and construction phase schedule and the Owner shall be responsible for prompt decisions and approvals so as to maintain the approved schedule. Any design, engineering, architectural, or other professional service required to be performed under this Agreement shall be performed by duly licensed personnel.

2.1.2 The Contractor shall prepare and the Owner approve a design phase schedule as follows: PHASE 1: Based upon the Owner's Project requirements, schematic Design Studies will be prepared by the Architect/Engineer. These Schematics are for the purpose of assisting the Owner in determining the feasibility of the project. PHASE 2: Upon approval of Schematic Designs and authorization from the Owner to proceed, the Architect/Engineer shall prepare Design Development documents to fix the size and character of the Project as to structural, mechanical and electrical systems, materials and other appropriate essential items in the Project. These Development Documents are the basis for the design and construction of the Project. PHASE 3: From approved Design Development Documents the Architect/Engineer will prepare working Drawings and Specifications setting forth in detail the requirements for the construction of the Project, and based upon codes, laws or regulations which have been enacted at the time of their preparation.

2.1.3 The Contractor, the Architect/Engineer and the Owner will work closely together to monitor the design in accordance with prior approvals so as to ensure that the Project can be constructed within the Lump Sum as defined in Article 6. As these working Drawings and Specifications are being completed, the Contractor will keep the Owner advised of the effects of any Owner requested changes on the Contract Time Schedule and/or the Lump Sum. *Construction of the Project shall be in accordance with these Drawings and Specifications as approved by the Owner. The Drawings and Specifications shall remain the property of the Contractor and are not to be used by the Owner on this or other projects without the written consent of the Contractor.*

2.1.4 After the completion of any Phase as set forth in Article 2.1.2, if the Project is no longer feasible from the standpoint of the Owner, the Owner may terminate this Agreement and pay the Contractor pursuant to Article 10.3.1.

2.1.5 The Contractor will assist the Owner in securing permits necessary for the construction of the Project.

©1986 Associated General Contractors of America

SAMPLE

2.2 Responsibilities With Respect to Construction

2.2.1 The Contractor will provide all construction supervision, inspection, labor, materials, tools, construction equipment and subcontracted items necessary for the execution and completion of the Project.

2.2.2 The Contractor will pay all sales, use, gross receipts and similar taxes related to the Work provided by the Contractor which have been legally enacted at the time of execution of this Agreement and for which the Contractor is liable.

2.2.3 The Contractor will prepare and submit for the Owner's approval an estimated progress schedule for the Project. This schedule shall indicate the dates for the starting and completion of the various stages of the design and construction. It shall be revised as required by the conditions of the Work and those conditions and events which are beyond the Contractor's control.

2.2.4 The Contractor shall at all times keep the premises free from the accumulation of waste materials or rubbish caused by his operations. At the completion of the Work, he shall remove all of his waste material and rubbish from and around the Project as well as all his tools, construction equipment, machinery and surplus materials.

2.2.5 The Contractor will give all notices and comply with all laws and ordinances legally enacted at the date of execution of the Agreement, which govern the proper execution of the Work.

2.2.6 The Contractor shall take necessary precautions for the safety of his employees on the Work, and shall comply with all applicable provisions of federal, state and municipal safety laws to prevent accidents or injury to persons on, about or adjacent to the Project site. He shall erect and properly maintain, at all times, as required by the conditions and progress of Work, necessary safeguards for the protection of workmen and the public. It is understood and agreed, however, that the Contractor shall have no responsibility for the elimination or abatement of safety hazards created or otherwise resulting from Work at the job site carried on by other persons or firms directly employed by the Owner as separate contractors or by the Owner's tenants, and the Owner agrees to cause any such separate contractors and tenants to abide by and fully adhere to all applicable provisions of federal, state and municipal safety laws and regulations and to comply with all reasonable requests and directions of the Contractor for the elimination or abatement of any such safety hazards at the job site.

2.2.7 The Contractor shall keep such full and detailed accounts as may be necessary for proper financial management under this Agreement. The system shall be satisfactory to the Owner, who shall be afforded access to all the Contractor's records, books, correspondence, instructions, drawings, receipts, vouchers, memoranda and similar data relating to this Agreement. The Contractor shall preserve all such records for a period of three years after the final payment or longer where required by law.

2.3 Royalties and Patents

2.3.1 The Contractor shall pay all royalties and license fees for materials, methods and systems incorporated in the work. He shall defend all suits or claims for infringement of any patent rights and shall save the Owner harmless from loss on account thereof except when a particular design, process or product is specified by the Owner. In such case the Contractor shall be responsible for such loss only if he has reason to believe that the design, process or product so specified is an infringement of a patent, and fails to give such information promptly to the Owner.

2.4 Warranties and Completion

2.4.1 The Contractor warrants to the Owner that all materials and equipment furnished under this Agreement will be new, unless otherwise specified, and that all Work will be of good quality, free from improper workmanship and defective materials and in conformance with the Drawings and Specifications. The Contractor agrees to correct all Work performed by him under this Agreement which proves to be defective in material and workmanship within a period of one year from the Date of Substantial Completion as defined in Paragraph 5.2, or for such longer periods of time as may be set forth with respect to specific warranties contained in the Specifications. This warranty is expressly in lieu of all other rights and remedies at law or in equity.

2.4.2 The Contractor will secure required certificates of inspection, testing or approval and deliver them to the Owner.

2.4.3 The Contractor will collect all written warranties and equipment manuals and deliver them to the Owner.

2.4.4 The Contractor, with the assistance of the Owner's maintenance personnel, will direct the checkout of utilities and operations of systems and equipment for readiness, and will assist in their initial start-up and testing.

SAMPLE

©1986 Associated General Contractors of America

2.5 Additional Services

2.5.1 The Contractor will provide the following additional services upon the request of the Owner. A written agreement between the Owner and Contractor shall define the extent of such additional services and the amount and manner in which the Contractor will be compensated for such additional services.

2.5.2 Services related to investigation, appraisals or evaluations of existing conditions, facilities or equipment, or verification of the accuracy of existing drawings or other Owner-furnished information.

2.5.3 Services related to Owner-furnished equipment, furniture and furnishings which are not a part of this Agreement.

2.5.4 Services for tenant or rental spaces not a part of this Agreement.

2.5.5 Obtaining and training maintenance personnel or negotiating maintenance service contracts.

ARTICLE 3

Owner's Responsibilities

3.1 The Owner shall provide full information regarding his requirements for the Project.

3.2 The Owner shall designate a representative who shall be fully acquainted with the Project, and has authority to approve changes in the scope of the Project, render decisions promptly, and furnish information expeditiously and in time to meet the dates set forth in Subparagraph 2.2.3.

3.3 The Owner shall furnish for the site of the Project all necessary surveys describing the physical characteristics, soils reports and subsurface investigations, legal limitations, utility locations, and a legal description.

3.4 The Owner shall secure and pay for all necessary approvals, easements, assessments and charges required for the construction, use, or occupancy of permanent structures or for permanent changes in existing facilities.

3.5 The Owner shall furnish such legal services as may be necessary for providing the items set forth in Paragraph 3.4, and such auditing services as he may require.

3.6 If the Owner becomes aware of any fault or defect in the Project or non-conformance with the Drawings or Specifications, he shall give prompt written notice thereof to the Contractor.

3.7 The Owner shall provide the insurance for the Project as provided in Paragraph 9.4.

3.8 The Owner shall bear the costs of any bonds that may be required.

3.9 The services and information required by the above paragraphs shall be furnished with reasonable promptness at the Owner's expense and the Contractor shall be entitled to rely upon the accuracy and the completeness thereof.

3.10 The Owner shall furnish reasonable evidence satisfactory to the Contractor, prior to commencing Work and at such future times as may be required, that sufficient funds are available and committed for the entire Cost of the Project. Unless such reasonable evidence is furnished, the Contractor is not required to commence or continue any Work, or may, if such evidence is not presented within a reasonable time, stop Work upon 15 days notice to the Owner. The failure of the Contractor to insist upon the providing of this evidence at any one time shall not be a waiver of the Owner's obligation to make payments pursuant to this Agreement, nor shall it be a waiver of the Contractor's right to request or insist that such evidence be provided at a later date.

3.11 The Owner shall have no contractual obligation to the Contractor's Subcontractors and shall communicate with such Subcontractors only through the Contractor.

SAMPLE

©1986 Associated General Contractors of America

ARTICLE 4

Subcontracts

4.1 All portions of the Work that the Contractor does not perform with his own forces shall be performed under subcontracts.

4.2 A Subcontractor is a person or entity who has a direct contract with the Contractor to perform any Work in connection with the Project. The term Subcontractor does not include any separate contractor employed by the Owner or the separate contractors' subcontractors.

4.3 No contractual relationship shall exist between the Owner and any Subcontractor. The Contractor shall be responsible for the management of the Subcontractors in the performance of their Work.

ARTICLE 5

Contract Time Schedule

5.1 The Work to be performed under this Agreement shall be commenced on or about and shall be substantially completed on or about

5.2 The Date of Substantial Completion of the Project or a designated portion thereof is the date when construction is sufficiently complete in accordance with the Drawings and Specifications so the Owner can occupy or utilize the Project or designated portion thereof for the use for which it is intended. Warranties called for by this Agreement or by the Drawings and Specifications shall commence on the Date of Substantial Completion of the Project or designated portion thereof. This date shall be established by a Certificate of Substantial Completion signed by the Owner and Contractor and shall state their respective responsibilities for security, maintenance, heat, utilities, damage to the Work and insurance. This Certificate shall also list the items to be completed or corrected and fix the time for their completion and correction.

5.3 If the Contractor is delayed at any time in the progress of the Project by any act or neglect of the Owner or by any separate contractor employed by the Owner, or by changes ordered in the Project, or by labor disputes, fire, unusual delay in transportation, adverse weather conditions not reasonably anticipated, unavoidable casualties, or any causes beyond the Contractor's control, or a delay authorized by the Owner pending arbitration, then the Date for Substantial Completion shall be extended by Change Order for the period caused by such delay.

ARTICLE 6

Lump Sum Price

6.1 The Lump Sum price for the Project is ($).

6.2 The Lump Sum is based upon laws, codes, and regulations in existence at the date of its establishment and upon criteria, Drawings, and Specifications as set forth in this agreement.

6.3 The Lump Sum will be modified for delays caused by the Owner and for Changes in the Project, all pursuant to Article 7.

6.4 Allowances

6.4.1 Allowances included in the Lump Sum are as set forth below:

©1986 Associated General Contractors of America

SAMPLE

6.4.2 Whenever the cost is more than or less than the Allowance, the Lump Sum shall be adjusted by Change Order as provided in Article 7.

ARTICLE 7

Changes in the Project

7.1 The Owner, without invalidating this Agreement, may order Changes in the Project within the general scope of this Agreement consisting of additions, deletions or other revisions. The Lump Sum, and the Contract Time Schedule shall be adjusted accordingly. All such Changes in the Project shall be authorized by Change Order.

7.1.1 A Change Order is a written order to the Contractor signed by the Owner or his authorized agent and issued after the execution of this Agreement, authorizing a Change in the Project and/or an adjustment in the Lump Sum or the Contract Time Schedule.

7.1.2 The increase or decrease in the Lump Sum resulting from a Change in the Project shall be determined in one or more of the following ways:

7.1.2.1 by mutual acceptance of a lump sum properly itemized and supported by sufficient substantiating data to permit evaluation; or

7.1.2.2 by unit prices stated in this Agreement or subsequently agreed upon; or

7.1.2.3 If none of the methods set forth in articles 7.1.2.1 and 7.1.2.2 is agreed upon, the Contractor shall promptly proceed with the Work required by the Change in the Project provided the Contractor receives a written order to proceed signed by the Owner. The increase in the Lump Sum shall then be determined on the basis of the reasonable costs of such Work and savings of those performing the Work attributed to the Change in the Project including a reasonable increase in the Contractor's overhead and profit. The amount of decrease in the Lum Sum to be allowed by Contractor to the Owner for any deletion or Change in the Project with results in a net decrease in cost will be the amount of the actual net decrease only. When both increases and decreases in costs of the Work are involved in any one Change in the Project, the increase in overhead and profit shall be figured on the basis of the net increase in costs, if any. Under this article and articles 7.1.2.1 and 7.1.2.2, the Contractor shall keep and present, in such form as the Owner may prescribe, an itemized accounting together with appropriate supporting data of the effect on the Lump Sum. The increase or decrease in the Lump Sum under this article and articles 7.1.2.1 and 7.1.2.2 shall be authorized by Change Order signed by the Owner or its authorized agent.

7.1.3 If unit prices are stated in this agreement or subsequently agreed upon, and if the quantities originally contemplated are so changed in a proposed Change Order or as a result of several Change Orders that application of the agreed unit prices to the quantities of Work proposed will cause substantial inequity to the Owner or the Contractor, the applicable unit prices and the Lump Sum shall be equitably adjusted.

7.1.4 Should concealed conditions encountered in the performance of the Work below the surface of the ground or should concealed or unknown conditions in an existing structure be at variance with the conditions indicated by the Drawings, Specifications, or Owner-furnished information or should unknown physical conditions below the surface of the ground or should concealed or unknown conditions in an existing structure of an unusual nature, differing materially from those ordinarily encountered and generally recognized as inherent in work of the character provided for in this Agreement, be encountered, the Lump Sum and the Contract Time Schedule shall be equitably adjusted by Change Order upon claim by either party made within a reasonable time after the first observance of the conditions.

SAMPLE

©1986 Associated General Contractors of America

7.2 Claims for Additional Cost or Time

7.2.1 If the Contractor wishes to make a claim for an increase in the Lump Sum or an extension in the Contract Time Schedule, he shall give the Owner written notice thereof within a reasonable time after the occurrence of the event giving rise to such claim. This notice shall be given by the Contractor before proceeding to execute the Work, except in an emergency endangering life or property in which case the Contractor shall act, at his discretion, to prevent threatened damage, injury or loss. Claims arising from delay shall be made within a reasonable time after the delay. Increases based upon design and estimating costs with respect to possible changes requested by the Owner, shall be made within a reasonable time after the decision is made not to proceed with the change. No such claim shall be valid unless so made. If the Owner and the Contractor cannot agree on the amount of the adjustment in the Lump Sum, and the Contract Time Schedule, it shall be determined pursuant to the provisions of Article 13. Any change in the Lump Sum or Contract Time Schedule resulting from such claim shall be authorized by Change Order.

7.3 Minor Changes in the Project

7.3.1 The Owner will have authority to order minor Changes in the Work not involving an adjustment in the Lump Sum or an extension of the Contract Time Schedule and not inconsistent with the intent of the Drawings and Specifications. Such Changes may be effected by written order and shall be binding on the Owner and the Contractor.

7.4 Emergencies

7.4.1 In any emergency affecting the safety of persons or property, the Contractor shall act, at his discretion, to prevent threatened damage, injury or loss. Any increase in the Lump Sum or extension of time claimed by the Contractor on account of emergency work shall be determined as provided in this Article.

ARTICLE 8

Payments to the Contractor

8.1 Payments shall be made by the Owner to the Contractor according to the following procedure:

8.1.1 On or before the ______ day of each month after Work has commenced, the Contractor shall submit to the Owner an Application for Payment in such detail as may be required by the Owner based on the Work completed and materials stored on the site and/or at locations approved by the Owner for the period ending on the ______ day of the month.

8.1.2 Within ten (10) days after his receipt of each monthly Application for Payment, the Owner shall pay directly to the Contractor the appropriate amounts for which Application for Payment is made therein. This payment request shall deduct the aggregate of amounts previously paid by the Owner.

8.1.3 If the Owner should fail to pay the Contractor at the time the payment of any amount becomes due, then the Contractor may, at any time thereafter, upon serving written notice that he will stop Work within seven (7) days after receipt of the notice by the Owner, and after such seven (7) day period, stop the Project until payment of the amount owing has been received. Written notice shall be deemed to have been duly served if sent by certified mail to the last known business address of the Owner.

8.1.4 Payments due but unpaid shall bear interest at the rate of two percentage points above the prime interest rate prevailing from time to time at the location of the Project.

8.2 The Contractor warrants and guarantees that title to all Work, materials and equipment covered by an Application for Payment whether incorporated in the Project or not, will pass to the Owner upon receipt of such payment by the Contractor free and clear of all liens, claims, security interests or encumbrances hereinafter referred to as Liens.

8.3 No Progress Payment nor any partial or entire use or occupancy of the Project by the Owner shall constitute an acceptance of any Work not in accordance with the Drawings and Specifications.

©1986 Associated General Contractors of America

SAMPLE

8.4 Final payment constituting the unpaid balance of the Project shall be due and payable when the Project is delivered to the Owner, ready for beneficial occupancy, or when the Owner occupies the Project, whichever event first occurs, provided that the Project be then substantially completed and this Agreement substantially performed. If there should remain minor items to be completed, the Contractor and the Owner shall list such items and the Contractor shall deliver, in writing, his guarantee to complete said items within a reasonable time thereafter. The Owner may retain a sum equal to 150 percent of the estimated cost of completing any unfinished items, provided that said unfinished items are listed separately and the estimated cost of completing any unfinished items is likewise listed separately. Thereafter, the Owner shall pay to the Contractor, monthly, the amount retained for incomplete items as each of said items is completed.

8.5 Before issuance of Final Payment, the Owner may request satisfactory evidence that all payrolls, materials bills and other indebtness connected with the Project have been paid or otherwise satisfied.

8.6 The making of Final Payment shall constitute a waiver of all claims by the Owner except those rising from: unsettled liens; improper workmanship or defective materials appearing within one year after the Date of Substantial Completion; and terms of any special guarantees required by the Drawings and Specifications.
8.7 The acceptance of Final Payment shall constitute a waiver of all claims by the Contractor except those previously made in writing and unsettled.

ARTICLE 9

Insurance, Indemnity and Waiver of Subrogation

9.1 Indemnity

9.1.1 The Contractor agrees to indemnify and hold the Owner harmless from all claims for bodily injury and property damage (other than the Work itself and other property insured under Paragraph 9.4) that may arise from the Contractor's operations under this Agreement.

9.1.2 The Owner shall cause any other contractor who may have a contract with the Owner to perform work in the areas where Work will be performed under this Agreement, to agree to indemnify the Owner and the Contractor and hold them harmless from all claims for bodily injury and property damage (other than property insured under Paragraph 9.4) that may arise from that contractor's operations. Such provisions shall be in a form satisfactory to the Contractor.

9.2 Contractor's Liability Insurance

9.2.1 The Contractor shall purchase and maintain such insurance as will protect him from the claims set forth below which may arise out of or result from the Contractor's operations under this Agreement whether such operations be by himself or by any Subcontractor or by anyone directly or indirectly employed by any of them, or by anyone for whose acts any of them may be liable:

9.2.1.1 Claims under workers' compensation, disability benefit and other similar employee benefit acts which are applicable to the Work to be performed;

9.2.1.2 Claims for damages because of bodily injury, occupational sickness or disease, or death of his employees under any applicable employer's liability law;

9.2.1.3 Claims for damages because of bodily injury, or death of any person other than his employees;

9.2.1.4 Claims for damages insured by usual personal injury liability coverage which are sustained (1) by any person as a result of an offense directly or indirectly related to the employment of such person by the Contractor or (2) by any other person;

9.2.1.5 Claims for damages, other than to the Work itself, because of injury to or destruction of tangible property, including loss of use therefrom;

SAMPLE

©1986 Associated General Contractors of America

9.2.1.6 Claims for damages because of bodily injury or death of any person or property damage arising out of the ownership, maintenance or use of any motor vehicle.

9.2.2 The Comprehensive General Liability Insurance shall include premises-operations (including explosion, collapse and underground coverage) elevators, independent contractors, completed operations, and blanket contractual liability on all written contracts, all including broad form property damage coverage.

9.2.3 The Contractor's Comprehensive General and Automobile Liability Insurance, as required by Subparagraphs 9.2.1 and 9.2.2 shall be written for not less than limits of liability as follows:

a. Comprehensive General Liability
   1. Bodily Injury $ ______________ Each Occurrence
      (Completed Operations)
      $ ______________ Aggregate
   2. Property Damage $ ______________ Each Occurrence
      $ ______________ Aggregate

b. Comprehensive Automobile Liability
   1. Bodily Injury $ ______________ Each Person
      $ ______________ Each Occurrence
   2. Property Damage $ ______________ Each Occurrence

9.2.4 Comprehensive General Liability Insurance may be arranged under a single policy for the full limits required or by a combination of underlying policies with the balance provided by an Excess or Umbrella Liability policy.

9.2.5 The foregoing policies shall contain a provision that coverages afforded under the policies will not be cancelled or not renewed until at least sixty (60) days' prior written notice has been given to the Owner. Certificates of Insurance showing such coverages to be in force shall be filed with the Owner prior to commencement of the Work.

9.3 Owner's Liability Insurance

9.3.1 The Owner shall be responsible for purchasing and maintaining his own liability insurance and, at his option, may purchase and maintain such insurance as will protect him against claims which may arise from operations under this Agreement.

9.4 Insurance to Protect Project

9.4.1 The Owner shall purchase and maintain property insurance in a form acceptable to the Contractor upon the entire Project for the full cost of replacement at the time of any loss. This insurance shall include as named insureds the Owner, the Contractor, Subcontractors and Subsubcontractors and shall insure against loss from the perils of Fire, Extended Coverage, and shall include "All Risk" insurance for physical loss or damage including, without duplication of coverage, at least theft, vandalism, malicious mischief, transit, collapse, flood, earthquake, testing, and damage resulting from defective design, workmanship or material. The Owner will increase limits of coverage, if necessary, to reflect estimated replacement cost. The Owner will be responsible for any co-insurance penalties or deductibles. If the Project covers an addition to or is adjacent to an existing building, the Contractor, Subcontractors and Subsubcontractors shall be named as additional insureds under the Owner's Property Insurance covering such building and its contents.

9.4.1.1 If the Owner finds it necessary to occupy or use a portion or portions of the Project prior to Substantial Completion thereof, such occupany shall not commence prior to a time mutually agreed to by the Owner and the Contractor and to which the insurance company or companies providing the property insurance have consented by endorsement to the policy or policies. This insurance shall not be cancelled or lapsed on account of such partial occupancy. Consent of the Contractor and of the insurance company or companies to such occupancy or use shall not be unreasonably withheld.

9.4.2 The Owner shall purchase and maintain such boiler and machinery insurance as may be required or necessary. This insurance shall include the interests of the Owner, the Contractor, Subcontractors and Subsubcontractors in the Work.

©1986 Associated General Contractors of America

SAMPLE

9.4.3 The Owner shall purchase and maintain such insurance as will protect the Owner and the Contractor against loss of use of Owner's property due to those perils insured pursuant to Subparagraph 9.4.1. Such policy will provide coverage for expenses of expediting materials, continuing overhead of the Owner and the Contractor, necessary labor expense including overtime, loss of income by the Owner and other determined exposures. Exposures of the Owner and the Contractor shall be determined by mutual agreement and separate limits of coverage fixed for each item.

9.4.4 The Owner shall file a copy of all policies with the Contractor before an exposure to loss may occur. Copies of any subsequent endorsements will be furnished to the Contractor. The Contractor will be given sixty (60) days notice of cancellation, non-renewal, or any endorsements restricting or reducing coverage. If the Owner does not intend to purchase such insurance, he shall inform the Contractor in writing prior to the commencement of the Work. The Contractor may then effect insurance which will protect the interests of himself, the Subcontractors and their Subsubcontractors in the Project, the cost of which shall be added to the Lump Sum by Change Order. If the Contractor is damaged by failure of the Owner to purchase or maintain such insurance or to so notify the Contractor, the Owner shall bear all reasonable costs properly attributable thereto.

9.5 Property Insurance Loss Adjustment

9.5.1 Any insured loss shall be adjusted with the Owner and the Contractor and made payable to the Owner and Contractor as trustees for the insureds, as their interests may appear, subject to any applicable mortgagee clause.

9.5.2 Upon the occurrence of an insured loss, monies received will be deposited in a separate account and the trustees shall make distribution in accordance with the agreement of the parties in interest, or in the absence of such agreement, in accordance with an arbitration award pursuant to Article 13. If the trustees are unable to agree between themselves on the settlement of the loss, such dispute shall also be submitted to arbitration pursuant to Article 13.

9.6 Waiver of Subrogation

9.6.1 The Owner and Contractor waive all rights against each other, the Architect/Engineer, Subcontractors and Subsubcontractors for damages caused by perils covered by insurance provided under Paragraph 9.4, except such rights as they may have to the proceeds of such insurance held by the Owner and Contractor as trustees. The Contractor shall require similar waivers from all Subcontractors and Subsubcontractors.

9.6.2 The Owner and Contractor waive all rights against each other and the Architect/Engineer, Subcontractors and Subsubcontractors for loss or damage to any equipment used in connection with the Project which loss is covered by any property insurance. The Contractor shall require similar waivers from all Subcontractors and Subsubcontractors.

9.6.3 The Owner waives subrogation against the Contractor, Architect/Engineer, Subcontractors, and Subsubcontractors on all property and consequential loss policies carried by the Owner on adjacent properties and under property and consequential loss policies purchased for the Project after its completion.

9.6.4 If the policies of insurance referred to in this Paragraph require an endorsement to provide for continued coverage where there is a waiver of subrogation, the owners of such policies will cause them to be so endorsed.

SAMPLE

ARTICLE 10

TERMINATION OF THE AGREEMENT AND OWNER'S RIGHT TO PERFORM CONTRACTOR'S OBLIGATIONS

10.1 Termination by the Contractor

10.1.1 If the Project is stopped for a period of thirty (30) days under an order of any court or other public authority having jurisdiction, or as a result of an act of government, such as a declaration of a national emergency making materials unavailable, through no act or fault of the Contractor or if the Project should be stopped for a period of thirty (30) days by the Contractor for the Owner's failure to make payment thereon, then the Contractor may, upon seven days' written notice to the Owner, terminate this Agreement and recover from the Owner payment for all Work executed, the Lump Sum earned to date, and for any proven loss sustained upon any materials, equipment, tools, construction equipment and machinery, including reasonable profit and damages.

©1986 Associated General Contractors of America

10.2 Owner's Right to Perform Contractor's Obligations and Termination by the Owner for Cause

10.2.1 If the Contractor fails to perform any of his obligations under this Agreement, including any obligation he assumes to perform Work with his own forces, the Owner may, after seven days' written notice, during which period the Contractor fails to perform such obligation, make good such deficiencies. The Lump Sum, if any, shall be reduced by the cost to the Owner of making good such deficiencies.

10.2.2 If the Contractor is adjudged a bankrupt, or if he makes a general assignment for the benefit of his creditors, or if a receiver is appointed on account of his insolvency, or if he persistently or repeatedly refuses or fails, except in cases for which extension of time is provided, to supply enough properly skilled workmen or proper materials, or if he fails to make proper payment to Subcontractors or for materials or labor, or persistently disregards laws, ordinances, rules, regulations or orders of any public authority having jurisdiction, or otherwise is guilty of a substantial violation of a provision of this Agreement, then the Owner may, without prejudice to any right or remedy and after giving the Contractor and his surety, if any, seven (7) days' written notice, during which period the Contractor fails to cure the violation, terminate the employment of the Contractor and take possession of the site and of all materials, equipment, tools, construction equipment and machinery thereon owned by the Contractor and may finish the Work by whatever reasonable method he may deem expedient. In such case, the Contractor shall not be entitled to receive any further payment until the Work is finished nor shall he be relieved from his obligations assumed under Article 6.

10.3 Termination by Owner Without Cause

10.3.1 If the Owner terminates the Agreement other than pursuant to 10.2.2, he shall pay the Contractor the total of: (a.) Costs incurred by the Contractor in performing the Project, including initial costs and preparatory expenses; (b.) Costs incurred in settling and paying termination claims under terminated subcontracts; (c.) Accounting, legal, clerical and other expenses incurred as a result of the termination; (d.) Storage, transportation, demobilization and other costs incurred for the preservation, protection or disposition of material and equipment on the Project; (e.) Any other necessary and reasonable costs incurred by the Contractor as a result of the Owner's termination of this Agreement; (f.) Overhead at ten percent (10%) of the total amount of (a.) thorugh (e.) above; profit at ten percent (10%) of the total amount of (a) through (f) above, as adjusted pursuant to Articles 6 and 7. In calculating the amount due the Contractor under this clause, a deduction shall be made for all payments to the Contractor under this Agreement.

ARTICLE 11

Assignment and Governing Law

11.1 Neither the Owner nor the Contractor shall assign his interest in this Agreement without the written consent of the other except as to the assignment of proceeds.

11.2 This Agreement shall be governed by the law in effect at the location of this Project.

ARTICLE 12

Miscellaneous Provisions

SAMPLE

©1986 Associated General Contractors of America

ARTICLE 13

Arbitration

13.1 *AGREEMENT TO ARBITRATE:* All claims, disputes and matters in question arising out of, or relating to this Agreement or the breach thereof, except for claims which have been waived by the making or acceptance of final payment, and the claims described in Article 13.7, shall be decided by arbitration in accordance with the Construction Industry Arbitration Rules of the American Arbitration Association then in effect unless the parties mutually agree otherwise. This agreement to arbitrate shall be specifically enforceable under the prevailing arbitration law.

13.2 *NOTICE OF DEMAND:* Notice of the demand for arbitration shall be filed in writing with the other party to this Agreement and with the American Arbitration Association. The demand for arbitration shall be made within a reasonable time after written notice of the claim, dispute or other matter in question has been given, and in no event shall it be made after the date of final acceptance of the Work by the Owner or when institution of legal or equitable proceedings based on such claim, dispute or other matter in question would be barred by the applicable statute of limitations, whichever shall first occur. The location of the arbitration proceedings shall be the city of the Contractor's headquarters or ______________________________

____

____

13.3 *AWARD:* The award rendered by the arbitrator(s) shall be final and judgment may be entered upon it in accordance with applicable law in any court having jurisdiction.

13.4 *WORK CONTINUATION AND PAYMENT:* Unless otherwise agreed in writing, the Contractor shall carry on the Work and maintain the Schedule of Work pending arbitration, and, if so, the Owner shall continue to make payments in accordance with this Agreement.

13.5 *NO LIMITATION OF RIGHTS OR REMEDIES:* Nothing in this Article shall limit any rights or remedies not expressly waived by the Contractor which the Contractor may have under lien laws or payment bonds.

13.6 *SAME ARBITRATORS:* To the maximum extent permitted by law, all claims which are related to or dependent upon each other, shall be heard by the same arbitrator or arbitrators even through the parties are not the same.

13.7 *EXCEPTIONS:* This agreement to arbitrate shall not apply to any claim of contribution or indemnity asserted by one party to this Agreement against the other party and arising out of any action brought in a state or federal court or in arbitration by a person who is under no obligation to arbitrate the subject matter of such action with either of the parties hereto. In any dispute arising over the application of this Article 13.7, the question of arbitrability shall be decided by the appropriate court and not by arbitration.

Attest: ______________________________ Owner: ______________________________

Attest: ______________________________ Contractor: ______________________________

SAMPLE

©1986 Associated General Contractors of America

THE ASSOCIATED GENERAL CONTRACTORS OF AMERICA

SUBCONTRACT FOR BUILDING CONSTRUCTION

SAMPLE

TABLE OF ARTICLES

REPRESENTED WITH THE PERMISSION OF THE ASSOCIATED GENERAL CONTRACTORS OF AMERICA (AGC). COPIES OF CURRENT FORMS MAY BE OBTAINED FROM AGC'S PUBLICATIONS DEPARTMENT, 1957 E ST. N.W., WASHINGTON, D.C. 20006.

This Agreement has important legal and insurance consequences. Consultation with an attorney and insurance consultant is encouraged with respect to its completion or modification and particularly when used with other than AIA A201 General Conditions of the Contract for Construction, August 1976 edition.

©Associated General Contractors of America

TABLE OF CONTENTS

SAMPLE

©Associated General Contractors of America

SUBCONTRACT FOR BUILDING CONSTRUCTION

ARTICLE 1

AGREEMENT

This Agreement made this ________ day of ____________, 19____, and effective the ________ day of ____________, 19____, by and between ____,

____

hereinafter called the Contractor and ____,

____

hereinafter called the Subcontractor, to perform part of the Work on the following Project:

PROJECT:

OWNER:

ARCHITECT:

CONTRACTOR:

SUBCONTRACTOR:

CONTRACT PRICE:

SAMPLE

Notice to the parties shall be given at the above addresses.

ARTICLE 2

SCOPE OF WORK

2.1 SUBCONTRACTOR'S WORK. The Contractor employs the Subcontractor as an independent contractor, to perform the work described in Article 16. The Subcontractor shall perform such work (hereinafter called the "Subcontractor's Work") under the general direction of the Contractor and in accordance with this Agreement and the Contract Documents.

2.2 CONTRACT DOCUMENTS. The Contract Documents which are binding on the Subcontractor are as set forth in Article 16.5.

Upon the Subcontractor's request the Contractor shall furnish a copy of any part of these documents.

2.3 CONFLICTS. In the event of a conflict between this Agreement and the Contract Documents, this Agreement shall govern, except as follows:

©Associated General Contractors of America

ARTICLE 3

SCHEDULE OF WORK

3.1 TIME IS OF ESSENCE. Time is of the essence for both parties, and they mutually agree to see to the performance of their respective work and the work of their subcontractors so that the entire Project may be completed in accordance with the Contract Documents and the Schedule of Work. The Contractor shall prepare the Schedule of Work and revise such schedule as the Work progresses.

3.2 DUTY TO BE BOUND. Both the Contractor and the Subcontractor shall be bound by the Schedule of Work. The Subcontractor shall provide the Contractor with any requested scheduling information for the Subcontractor's Work. The Schedule of Work and all subsequent changes thereto shall be submitted to the Subcontractor in advance of the required performance.

3.3 SCHEDULE CHANGES. The Subcontractor recognizes that changes will be made in the Schedule of Work and agrees to comply with such changes subject to a reservation of rights arising hereunder.

3.4 PRIORITY OF WORK. The Contractor shall have the right to decide the time, order and priority in which the various portions of the Work shall be performed and all other matters relative to the timely and orderly conduct of the Subcontractor's Work.

The Subcontractor shall commence its work within ____ days of notice to proceed from the Contractor and if such work is interrupted for any reason the Subcontractor shall resume such work within two working days from the Contractor's notice to do so.

ARTICLE 4

CONTRACT PRICE

The Contractor agrees to pay to the Subcontractor for the satisfactory performance of the subcontractor's Work the sum of ______________________________________

____

Dollars ($ ____________________) in accordance with Article 5, subject to additions or deductions per Article 6.

ARTICLE 5

PAYMENT

5.1 GENERAL PROVISIONS.

5.1.1 SCHEDULE OF VALUES. The Subcontractor shall provide a schedule of values satisfactory to the Contractor and the Owner no more than fifteen (15) days from the date of execution of this Agreement.

5.1.2 ARCHITECT VERIFICATION. Upon request the Contractor shall give the Subcontractor written authorization to obtain directly from the Architect the percentage of completion certified for the Subcontractor's Work.

5.1.3. PAYMENT USE RESTRICTION. No payment received by the Subcontractor shall be used to satisfy or secure any indebtedness other than one owed by the Subcontractor to a person furnishing labor or materials for use in performing the Subcontractor's Work.

5.1.4. PAYMENT USE VERIFICATION. The Contractor shall have the right at all times to contact the Subcontractor's subcontractors and suppliers to ensure that the same are being paid by the Subcontractor for labor or materials furnished for use in performing the Subcontractor's Work.

5.1.5 PARTIAL LIEN WAIVERS AND AFFIDAVITS. When required by the Contractor, and as a prerequisite for payment, the Subcontractor shall provide, in a form satisfactory to the Owner and the Contractor, partial lien or claim waivers and affidavits from the Subcontractor, and its sub-subcontractors and suppliers for the completed Subcontractor's Work. Such waivers may be made conditional upon payment.

5.1.6 SUBCONTRACTOR PAYMENT FAILURE. In the event the Contractor has reason to believe that labor, material or other obligations incurred in the performance of the Subcontractor's Work are not being paid, the Contractor shall give written notice of such claim or lien to the Subcontractor and may take any steps deemed necessary to insure that any progress payment shall be utilized to pay such obligations.

If upon receipt of said notice, the Subcontractor does not:

(a) supply evidence to the satisfaction of the Contractor that the monies owing to the claimant have been paid; or

(b) post a bond indemnifying the Owner, the Contractor, the Contractor's surety, if any, and the premises from such claim or lien;

then the Contractor shall have the right to retain out of any payments due or to become due to the Subcontractor a reasonable amount to protect the Contractor from any and all loss, damage or expense including attorney's fees arising out of or relating to any such claim or lien until the claim or lien has been satisfied by the Subcontractor.

5.1.7 PAYMENT NOT ACCEPTANCE. Payment to the Subcontractor is specifically agreed not to constitute or imply acceptance by the Contractor or the Owner of any portion of the Subcontractor's Work.

SAMPLE

©Associated General Contractors of America

5.2 PROGRESS PAYMENTS

5.2.1 APPLICATION. The Subcontractor's progress payment application for work performed in the preceding payment period shall be submitted to the Contractor per the terms of this Agreement and specifically Articles 5.1.1, 5.2.2, 5.2.3, and 5.2.4 for approval of the Contractor and ____

____

The Contractor shall forward, without delay, the approved value to the Owner for payment.

5.2.2 RETAINAGE/SECURITY. The rate of retainage shall not exceed the percentage retained from the Contractor's payment by the Owner for the Subcontractor's Work provided the Subcontractor furnishes a bond or other security to the satisfaction of the Contractor.

If the Subcontractor has furnished such bond or security; its work is satisfactory and the Contract Documents provide for reduction of retainage at a specified percentage of completion, the Subcontractor's retainage shall also be reduced when the Subcontractor's Work has attained the same percentage of completion and the Contractor's retainage for the Subcontractor's Work has been so reduced by the Owner.

However, if the Subcontractor does not provide such bond or security, the rate of retainage shall be ______%

5.2.3 TIME OF APPLICATION. The Subcontractor shall submit progress payment applications to the Contractor no later than the ________ day of each payment period for work performed up to and including the ________ day of the payment period indicating work completed and, to the extent allowed under Article 5.2.4, materials suitably stored during the preceding payment period.

5.2.4. STORED MATERIALS. Unless otherwise provided in the Contract Documents, and if approved in advance by the Owner, applications for payment may include materials and equipment not incorporated in the Subcontractor's Work but delivered and suitably stored at the site or at some other location agreed upon in writing. Approval of payment application for such stored items on or off the site shall be conditioned upon submission by the Subcontractor of bills of sale and applicable insurance or such other procedures satisfactory to the Owner and Contractor to establish the Owner's title to such materials and equipment or otherwise protect the Owner's and Contractor's interest therein, including transportation to the site.

5.2.5 TIME OF PAYMENT. Progress payments to the Subcontractor for satisfactory performance of the Subcontractor's Work shall be made no later than seven (7) days after receipt by the Contractor of payment from the Owner for such Subcontractor's Work.

5.2.6 PAYMENT DELAY. If for any reason not the fault of the Subcontractor, the Subcontractor does not receive a progress payment from the Contractor within seven (7) days after the date such payment is due, as defined in Article 5.2.5, then the Subcontractor, upon giving an additional seven (7) days written notice to the Contractor, and without prejudice to and in addition to any other legal remedies, may stop work until payment of the full amount owing to the Subcontractor has been received. To the extent obtained by the Contractor under the Contract Documents, the contract price shall be increased by the amount of the Subcontractor's reasonable costs of shut-down, delay, and start-up, which shall be effected by appropriate Change Order.

If the Subcontractor's Work has been stopped for thirty (30) days because the Subcontractor has not recieved progress payments as required hereunder, the Subcontractor my terminate this Agreement upon giving the Contractor an additional seven (7) days written notice.

5.3 FINAL PAYMENT.

5.3.1 APPLICATION. Upon acceptance of the Subcontractor's Work by the Owner, the Contractor, and if necessary, the Architect; and upon the Subcontractor furnishing evidence of fulfillment of the Subcontractor's obligations in accordance with the Contract Documents and Article 5.3.2, the Contractor shall forward the Subcontractor's application for final payment without delay.

5.3.2 REQUIREMENTS. Before the Contractor shall be required to forward the Subcontractor's application for final payment to the Owner, the Subcontractor shall submit to the Contractor:

(a) an affidavit that all payrolls, bills for materials and equipment, and other indebtedness connected with the Subcontractor's Work for which the Owner or his property or the Contractor or the Contractor's surety might in any way be liable, have been paid or otherwise satisfied;

(b) consent of surety to final payment, if required;

(c) satisfaction of required closeout procedures, and

(d) other data if required by the Contractor or Owner, such as receipts, releases, and waivers of liens to the extent and in such form as may be designated by the Contractor or Owner.

Final payment shall constitute a waiver of all claims by the Subcontractor relating to the Subcontractor's Work, but shall in no way relieve the Subcontractor of liability for the obligations assumed under Article 9.10 hereof, or for faulty or defective work appearing after final payment.

SAMPLE

©Associated General Contractors of America

5.3.3 TIME OF PAYMENT. Final payment of the balance due of the contract price shall be made to the Subcontractor:

(a) upon receipt of the Owner's waiver of all claims related to the Subcontractor's Work except for unsettled liens, unknown defective work, and non-compliance with the Contract Documents or warranties; and

(b) within seven (7) days after receipt by the Contractor of final payment from the Owner for such Subcontractor's Work.

5.3.4 FINAL PAYMENT DELAY. If the Owner or its designated agent does not issue a certificate for Final Payment or the Contractor does not receive such payment for any cause which is not the fault of the Subcontractor, the Contractor shall promptly inform the Subcontractor in writing. The Contractor shall also diligently pursue, with the assistance of the Subcontractor, the prompt release by the Owner of the final payment due for the Subcontractor's Work. At the Subcontractor's request and joint expense, to the extent agreed upon in writing, the Contractor shall institute all reasonable legal remedies to mitigate the damages and pursue full payment of the Subcontractor's application for final payment including interest thereon.

5.4 LATE PAYMENT INTEREST. To the extent obtained by the Contractor under the Contract Documents, progress payments or final payment due and unpaid under this Agreement shall bear interest from the date payment is due at the rate provided in the Contract Documents, or, in the absence thereof, at the legal rate prevailing at the place of the Project.

ARTICLE 6
CHANGES, CLAIMS AND DELAYS

6.1 CHANGES. When the Contractor so orders in writing, the Subcontractor, without nullifying this Agreement, shall make any and all changes in the Work which are within the general scope of this agreement.

Adjustments in the contract price or contract time, if any, resulting from such changes shall be set forth in a Subcontract Change Order pursuant to the Contract Documents.

No such adjustments shall be made for any such changes performed by the Subcontractor that have not been so ordered by the Contractor.

6.2 CLAIMS RELATING TO OWNER. The Subcontractor agrees to make all claims for which the Owner is or may be liable in the manner provided in the Contract Documents for like claims by the Contractor upon the Owner.

Notice of such claims shall be given by the Subcontractor to the Contractor within one (1) week prior to the beginning of the Subcontractor's Work or the event for which such claim is to be made, or immediately upon the Subcontractor's first knowledge of the event, whichever shall first occur; otherwise, such claims shall be deemed waived.

The Contractor agrees to permit the Subcontractor to prosecute and claim, in the name of the Contractor, for the use and benefit of the Subcontractor in the manner provided in the Contract Documents for like claims by the Contractor upon the Owner.

6.3 CLAIMS RELATING TO CONTRACTOR. The Subcontractor shall give the Contractor written notice of all claims not included in Article 6.2 within five (5) days of the beginning of the event for which claim is made; otherwise, such claims shall be deemed waived.

All unresolved claims, disputes and other matters in question between the Contractor and the Subcontractor not relating to claims included in Article 6.2 shall be resolved in the manner provided in Article 14 herein.

6.4 DELAY. If the progress of the Subcontractor's Work is substantially delayed without the fault or responsibility of the Subcontractor, then the time for the Subcontractor's Work shall be extended by Change Order to the extent obtained by the Contractor under the Contract Documents and the Schedule of Work shall be revised accordingly.

The Contractor shall not be liable to the Subcontractor for any damages or additional compensation as a consequence of delays caused by any person not a party to this Agreement unless the Contractor has first recovered the same on behalf of the Subcontractor from said person, it being understood and agreed by the Subcontractor that, apart from recovery from said person, the Subcontractor's sole and exclusive remedy for delay shall be an extension in the time for performance of the Subcontractor's Work.

6.5 LIQUIDATED DAMAGES. If the Contract Documents provide for liquidated or other damages for delay beyond the completion date set forth in the Contract Documents, and are so assessed, then the Contractor may assess same against the Subcontractor in proportion to the Subcontractor's share of the responsibility for such delay. However the amount of such assessment shall not exceed the amount assessed against the Contractor.

ARTICLE 7
CONTRACTOR'S OBLIGATIONS

7.1 OBLIGATIONS DERIVATIVE. The Contractor binds itself to the Subcontractor under this agreement in the same manner as the Owner is bound to the Contractor under the Contract Documents.

7.2 AUTHORIZED REPRESENTATIVE. The Contractor shall designate one or more persons who shall be the Contractor's authorized representative(s) a) on-site and b)

©Associated General Contractors of America

off-site. Such authorized representative(s) shall be the only person(s) the Subcontractor shall look to for instructions, orders and/or directions, except in an emergency.

7.3 STORAGE APPLICATIONS. The Contractor shall allocate adequate storage areas, if available, for the Subcontractor's materials and equipment during the course of the Subcontractor's Work.

7.4 TIMELY COMMUNICATIONS. The Contractor shall transmit, with reasonable promptness, all submittals, transmittals, and written approvals relating to the Subcontractor's Work.

7.5 NON-CONTRACTED SERVICES. The Contractor agrees, except as otherwise provided in this Agreement, that no claim for non-contracted construction services rendered or materials furnished shall be valid unless the Contractor provides the Subcontractor notice:

(a) prior to furnishing of the services and materials, except in an emergency affecting the safety of persons or property;

(b) in writing of such claim within three days of first furnishing such services or materials; and

(c) the written charges for such services or materials no later than the fifteenth (15th) day of the calendar month following that in which the claim originated.

ARTICLE 8
SUBCONTRACTOR'S OBLIGATIONS

8.1 OBLIGATIONS DERIVATIVE. The Subcontractor binds itself to the Contractor under this Agreement in the same manner as the Contractor is bound to the Owner under the Contract Documents.

8.2 RESPONSIBILITIES. The Subcontractor shall furnish all of the labor, materials, equipment, and services, including, but not limited to, competent supervision, shop drawings, samples, tools, and scaffolding as are necessary for the proper performance of the Subcontractor's Work.

The Subcontractor shall provide a list of proposed subcontractors, and suppliers, be responsible for taking field dimensions, providing tests, ordering of materials and all other actions as required to meet the Schedule of Work.

8.3 TEMPORARY SERVICES. The Subcontractor shall furnish all temporary services and/or facilities necessary to perform its work, except as provided in Article 16. Said article also identifies those common temporary services (if any) which are to be furnished by this subcontractor.

8.4 COORDINATION. The Subcontractor shall:

(a) cooperate with the Contractor and all others whose work may interfere with the Subcontractor's Work;

(b) specifically note and immediately advise the Contractor of any such interference with the Subcontractor's Work; and

(c) participate in the preparation of coordination drawings and work schedules in areas of congestion.

8.5 AUTHORIZED REPRESENTATIVE. The Subcontractor shall designate one or more persons who shall be the authorized Subcontractor's representative(s) a) on-site and b) off-site. Such authorized representative(s) shall be the only person(s) to whom the Contractor shall issue instructions, orders or directions, except in an emergency.

8.6 PROVISION FOR INSPECTION. The Subcontractor shall notify the Contractor when portions of the Subcontractor's Work are ready for inspection. The Subcontractor shall at all times furnish the Contractor and its representatives adequate facilities for inspecting materials at the site or any place where materials under this Agreement may be in the course of preparation, process, manufacture or treatment.

The Subcontractor shall furnish to the Contractor in such detail and as often as required, full reports of the progress of the Subcontractor's Work irrespective of the location of such work.

8.7 SAFETY AND CLEANUP. The Subcontractor shall follow the Contractor's clean-up and safety directions, and

(a) at all times keep the building and premises free from debris and unsafe conditions resulting from the Subcontractor's Work; and

(b) broom clean each work area prior to discontinuing work in the same.

If the Subcontractor fails to immediately commence compliance with such safety duties or commence clean-up duties within 24 hours after receipt from the Contractor of written notice of noncompliance, the Contractor may implement such safety or cleanup measures without further notice and deduct the cost thereof from any amounts due or to become due the Subcontractor.

8.8 PROTECTION OF THE WORK. The Subcontractor shall take necessary precautions to properly protect the Subcontractor's Work and the work of others from damage caused by the Subcontractor's operations. Should the Subcontractor cause damage to the Work or property of the Owner, the Contractor or others, the Subcontractor shall promptly remedy such damage to the satisfaction of the Contractor, or the Contractor may so remedy and deduct the cost thereof from any amounts due or to become due the Subcontractor.

8.9 PERMITS, FEES AND LICENSES. The Subcontractor shall give adequate notices to authorities pertaining to the Subcontractor's Work and secure and pay for all

SAMPLE

©Associated General Contractors of America

permits, fees, licenses, assessments, inspections and taxes necessary to complete the Subcontractor's Work in accordance with the Contract Documents.

To the extent obtained by the Contractor under the Contract Documents, the Subcontractor shall be compensated for additional costs resulting from laws, ordinances, rules, regulations and taxes enacted after the date of the Agreement.

8.10 ASSIGNMENT. The Subcontractor shall not assign this Agreement not its proceeds nor subcontract the whole nor any part of the Subcontractor's Work without prior written approval of the Contractor which shall not be unreasonably withheld. See Article 16.4 for sub-subcontractors and suppliers previously approved by the Contractor.

8.11 NON-CONTRACTED SERVICES. The Subcontractor agrees, except as otherwise provided in this Agreement, that no claim for non-contracted construction services rendered or materials furnished shall be valid unless the Subcontractor provides the Contractor notice:

(a) prior to furnishing of the services or materials, except in an emergency affecting the safety of persons or property;

(b) in writing of such claim within three days of first furnishing such services or materials; and

(c) the written charge for such services or materials no later than the fifteenth (15th) day of the calendar month following that in which the claim originated.

ARTICLE 9
SUBCONTRACT PROVISIONS

9.1 LAYOUT RESPONSIBILITY AND LEVELS. The Contractor shall establish principal axis lines of the building and site whereupon the Subcontractor shall lay out and be strictly responsible for the accuracy of the Subcontractor's Work and for any loss or damage to the Contractor or others by reason of the Subcontractor's failure to set out or perform its work correctly. The Subcontractor shall exercise prudence so that the actual final conditions and details shall result in perfect alignment of finish surfaces.

9.2 WORKMANSHIP. Every part of the Subcontractor's Work shall be executed in strict accordance with the Contract Documents in the most sound, workmanlike, and substantial matter. All workmanship shall be of the best of its several kinds, and all materials used in the Subcontractor's Work shall be furnished in ample quantities to facilitate the proper and expeditious execution of the work, and shall be new except such materials as may be expressly provided in the Contract Documents to be otherwise.

9.3 MATERIALS FURNISHED BY OTHERS. In the event the scope of the Subcontractor's Work includes installation of materials or equipment furnished by others, it shall be the responsibility of the Subcontractor to examine the items so provided and thereupon handle, store and install the items with such skill and care as to ensure a satisfactory and proper installation. Loss or damage due to acts of the Subcontractor shall be deducted from any amounts due or to become due the Subcontractor.

9.4 SUBSTITUTIONS. No substitutions shall be made in the Subcontractor's Work unless permitted in the Contract Documents and only then upon the Subcontractor first receiving all approvals required under the Contract Documents for substitutions. The Subcontractor shall indemnify the Contractor as a result of such substitutions, whether or not the Subcontractor has obtained approval thereof.

9.5 USE OF CONTRACTOR'S EQUIPMENT. The Subcontractor, its agents, employees, subcontractors or suppliers shall not use the Contractor's equipment without the express written permission of the Contractor's designated representative.

If the Subcontractor or any of its agents, employees, suppliers or lower tier subcontractors utilize any machinery, equipment, tools, scaffolding, hoists, lifts or similar items owned, leased, or under the control of the Contractor, the Subcontractor shall be liable to the Contractor as provided in Article 12 for any loss or damage (including personal injury or death) which may arise from such use, except where such loss or damage shall be found to have been due solely to the negligence of the Contractor's employees operating such equipment.

9.6 CONTRACT BOND REVIEW. The Contractor's Payment Bond for the Project, if any, may be reviewed and copied by the Subcontractor.

9.7 OWNER ABILITY TO PAY. The Subcontractor shall have the right to receive from the Contractor information relative to the Owner's financial ability to pay for the Work.

9.8 PRIVITY. Until final completion of the Project, the Subcontractor agrees not to perform any work directly for the Owner or any tenants thereof, or deal directly with the Owner's representatives in connection with the Project, unless otherwise directed in writing by the Contractor. All work for this Project performed by the Subcontractor shall be processed and handled exclusively by the Contractor.

9.9 SUBCONTRACT BOND. If a Performance and Payment Bond is not required of the Subcontractor under Article 16, then within the duration of this Agreement, the Contractor may require such bonds and the Subcontractor

SAMPLE

©Associated General Contractors of America

shall provide the same.

Said bonds shall be in the full amount of this Agreement in a form and by a surety satisfactory to the Contractor.

The Subcontractor shall be reimbursed without retainage for cost of same simultaneously with the first progress payment hereunder.

The reimbursement amount for the bonds shall not exceed the manual rate for such subcontractor work.

Retainage reduction provisions of Article 5.2.2 shall not apply when bonds are furnished under the terms of the Article.

In the event the Subcontractor shall fail to promptly provide such requested bonds, the Contractor may terminate this Agreement and re-let the work to another Subcontractor and all Contractor costs and expenses incurred thereby shall be paid by the Subcontractor.

9.10 WARRANTY. The Subcontractor warrants its work against all deficiencies and defects in materials and/or workmanship and as called for in the Contract Documents.

The Subcontractor agrees to satisfy such warranty obligations which appear within the guarantee or warranty period established in the Contract Documents without cost to the Owner or the Contractor.

If no guarantee or warranty is required of the Contractor in the Contract Documents, then the Subcontractor shall guarantee or warranty its work as described above for the period of one year from the date(s) of substantial completion of all or a designated portion of the Subcontractor's Work or acceptance or use by the Contractor or Owner of designated equipment, whichever is sooner.

The Subcontractor further agrees to execute any special guarantees or warranties that shall be required for the Subcontractor's Work prior to final payment.

ARTICLE 10
RECOURSE BY CONTRACTOR

10.1 FAILURE OF PERFORMANCE

10.1.1 NOTICE TO CURE. If the Subcontractor refuses or fails to supply enough properly skilled workers, proper materials, or maintain the Schedule of Work, or it fails to make prompt payment for its workers, sub-subcontractors or suppliers, disregards laws, ordinances, rules, regulations or orders of any public authority having jurisdiction, or otherwise is guilty of a material breach of a provision of this Agreement, and fails within three (3) working days after receipt of written notice to commence and continue satisfactory correction of such default with diligence and promptness, then the Contractor, without prejudice to any rights or remedies, shall have the right to any or all of the following remedies:

(a) supply such number of workers and quantity of materials, equipment and other facilities as the Contractor deems necessary for the completion of the Subcontractor's Work; or any part thereof which the Subcontractor has failed to complete or perform after the aforesaid notice, and charge the cost thereof to the Subcontractor, who shall be liable for the payment of same including reasonable overhead, profit and attorney's fees;

(b) contract with one or more additional contractors to perform such part of the Subcontractor's Work as the Contractor shall determine will provide the most expeditious completion of the total Work and charge the cost thereof to the Subcontractor;

(c) withhold payment of any monies due the Subcontractor pending corrective action to the extent required by and to the satisfication of the contractor and ________________________________ ; and

(d) in the event of an emergency affecting the safety of persons or property, the Contractor may proceed as above without notice.

10.1.2 TERMINATION BY CONTRACTOR. If the Subcontractor fails to commence and satisfactorily continue correction of a default within three (3) working days after receipt by the Subcontractor of the notice issued under Article 10.1.1, then the Contractor may, in lieu of or in addition to Article 10.1.1, issue a second written notice, by certified mail, to the Subcontractor and its surety, if any. Such notice shall state that if the Subcontractor fails to commence and continue correction of a default within seven (7) working days after receipt by the Subcontractor of the notice, the Contractor may terminate this Agreement and use any materials, implements, equipment, appliances or tools furnished by or belonging to the Subcontractor to complete the Subcontractor's Work. The Contractor also may furnish those materials, equipment and/or employ such workers or Subcontractors as the Contractor deems necessary to maintain the orderly progress of the Work.

All of the costs incurred by the Contractor in so performing the Subcontractor's Work, including reasonable overhead, profit and attorney's fees, shall be deducted from any monies due or to become due the Subcontractor. The Subcontractor shall be liable for the payment of any amount by which such expense may exceed the unpaid balance of the subcontract price.

10.1.3 USE OF SUBCONTRACTOR'S EQUIPMENT. If the Contractor performs work under this Article or sublets such work to be so performed, the Contractor and/or the persons to whom work has been sublet shall have the right to take and use any materials, implements, equipment, appliances or tools furnished by, belonging or delivered to the Subcontractor and located at the Project.

SAMPLE

©Associated General Contractors of America

10.2 BANKRUPTCY

10.2.1 TERMINATION ABSENT CURE. Upon the appointment of a receiver for the Subcontractor or upon the Subcontractor making an assignment for the benefit of creditors, the Contractor may terminate this Agreement upon giving three (3) working days written notice, by certified mail, to the Subcontrator and its surety, if any. If an order for relief is entered under the bankruptcy code with respect to the Subcontractor, the Contractor may terminate this Agreement by giving three (3) working days written notice, by certified mail, to the Subcontractor, its trustee, and its surety, if any, unless the Subcontractor, the surety, or the trustee:

(a) promptly cures all defaults;
(b) provides adequate assurances of future performance;
(c) compensates the Contractor for actual pecuniary loss resulting from such defaults; and
(d) assumes the obligations of the Subcontractor within the statutory time limits.

10.2.2 INTERIM REMEDIES. If the Subcontractor is not performing in accordance with the Schedule of Work at the time of entering an order of relief, or at any subsequent time, the Contractor, while awaiting the decision of the Subcontractor or its trustee to reject or to accept this Agreement and provide adequate assurance of its ability to perform hereunder, may avail itself of such remedies under this Article as are reasonably necessary to maintain the Schedule of Work.

The Contractor may offset against any sums due or to become due the Subcontractor all costs incurred in pursuing any of the remedies provided hereunder, including, but not limited to, reasonable overhead, profit and attorney's fees.

The Subcontractor shall be liable for the payment of any amount by which such expense may exceed the unpaid balance of the contact price.

10.2.3 SUSPENSION BY OWNER. Should the Owner suspend the Prime Contract or any part of the Prime Contract which includes the Subcontractor's Work, the Contractor shall so notify the Subcontractor in writing and upon receipt of said notice the Subcontractor shall immediately suspend the Subcontractor's Work.

In the event of such Owner suspension, the Contractor's liability to the Subcontractor is limited to the extent of the Contractor's recovery on the Subcontractor's behalf under the Contract Documents. The Contractor agrees to cooperate with the Subcontractor, at the Subcontractor's expense, in the prosecution of any Subcontractor claim arising out of an Owner suspension and to permit the Subcontractor to prosecute said claim, in the name of the Contractor, for the use and benefit of the Subcontractor.

10.4 TERMINATION BY OWNER. Should the Owner terminate the Prime Contract or any part of the Prime Contract which includes the Subcontractor's Work, the Contractor shall so notify the Subcontractor in writing and upon receipt of said notice, this Agreement shall be terminated and the Subcontractor shall immediately stop the Subcontractor's Work.

In the event of such Owner termination, the Contractor's liability to the Subcontractor is limited to the extent of the Contractor's recovery on the Subcontractor's behalf under the Contract Documents.

The Contractor agrees to cooperate with the Subcontractor, at the Subcontractor's expense, in the prosecution of any Subcontractor claim arising out of the Owner termination and to permit the Subcontractor to prosecute said claim, in the name of the Contractor, for the use and benefit of the Subcontractor, or assign the claim to the Subcontractor.

10.5 TERMINATION FOR CONVENIENCE. The Contractor may order the Subcontractor in writing to suspend, delay, or interrupt all or any part of the Subcontractor's Work for such period of time as may be determined to be appropriate for the convenience of the Contractor.

The Subcontractor shall notify the Contractor in writing within ten (10) working days after receipt of the Contractor's order of the effect of such order upon the Subcontractor's Work, and the contract price or contract time shall be adjusted by Subcontract Change Order for any increase in the time or cost of performance of this Agreement caused by such suspension, delay, or interruption.

No claim under this Article shall be allowed for any costs incurred more than ten (10) working days prior to the Subcontractor's notice to the Contractor.

Neither the contract price nor the contract time shall be adjusted under this Article for any suspension, delay or interruption to the extent that performance would have been so suspended, delayed, or interrupted by the fault or negligence of the Subcontractor.

10.6 WRONGFUL EXERCISE. If the Contractor wrongfully exercises any option under this Article, the Contractor shall be liable to the Subcontractor solely for the reasonable value of work performed by the Subcontractor prior to the Contractor's wrongful action, including reasonable overhead and profit, less prior payments made, and attorney's fees.

©Associated General Contractors of America

ARTICLE 11

LABOR RELATIONS

(Insert here any conditions, obligations or requirements relative to labor relations and their effect on the project. Legal counsel is recommended.)

SAMPLE

©Associated General Contractors of America

ARTICLE 12

INDEMNIFICATION

12.1 SUBCONTRACTOR'S PERFORMANCE. To the fullest extent permitted by law, the Subcontractor shall indemnify and hold harmless the Owner, the Architect, the Contractor (including the affiliates, parents and subsidiaries) and other contractors and subcontractors and all of their agents and employees from and against all claims, damages, loss and expenses, including but not limited to attorney's fees, arising out of or resulting from the performance of the Subcontractor's Work provided that

(a) any such claim, damage, loss, or expense is attributable to bodily injury, sickness, disease, or death, or to injury to or destruction of tangible property (other than the Subcontractor's Work itself) including the loss of use resulting therefrom, to the extent caused or alleged to be caused in whole or in part by any negligent act or omission of the Subcontractor or anyone directly or indirectly employed by the Subcontractor or for anyone for whose acts the Subcontractor may be liable, regardless of whether it is caused in part by a party indemnified hereunder.

(b) such obligation shall not be construed to negate, or abridge, or otherwise reduce any other right or obligation of indemnity which would otherwise exist as to any party or person described in this Article 12.

12.2 NO LIMITATION UPON LIABILITY. In any and all claims against the Owner, the Architect, the Contractor (including its affiliates, parents and subsidiaries) and other contractors or subcontractors, or any of their agents or employees, by any employee of the Subcontractor, anyone directly or indirectly employed by the Subcontractor or anyone for whose acts the Subcontractor may be liable, the indemnification obligation under this Article 12 shall not be limited in any way by any limitation on the amount or type of damages, compensation or benefits payable by or for the Subcontractor under worker's or workmen's compensation acts, disability benefit acts or other employee benefit acts.

12.3 ARCHITECT EXCLUSION. The obligation of the Subcontractor under this Article 12 shall not extend to the liability of the Architect, its agents or employees, arising out of (a) the preparation or approval of maps, drawings, opinions, reports, surveys, Change Orders, designs or specifications, or (b) the giving of or the failure to give directions or instructions by the Architect, its agents or employees provided such giving or failure to give is the primary cause of the injury or damage.

12.4 COMPLIANCE WITH LAWS. The Subcontractor agrees to be bound by, and at its own cost, comply with all federal, state and local laws, ordinances and regulations (hereinafter collectively referred to as "laws") applicable to the Subcontractor's Work including, but not limited to, equal employment opportunity, minority business enterprise, women's business enterprise, disadvantaged business enterprise, safety and all other laws with which the Contractor must comply according to the Contract Documents.

The Subcontractor shall be liable to the Contractor and the Owner for all loss, cost and expense attributable to any acts of commission or omission by the Subcontractor, its employees and agents resulting from the failure to comply therewith, including, but not limited to, any fines, penalties or corrective measures.

12.5 PATENTS. Except as otherwise provided by the Contract Documents, the Subcontractor shall pay all royalties and license fees which may be due on the inclusion of any patented materials in the Subcontractor's Work. The Subcontractor shall defend all suits for claims for infringement of any patent rights arising out of the Subcontractor's Work, which may be brought against the Contractor or Owner, and shall be liable to the Contractor and Owner for all loss, including all costs, expenses, and attorney's fees.

ARTICLE 13

INSURANCE

13.1 SUBCONTRACTOR'S INSURANCE. Prior to start of the Subcontractor's Work, the Subcontractor shall procure for the Subcontractor's Work and maintain in force Worker's Compensation Insurance, Employer's Liability Insurance, Comprehensive General Liability Insurance, and all insurance required of the Contractor under the Contract Documents except as follows:

The Contractor, Owner and Architect shall be named as additional insureds on each of these policies except for Worker's Compensation.

This insurance shall include contractual liability insurance covering the Subcontractor's obligations under Article 12.

13.2 MINIMUM LIMITS OF LIABILITY. The Subcontractor's Comprehensive General and Automobile Liability Insurance, as required by Article 13.1, shall be written with limits of liability not less than the following:

A. Comprehensive General Liability including completed operations

1. Bodily Injury $________ Each Occurrence
   $________ Aggregate
2. Property Damage $________ Each Occurrence
   $________ Aggregate

B. Comprehensive Automobile Liability

1. Bodily Injury $________ Each Person
   $________ Each Occurrence
2. Property Damage $________ Each Occurrence

©Associated General Contractors of America

SAMPLE

13.3 NUMBER OF POLICIES. Comprehensive General Liability Insurance and other liability insurance may be arranged under a single policy for the full limits required or by a combination of underlying policies with the balance provided by an Excess or Umbrella Liability Policy.

13.4 CANCELLATION, RENEWAL OR MODIFICATION. The Subcontractor shall maintain in effect all insurance coverage required under this Agreement at the Subcontractor's sole expense and with insurance companies acceptable to the Contractor.

All insurance policies shall contain a provision that the coverages afforded thereunder shall not be cancelled or not renewed, nor restrictive modifications added, until at least thirty (30) days prior written notice has been given to the Contractor unless otherwise specifically required in the Contract Documents.

Certificate of Insurance, or certified copies of policies acceptable to the Contractor, shall be filed with the Contractor prior to the commencement of the Subcontractor's Work.

In the event the Subcontractor fails to obtain or maintain any insurance coverage required under this agreement, the Contractor may purchase such coverage and charge the expense thereof to the Subcontractor, or terminate this Agreement.

13.5 WAIVER OF RIGHTS. The Contractor and Subcontractor waive all rights against each other and the Owner, the Architect, separate contractors, and all other subcontractors for loss or damage to the extent covered by Builder's Risk or any other property or equipment insurance, except such rights as they may have to the proceeds of such insurance; provided, however, that such waiver shall not extend to the acts of the Architect listed in Article 12.3.

Upon written request of the Subcontractor, the Contractor shall provide the Subcontractor with a copy of the Builder's Risk policy of insurance or any other equipment insurance in force for the Project and procured by the Contractor. The Subcontractor shall satisfy itself to the existence and extent of such insurance prior to commencement of the Subcontractor's Work.

If the Owner or Contractor have not purchased Builder's Risk insurance for the full insurable value of the Subcontractor's Work less a reasonable deductible, then the Subcontractor may procure such insurance as will protect the interests of the Subcontractor, its subcontractors and their subcontractors in the Work, and, by appropriate Subcontractor Change Order, the cost of such additional insurance shall be reimbursed to the Subcontractor.

If not covered under the Builder's Risk policy of insurance or any other property or equipment insurance required by the Contract Documents, the Subcontractor shall procure and maintain at the Subcontractor's own expense property and equipment insurance for portions of the Subcontractor's Work stored off the site or in transit, when such portions of the Subcontractor's Work are to be included in an application for payment under Article 5.

13.6 ENDORSEMENT. If the policies of insurance referred to in this Article require an endorsement to provide for continued coverage where there is a waiver of subrogation, the owners of such policies will cause them to be so endorsed.

ARTICLE 14

ARBITRATION

14.1 AGREEMENT TO ARBITRATE. All claims, disputes and matters in question arising out of, or relating to, this Agreement or the breach thereof, except for claims which have been waived by the making or acceptance of final payment, and the claims described in Article 14.7, shall be decided by arbitration in accordance with the Construction Industry Arbitration Rules of the American Arbitration Association then in effect unless the parties mutually agree otherwise. This agreement to arbitrate shall be specifically enforceable under the prevailing arbitration law.

14.2 NOTICE OF DEMAND. Notice of the demand for arbitration shall be filed in writing with the other party to this Agreement and with the American Arbitration Association. The demand for arbitration shall be made within a reasonable time after written notice of the claim, dispute or other matter in question has been given, and in no event shall it be made after the date of final acceptance of the Work by the Owner or when institution of legal or equitable proceedings based on such claim, dispute or other matter in question would be barred by the applicable statute of limitations, whichever shall first occur. The location of the arbitration proceedings shall be the city of the Contractor's headquarters or

____

____

14.3 AWARD. The award rendered by the arbitrator(s) shall be final and judgment may be entered upon it in accordance with applicable law in any court having jurisdiction.

14.4 WORK CONTINUATION AND PAYMENT. Unless otherwise agreed in writing, the Subcontractor shall carry on the Work and maintain the Schedule of Work pending arbitration, and if so, the Contractor shall continue to make payments in accordance with this Agreement.

14.5 NO LIMITATION OF RIGHTS OR REMEDIES. Nothing in this Article shall limit any rights or remedies not expressly waived by the Subcontractor which the Subcontractor may have under lien laws or payment bonds.

©Associated General Contractors of America

SAMPLE

14.6 SAME ARBITRATORS. To the extent not prohibited by their contracts with others, the claims and disputes of the Owner, Contractor, Subcontractor and other subcontractors involving a common question of fact or law shall be heard by the same arbitrator(s) in a single proceeding.

14.7 EXCEPTIONS. This agreement to arbitrate shall not apply to any claim:

(a) of contribution or indemnity asserted by one party to this Agreement against the other party and arising out of an action brought in a state or federal court or in arbitration by a person who is under no obligation to arbitrate the subject matter of such action with either of the parties hereto; or does not consent to such arbitration; or

(b) asserted by the Subcontractor against the Contractor if the Contractor asserts said claim, either in whole or part, against the Owner and the contract between the Contractor and Owner does not provide for binding arbitration, or does so provide but the two arbitration proceedings are not consolidated, or the Contractor and Owner have not subsequently agreed to arbitrate said claim, in either case of which the parties hereto shall so notify each other either before or after demand for arbitration is made.

In any dispute arising over the application of this Article 14.7, the question of arbitrability shall be decided by the appropriate court and not by arbitration.

ARTICLE 15

CONTRACT INTERPRETATION

15.1 INCONSISTENCIES AND OMISSIONS. Should inconsistencies or omissions appear in the Contract Documents, it shall be the duty of the Subcontractor to so notify the Contractor in writing within three (3) working days of the Subcontractor's discovery thereof. Upon receipt of said notice, the Contractor shall instruct the Subcontractor as to the measures to be taken and the Subcontractor shall comply with the Contractor's instructions.

15.2 LAW AND EFFECT. This Agreement shall be governed by the law of the state of ______________________

15.3 SEVERABILITY AND WAIVER. The partial or complete invalidity of any one or more provisions of this Agreement shall not affect the validity or continuing force and effect of any other provision. The failure of either party hereto to insist, in any one or more instances, upon the performance of any of the terms, covenants or conditions of this Agreement, or to exercise any right herein, shall not be construed as a waiver or relinquishment of such term, covenant, condition or right as respects further performance.

15.4 ATTORNEY'S FEES. Should either party employ an attorney to institute suit or demand arbitration to enforce any of the provisions hereof, to protect its interest in any matter arising under this Agreement, or to collect damages for the breach of the Agreement or to recover on a surety bond given by a party under this Agreement, the prevailing party shall be entitled to recover reasonable attorney's fees, costs, charges, and expenses expended or incurred therein.

15.5 TITLES. The titles given to the Articles of this Agreement are for ease of reference only and shall not be relied upon or cited for any other purpose.

15.6 ENTIRE AGREEMENT. This Agreement is solely for the benefit of the signatories hereto and represents the entire and integrated agreement between the parties hereto and supersedes all prior negotiations, representations, or agreements, either written or oral.

ARTICLE 16

SPECIAL PROVISIONS

16.1 PRECEDENCE. It is understood the work to be performed under this Agreement, including the terms and conditions thereof, is as described in Articles 1 through 16 herein together with the following Special Provisions, which are intended to complement same. However, in the event of any inconsistency, these Special Provisions shall govern.

16.2 SCOPE OF WORK. All work necessary or incidental to complete the ______________________

____

____

____

Work for the Project in strict accordance with the Contract Documents and as more particularly, though not exclusively, specified in: ______________________

____

____

____

____

____

with the following additions or deletions:

SAMPLE

©Associated General Contractors of America

16.3 COMMON TEMPORARY SERVICES. The following "Project" common temporary services and/or facilities are for use of all project personnel and shall be furnished as herein below noted:

By this subcontractor;

By others;

16.4 OTHER SPECIAL PROVISIONS. (Insert here any special provisions required by this subcontract.)

16.5 CONTRACT DOCUMENTS. (List applicable contract documents including specifications, drawings, addenda, modifications and exercised alternates. Identify with general description, sheet numbers and latest date including revisions.

SAMPLE

IN WITNESS WHEREOF, the parties hereto have executed this Agreement under seal, the day and year first above written.

____
Subcontractor

____
Contractor

By ______________________________
(Title)

By ______________________________
(Title)

©Associated General Contractors of America

THE ASSOCIATED GENERAL CONTRACTORS OF AMERICA

SUBCONTRACT (SHORT FORM)

TABLE OF ARTICLES

1. CONTRACT PAYMENT
2. SCOPE OF WORK
3. SCHEDULE OF WORK
4. CHANGES
5. FAILURE OF PERFORMANCE
6. INSURANCE
7. INDEMNIFICATION
8. WARRANTY
9. SPECIAL PROVISIONS

SAMPLE

REPRESENTED WITH THE PERMISSION OF THE ASSOCIATED GENERAL CONTRACTORS OF AMERICA (AGC). COPIES OF CURRENT FORMS MAY BE OBTAINED FROM AGC'S PUBLICATIONS DEPARTMENT, 1957 E ST. N.W., WASHINGTON, D.C. 20006.

This document conforms to the high standards for AGC documents and is recommended for use only where the covered work is of small dollar amount and will be completed within a relatively short period of time. For all other work, AGC recommends its comprehensive *Subcontract for Building Construction* (AGC-600).

This document has important legal and insurance consequences. AGC encourages consultation with an attorney and insurance consultant when completing or modifying this document.

AGC DOCUMENT NO. 603 • SUBCONTRACT (SHORT FORM) • 1987 MASTER INDEX NO. 1-1-0-37.4

© Copyright 1987, The Associated General Contractors of America

Special Instructions

AGREEMENT: In the Agreement section above Article 1, the first blanks should be filled in with the current day, month, and year. The second set of date blanks should be filled in with the day, month, and year when the subcontract is to become effective. Both sets of blanks should be filled in, even if the two dates are identical, so there is no confusion over what the parties intend with respect to the effective date.

The legal name of the contractor firm should be filled in the blank before "(Contractor)," and the legal name of the subcontractor firm in the following blank.

After "Project," fill in the name and address of the project where the subcontract work will be performed. After "Owner," fill in the owner's legal business name and address. Similarly, after "Architect," "Contractor," and "Subcontractor," fill in the names of the business firms, the firms' addresses, and the telephone numbers.

ARTICLE 1: Fill in the dollar amount in words in the blank before the word "Dollars." Use numbers to fill in the same amount after the "$" in parentheses. The rate of retainage should be filled in in numbers before the "%" symbol on the next line.

ARTICLE 2: Fill in the first blank with the general term for the work to be performed (for example, roofing or mechanical work). In the second blank, fill in the titles of specific documents containing the description.

ARTICLE 9: Insert any other special requirements agreed to by the parties. For example: bonds furnished by subcontractor; liquidated damages; governmental authority requirements; termination provisions; project work conditions or labor relations; dispute resolution procedures (including attorneys fee provisions); time limits; lien waivers; payment affidavits; or insurance requirements.

SIGNATURES: The top-level lines should be filled in with the legal names of the contractor and subcontractor business firms. The second-level lines should be filled in with the signatures of the persons representing each firm, with each person's name and business title typed or printed below the signature line. The subcontractor's federal tax identification number should be typed or printed in numbers on the last line.

SAMPLE

SKILL RESPONSIBILITY INTEGRITY
THE ASSOCIATED GENERAL CONTRACTORS OF AMERICA

JOB NO.: ______________________ ACCOUNT CODE: ______________________

DATE: ______________________

SUBCONTRACT
(SHORT FORM)

This agreement is made this ______ day of ______________ 19____, and effective the ______________ day of ______________ 19____, by and between ____ (Contractor) and ____ (Subcontractor) to perform the Work identified in Article 2 in accordance with the Project's Contract Documents.

PROJECT:
OWNER:
ARCHITECT:
CONTRACTOR:
SUBCONTRACTOR:

ARTICLE 1

CONTRACT PAYMENT. The Contractor agrees to pay Subcontractor for satisfactory performance of Subcontractor's Work the sum of ____ Dollars ($__________).

Progress payments, less retainage of ________ %, shall be made to Subcontractor for Work satisfactorily performed no later than seven (7) days after receipt by Contractor of payment from Owner for Subcontractor's Work. Final payment of the balance due shall be made to Subcontractor no later than seven (7) days after receipt by Contractor of final payment from Owner for Subcontractor's Work. These payments are subject to receipt of such lien waivers, affidavits, warranties and guarantees required by the Contract Documents or Contractor.

ARTICLE 2

SCOPE OF WORK. Subcontractor agrees to commence Subcontractor's Work herein described upon notification by Contractor, and to perform and complete such Work in accordance with Contract Documents and under the general direction of Contractor in accord with Contractor's schedule. This shall include all work necessary or incidental to complete the:

____

Work for the Project as more particularly, though not exclusively specified in ______________________________

____

____

ARTICLE 3

SCHEDULE OF WORK. Time is of the essence. Subcontractor shall provide Contractor with any requested scheduling information of Subcontractor's Work. The Schedule of Work, including that of this Subcontract shall be prepared by Contractor and may be revised as the Work progresses.

Subcontractor recognizes that changes may be made in the Schedule of Work and agrees to comply with such changes without additional compensation.

Subcontractor shall coordinate its work with all other contractors, subcontractors, and suppliers on the Project so as not to delay or damage their performance, work, or the Project.

SAMPLE

ARTICLE 4

CHANGES. Contractor, without nullifying this Agreement, may direct Subcontractor in writing to make changes to Subcontractor's Work. Adjustment, if any, in the contract price or contract time resulting from such changes shall be set forth in a Subcontract Change Order pursuant to the Contract Documents.

ARTICLE 5

FAILURE OF PERFORMANCE. Should Subcontractor fail to satisfy contractual deficiencies within three (3) working days from receipt of Contractor's written notice, then the Contractor, without prejudice to any right or remedies, shall have the right to take whatever steps it deems necessary to correct said deficiencies and charge the cost thereof to Subcontractor, who shall be liable for payment of same, including reasonable overhead, profit and attorneys fees.

ARTICLE 6

INSURANCE. Prior to the start of Subcontractor's Work, Subcontractor shall procure and maintain in force for the duration of the Work, Worker's Compensation Insurance, Employer's Liability Insurance, Comprehensive General Liability Insurance and all insurance required of Contractor under the Contract Documents. Contractor, Owner and Architect shall be named as additional insureds on each of these policies, except for Worker's Compensation.

ARTICLE 7

INDEMNIFICATION. To the fullest extent permitted by law, Subcontractor shall indemnify and hold harmless Owner, Architect, Architect's consultants, and Contractor from all damages, losses, or expenses, including attorneys fees, from any claims or damages for bodily injury, sickness, disease, or death, or from claims for damage to tangible property, other than the Work itself. This indemnification shall extend to claims resulting from performance of this Subcontract and shall apply only to the extent that the claim or loss is caused in whole or in part by any negligent act or omission of Subcontractor or any of its agents, employees, or subcontractors. This indemnity shall be effective regardless of whether the claim or loss is caused in some part by a party to be indemnified. The obligation of Subcontractor under this Article shall not extend to claims or losses that are primarily caused by the Architect, or Architect's consultant's performance or failure to perform professional responsibilities.

ARTICLE 8

WARRANTY. Subcontractor warrants its work against all deficiencies and defects in materials and/or workmanship and agrees to satisfy same without cost to Owner or Contractor for a period of one (1) year from the date of Substantial Completion of the Project or per Contract Documents, whichever is longer.

ARTICLE 9

SPECIAL PROVISIONS. (Insert any special provisions required by this Subcontract).

SAMPLE

In witness whereof, the parties have executed this Agreement under Seal, the day and year first written above.

______________________________	______________________________
SUBCONTRACTOR (FIRM NAME)	CONTRACTOR (FIRM NAME)
______________________________	______________________________
BY (Type or print signer's name and title)	BY (Type or print signer's name and title)

Subcontractor's Federal Tax ID Number: ____________________

© Copyright 1987, Associated General Contractors of America (AGC No. 603)

THE ASSOCIATED GENERAL CONTRACTORS

STANDARD DESIGN-BUILD SUBCONTRACT AGREEMENT WITH SUBCONTRACTOR NOT PROVIDING DESIGN

SAMPLE

This Document shall be used in conjunction with AGC Document 430 – Conditions Between Contractor and Subcontractor for Design-Build.

This Agreement made at ______________________________
this __________ day of ______________, 19_____, by and between ______________________________,
hereinafter referred to as the Contractor, and ______________________________, hereinafter referred to as the Subcontractor, to perform part of the Work on the following Project:

PROJECT:

OWNER:

REPRESENTED WITH THE PERMISSION OF THE ASSOCIATED GENERAL CONTRACTORS OF AMERICA (AGC). COPIES OF CURRENT FORMS MAY BE OBTAINED FROM AGC'S PUBLICATIONS DEPARTMENT, 1957 E ST. N.W., WASHINGTON, D.C. 20006.

Certain provisions of this document have been derived, with modifications, from the following document published by The American Institute of Architects: AIA Document A201, General Conditions, © 1976. Usage of AIA language, with the permission of AIA, does not imply AIA endorsement or approval of this document. Further reproduction of copyrighted AIA materials without separate written permission from AIA is prohibited.

AGC DOCUMENT NO. 450 • STANDARD DESIGN-BUILD SUBCONTRACT AGREEMENT WITH SUBCONTRACTOR PROVIDING DESIGN • January 1983

© 1986, Associated General Contractors of America.

ARTICLE 1

SCOPE OF WORK

1.1 The Contractor employs the Subcontractor as an independent contractor to construct a part of the Project for which the Contractor has contracted with the Owner. The Subcontractor's portion of the Project, hereinafter referred to as the "Work," is set out in Exhibit A attached hereto. The Subcontractor agrees to perform such Work under the general direction of the Contractor and subject to the final approval of the Owner, in accordance with the Subcontract Documents. This Agreement and the Subcontract Documents incorporated herein represent the entire agreement between the parties and supercede all prior negotiations, representations, or agreements.

1.2 The Subcontract Documents, hereinafter referred to as the "Subcontract," include this Agreement, the Conditions Between Contractor and Subcontractor for Design Build [AGC Document 430], and documents set forth therein, all of which are more specifically identified in Exhibit B attached hereto. If any provisions of such documents conflict with the terms of this Agreement, the terms of this Agreement shall control. The Subcontractor binds himself to the Contractor for the performance of Subcontractor's Work in the same manner as the Contractor is bound to the Owner for such performance under Contractor's contract with the Owner.

1.3 In the performance of his Work, Subcontractor will:

.1 Furnish all labor and materials, along with competent supervision, shop drawings and samples, tools, equipment, scaffolding, permits and fees necessary for the construction of the Work; and,

.2 Give all notices and comply with all applicable laws, building codes, ordinances, regulations and orders of any public authority bearing on the design and construction of the Work under this Agreement.

ARTICLE 2

PROJECT SCHEDULE

2.1 Subcontractor will commence, and thereafter prosecute his Work in accordance with the Project Schedule so as not to cause any delays or interference with the completion of the Project or in the obtaining of payments by the Contractor from the Owner or the final acceptance of the Project by the Owner. If the Subcontractor does not commence the Work in accordance with the Project Schedule, or if at any time the Work is not performed in accordance with such schedule, the Subcontractor agrees, upon three (3) days' written notice from the Contractor, to provide the necessary personnel and supply such equipment, materials, overtime work, workers and other devices and facilities as necessary so as to expedite the Work. Such notice, once given, shall continue in effect until the Work specified therein has been fully completed, even though the Subcontractor has initially acted under the notice but has failed to continue to do so until complete performance thereof. Subcontractor shall work overtime, at the direction of Contractor without additional cost to Contractor if such overtime work is necessary to cure delinquency in maintaining the Project Schedule and such delinquency is due to delays by Subcontractor.

ARTICLE 3

CHANGES IN THE WORK

3.1 The Contractor may order changes in the Work consistent with the provisions of the Conditions Between Contractor and Subcontractor for Design-Build and within the general scope of this Agreement, consisting of additions, deletions or other revisions.

3.2 No claims for extra work or changes will be recognized or paid unless prior approval has been obtained in writing from the Contractor.

ARTICLE 4

GENERAL PROVISIONS

4.1 The Work performed under this Agreement is subject to the approval of the Contractor and Owner.

© 1986, Associated General Contractors of America.

SAMPLE

4.2 If the Work, or any portion thereof is not acceptable, the Subcontractor shall be responsible for the cost of remedying unaccepted Work, whether such remedial Work is performed by the Subcontractor or by any other entity at the request of the Contractor or Owner.

ARTICLE 5

CONTRACT SUM AND PAYMENTS

5.1 The Contractor agrees to pay to the Subcontractor for the satisfactory completion of Subcontractor's Work the Contract Sum of ________________________ ($______________) in monthly payments of __________ percent of the work performed in any preceding month, in accordance with estimates prepared by the Subcontractor and approved by the Contractor. Payment of the approved portion of the Subcontractor's monthly estimate and final payment shall be conditioned upon receipt by the Contractor of his payment from the Owner.

5.2 Subcontractor shall provide with his monthly applications for payment completed lien waivers and affadavits from his subcontractors and suppliers in a form satisfactory to the Owner and Contractor. Approval and payment of Subcontractor's monthly estimate is specifically agreed not to constitute or imply acceptance by the Contractor or Owner of any portion of the Subcontractor's Work. Final payment shall not constitute acceptance of defective work.

5.3 The Subcontractor agrees and covenants that money received for the performance of this Agreement shall be used solely for the benefit of persons and firms supplying labor, materials, supplies, tools machines, equipment, plant or services exclusively for this Project in connection with this Agreement and having the right to assert liens or other claims against the land improvements, or funds involved in this Project or against any bond or other security posted by Contractor or Owner; that any money paid to the Subcontractor pursuant to this Agreement shall immediately become and constitute a trust fund for the benefit of said persons and firms, and shall not in any instance be diverted by the Subcontractor to any other purpose until all obligations and claims arising hereunder have been fully discharged.

5.4 The Contractor may deduct from any amounts due or to become due to the Subcontractor any sum or sums owing by the Subcontractor to the Contractor; and in the event of any breach by the Subcontractor of any provision or obligation of this Subcontract, or in the event of the assertion by other parties of any claim or lien against the Owner, the Contractor, Contractor's Surety, or the premises upon which the Work was performed, which claim or lien arises out of the Subcontractor's performance of this Agreement, the Contractor shall have the right, but is not required, to retain out of any payments due or to become due to the Subcontractor an amount sufficient to completely protect the Contractor from any and all loss, damage, or expense therefrom, until the claim or lien has been adjusted by the Subcontractor to the satisfaction of the Contractor. This paragraph shall be applicable even though the Subcontractor has posted a 100% labor and material payment bond and a performance bond.

5.5 Final payment will be made when the completed project is accepted by the Owner; the Subcontractor has submitted completed lien waivers and affidavits from his subcontractors and suppliers in a form and to the extent required by the Owner and Contractor; and the Contractor has received final payment from the Owner.

ARTICLE 6

INSURANCE

6.1 The Subcontractor shall, within __________ days of signing this Agreement, but before performing any Work, provide the Contractor with certificates of insurance indicating coverage for Comprehensive General Liability, Comprehensive Auto Liability, claims under workers compensation, disability benefit, and other similar employee benefit acts which are applicable to the work to be performed in accordance with the Conditions Between Contractor and Subcontractor for Design-Build [AGC Document 430] for the following limits:

Comprehensive General Liability		Comprehensive Auto	
Bodily Injury	$______________	Bodily Injury	$______________
Property Damage	$______________	Property Damage	$______________

Workers' Compensation $ Legal Limit

© 1986, Associated General Contractors of America.

SAMPLE

6.2 The Builders Risk Insurance contains a deductible of $____________. Each insured shall bear his loss within the deductible unless the Contractor's Agreement with the Owner provides that the Owner shall bear any such loss.

ARTICLE 7

MISCELLANEOUS PROVISIONS

7.1 Governing Law: This Agreement shall be governed by the law in effect at the location of the Project.

IN WITNESS WHEREOF the parties hereto have executed this Agreement under seal, the day and year first above written.

____
SUBCONTRACTOR

By________________________

ATTEST:

____

____
CONTRACTOR

By________________________
(Title)

ATTEST:

____

SAMPLE

© 1986, Associated General Contractors of America.

EXHIBIT A

The Subcontractor's Work on this Project shall consist of the following portions of the Project:

SAMPLE

© 1986, Associated General Contractors of America.

EXHIBIT B

The Contract Documents include:

1. The Agreement Between the Owner and Contractor dated _______________.
2. The General, Special and Supplementary Conditions identified as:
3. The Project Schedule dated _______________
4. The following drawings, specifications, and criteria:
5. The Contract Sum includes the following allowances:

SAMPLE

© 1986, Associated General Contractors of America.

Index

D

P

Q

R

S

T

V

W

94719

HD
9715
.A2
G73
1989

97-612-L

J. Edward.

tion paperwork

GASTON COLLEGE
MORRIS LIBRARY & MEDIA CENTER
DALLAS, NC 28034